Achieving Interoperability in Critical IT and Communication Systems

For a listing of recent titles in the
Artech House Mobile Communications Library,
turn to the back of this book.

Achieving Interoperability in Critical IT and Communication Systems

Robert I. Desourdis, Jr.
Peter J. Rosamilia
Christopher P. Jacobson
James E. Sinclair
James R. McClure

Editors

ARTECH HOUSE
BOSTON | LONDON
artechhouse.com

Library of Congress Cataloging-in-Publication Data
A catalog record of this book is available from the Library of Congress.

British Library Cataloguing in Publication Data
A catalogue record of this book is available from the British Library.

ISBN 978-1-59693-389-7

Cover design by Igor Valdman

685 Canton Street
Norwood, MA 02062

10 9 8 7 6 5 4 3 2 1

This work is dedicated to the men and women of public safety and emergency response, and particularly to the memory of those who sacrificed themselves for the lives, health, and property of others.

Greater love has no one than this, that he lay down his life for his friends.
—John 15:13 NIV

Contents

Foreword

Perhaps once or twice in a century a nation undergoes an experience so unexpected, so shocking, so traumatic that it becomes a turning point in history. One such event was the Japanese bombing of Pearl Harbor. In the perspective of history, the span of 67 years is only a flash, but it is a long time in the life of an individual. Those today who remember Pearl Harbor as a direct experience have obtained senior citizen status yet the subject continues to fascinate them and their descendants.

How could it have happened? How could the Japanese have sailed over 3,000 miles across the Pacific without being seen? How could they have developed a torpedo that would not sink in the mud of Pearl Harbor, (which was less then 35 feet deep)? How could they have developed a bomb weighing 1,720 pounds, which would penetrate the deck of a battle ship? How could they have avoided air detection when they came within 250 miles of their target? How could they have refueled in the cold Northern Pacific three times on the way to their target and not be seen? How could they have caught the American fleet napping at Pearl Harbor, the Gibraltar of Asia and sink or damage 8 battle ships, kill over 2,400 men, wound 1,178, destroy almost 300 aircraft and damage another 128 and not get caught? Their losses were only 29 aircraft, 129 men, 1 major submarine, and five midget submarines.

From 1942 to 1946 there were 9 separate investigations, all inconclusive, but all emphasizing that there were errors of judgment and omission. For the past 50 years this writer and his team have been studying what happened and why. After careful research, we came up with 25 indicators of what went wrong and the lessons that should have been learned.

Sixty years later, in another century, on September 11, 2001, another "day of infamy" happened. The same type of questions were asked and the answers written in the famous *9/11 Commission Report* as were analyzed of the Pearl Harbor attack. There was lack of imagination, a failure to respond properly, and a failure of the various agencies involved to coordinate their efforts. We seem never to learn from history. Robert Desourdis Jr., senior system architect and vice president for Technology of Science Applications International Corporation (SAIC), has borrowed from what we learned from the Pearl Harbor attack; using much of the research of Gordon Prange, Katherine V. Dillon, and myself along with the *9/11 Commission Report* to show that we did not learn. He has developed a model for one to use so that if history repeats itself one should know what to do. It is a bold attempt to qualify and quantify what happened, for as weapons get more deadly we cannot afford another 9/11 or another Pearl Harbor.

This book is intended for a worldwide audience and is in an effort to bring the contents of Pearl Harbor and the 9/11 attack reports and research to the public. There is an old adage that "smart people learn from other peoples mistakes; dumb people learn from their own." Using Desourdis' model, if we have a catastrophe like Katrina and 9/11 we will know what to do. After defining interoperability and its role in public safety and emergency in Chapter 1, Chapter 2 depicts the consequences of failed interoperability. Chapter 3 relates lessons not learned from Pearl Harbor, 9/11, and Katrina. Chapter 4 defines the process for establishing interoperability; Chapters 5 and 6 relate operational and system architects of the system; and Chapter 7 defines how one can achieve success using interoperability. Throughout the book examples are drawn from the 25 lessons learned at Pearl Harbor and largely ignored on 9/11. This book is important because it not only adds to the literature but presents a synopsis of the alleged sins of omission. It also does an excellent job in analyzing in succinct form: events, failures to act, messages sent, and who knew what and when. It is well documented.

"Pearl Harbor never dies and no living person has seen the end of this," so said Admiral Kimmel's lawyer Charles Rudd to my mentor Gordon W. Prange, and for forty years Prange, and for 50 years Katherine Dillon and myself, have been studying this subject. We know that this will not be the last word and that many of the same conclusions were reached by others; however, this book is unique and an excellent addition to the literature about Pearl Harbor and 9/11. It is not based upon hearsay, or would have, should have, could have material, but actual documents and testimony depicting what happened those fateful days in December 1941 and September 11, 2001. It will not change many peoples' minds as to who was responsible, but will serve as a model for future reference so that we can prevent such catastrophes—and if it does happen again, we will know what to do.

Donald M. Goldstein, Ph.D.
Professor
Pittsburgh, Pennsylvania
June 2009

Foreword

What a treat for me to be asked by Bob Desourdis (no exaggeration "the" wireless guru) to write a foreword to his new book. If he is the technical yin of wireless, I'm way over on the yang side of legal, regulatory; but I think there was some method in his madness in asking me to write these few words. He knows I share with him a passion about the subject of public safety communications in a crisis. Discussions about this topic over the last few years have been obscured by the overuse and misunderstanding of the word "interoperability." The best remedy I can describe to cure that deficiency is to read this book which explodes the myths, dissects the errors, and compiles a literal road map for policy makers who may want, once and for all, to fix what is seriously broken. One certainly could not say that this book is written down to the layman, so for the reader there is hard work ahead; but the structure is cleverly anecdotal, referential, and fascinating enough to be layman friendly.

As I read the chapters explaining failures of interoperability, I was reminded of an old parlor game that most of us have played at least once. A group of people sit in a circle and one person composes a short vignette of a few sentences and whispers it to the fellow on his right. That player then whispers it (exactly as it was given to him) to the lady on his right and so forth until the last player receives the whispered message and then repeats it out loud. Invariably, the content, tone, and meaning of the simple message has changed beyond recognition. The point of the game, just like the point of this book, is that human beings communicate very poorly, sometimes with dire consequences, even with the best of intentions and the simplest of environments.

Readers of this book will learn in an entertaining way just what happens when interoperability breaks down. The why of it. The how of it. Desourdis constructs first an analysis and then a structural solution for policy makers to begin the monumental task of bringing interoperability to a level where the next disaster either is averted altogether or, at least, the damage is mitigated by the effective working of his improved system.

So much for the glass half full of my reaction to Bob's excellent book. It detracts in no way from his intellect or skill as a writer to say that his book also gives me a pronounced glass half empty reaction as well. His intention I'm guessing is to do just that and it certainly worked with me. I can't help seeing and hearing the ghosts of the lives that have been lost by a repeated pattern of human incompetence, particularly at higher management levels, when national or manmade crises unfold. The pedagogic technique of the book is to slip back and forth between disasters like Pearl Harbor, 9/11, Katrina, and such. So much time has passed, so many lives have been lost, and yet so little has been learned. It shakes me to think that whatever next may be in store for us as a disaster will result yet again in the deaths of innocents because the guys in control have yet to learn anything meaningful about fixing the complex process of interoperability.

You will like this book. I think it will give you a much better understanding of why humans do such a poor job of communicating. And the next time you have to find the perfect gift for anyone in a position of responsibility or command, this is it.

Morgan E. O'Brien, attorney
Former cofounder and chairman of Nextel Communications
June 2009

Preface

In 2001, several coauthors and I described the worldwide science and technology of land mobile radio and commercial wireless communications for public safety and emergency response in the Artech House book entitled *Emerging Public Safety Wireless Communication Systems*. The book was meant as a purely technical reference. We created that book hoping its information would help improve communications interoperability in public safety and emergency response by providing technical support to the engineers and decision makers. In fact, the writing of the book and the events since that time have shown that the lack of *technical* interoperability was a symptom (or a result) of the real interoperability problem, and not the cause of the problem. In other words, having the right technology is a necessary but far from sufficient condition for true interoperability.

History has shown interoperability to be a people problem, that is, leaders and planners failing to use best practices to develop the right processes and tools (including standards) for sharing trusted information that drives predictable collaborative action. This is the heart of the *holistic* interoperability problem. This failure could continue to be ignored except for the fact that public safety, a primary responsibility of government, is always at stake for reasons that a priori have nothing to do with technology. Moreover, it is always our public safety and emergency responders and hospital-based first receivers (e.g., hospital staff) that must show great initiative and self-sacrifice because of these planning failures. It is this initiative and sacrifice that often masks the failure of leaders and planners to provide this holistic interoperability. Even today, many decision makers treat this understanding of the "people problem" as a new discovery, when this fact has been well known by students of history for many decades (such as Dr. Donald Goldstein, who contributed an excellent foreword to this book).

Thus "buying the right thing" is never by itself the solution of a critical IT or communications interoperability problem. However, this behavior is a natural reaction of leaders and planners who fail to see technical interoperability as a *systems* problem, and who neither know or care to follow best practices to develop and tailor a systems solution. In fact, technology (as I learned in my technology and policy graduate work at MIT) does what people plan for it to do, and it is well known to be the simplest part of establishing true interoperability. This understanding recalled my early days studying interoperability as the spectrum analysis lead on the first year of the Public Safety Wireless Network (PSWN) Program. This program was started by the U.S. government in the mid-1990s to finally fix the radio course by the quality control reviewer Rozlyn A. Reece from Science Applications International Corporation (SAIC). I thank Mary F. Cates of SAIC for her export control review. I also thank Anne Rosamilia and Patricia Y. Stover for their editing to clarify key points in several chapters, and Karen A. Benson and Linda L. Parker for their editorial support.

I want to thank Mark O. Puputti of Leominster, Massachusetts, for providing a key biblical reference in the Dedication, which immediately follows the dedication to our responders and receivers who sacrifice for us all. I also thank him for arranging an eloquent biblical interoperability dedication provided by his pastor Daryl N. Larson of Fitchburg, Massachusetts, as follows:

> Now the body is not made up of one part but of many. If the foot should say, "Because I am not a hand, I do not belong to the body," it would not for that reason cease to be part of the body. And if the ear should say, "Because I am not an eye, I do not belong to the body," it would not for that reason cease to be part of the body. If the whole body were an eye, where would the sense of hearing be? If the whole body were an ear, where would the sense of smell be? But in fact God has arranged the parts in the body, every one of them, just as he wanted them to be. If they were all one part, where would the body be? As it is, there are many parts, but one body.
>
> The eye cannot say to the hand, "I don't need you!" And the head cannot say to the feet, "I don't need you!" On the contrary, those parts of the body that seem to be weaker are indispensable, and the parts that we think are less honorable we treat with special honor. And the parts that are unpresentable are treated with special modesty, while our presentable parts need no special treatment. But God has combined the members of the body and has given greater honor to the parts that lacked it, so that there should be no division in the body, but that its parts should have equal concern for each other. If one part suffers, every part suffers with it; if one part is honored, every part rejoices with it.
>
> —*1 Corinthians 12:14-26 NIV*[1]

My understanding of holistic interoperability for critical IT and communication systems has evolved from the interaction with many unsung heroes in public safety and emergency response. I thank several professionals who helped solidify my understanding of the versatility, capabilities, and needs at the local and state levels. They have collectively managed the people problems at the heart of interoperability, following elements of several best planning practices—versus "just buying something"—and inspired me to create this book. Although there are many stars in the evolving Massachusetts interoperability firmament, some (not all) of the brightest ones in my direct experience are Chief Philip L. Mahoney (Woburn), C. Blair Sutherland (State Police), Edward F. Kelly (Melrose), Mark Foster (Beverly), Chief Terry Dun (Franklin), Lt. Mark J. Cady (Worcester), George Fosque (Cambridge), Russ Bond (Highway Department), Nancy Ridley (Health Department), Edward C.

1. Scripture taken from the HOLY BIBLE, NEW INTERNATIONAL VERSION®. Copyright © 1973, 1978, 1984 International Bible Society. Used by permission of Zondervan. All rights reserved. The "NIV" and "New International Version" trademarks are registered in the United States Patent and Trademark Office by International Bible Society. Use of either trademark requires the permission of International Bible Society.

McNamara (Worcester), Jonathan L. Epstein (Wakefield), Ronald W. Burson (Martha's Vineyard), Stephen A. Hooke (Holbrook), and Officer Scott Wilder (Brookline), in the great Commonwealth of Massachusetts.

I am also indebted to several individuals for there appreciation of true interoperability. Their duty to country, continued resolve for best practices, and insistence on stovepipe-independent interoperability despite the bureaucratic consequences, have eclipsed many in their respective domains. These individuals include C. Sean Thrash, CDR Pat DiBari, John M. Contestabile, and Charles E. Cape. Each of their histories in this regard would be a book unto itself—and somewhat more exciting. I thank Michael D. Paulette for providing useful national emergency planning documents.

This book would not have been realized if not for the belief and persistence of Mark E. Walsh, chief acquisitions editor of Artech House, to create a definitive work on interoperability. I also thank Lindsey C. Gendall, assistant acquisitions editor, for her help in coordinating the completion of the manuscript. Similarly, I thank Erin M. Donahue, production editor, for supporting this work to publication and Christopher D. Stanfa, web editor, for his help in preparing the book annex Web site.

Finally, I thank my mother, Margaret Gerinecz Desourdis (1921–2005), and my father, Robert Irving Desourdis, Sr., (1922–1991), for making it possible to create this work. May they forever rest in peace.

CHAPTER 1

Interoperability Defined

Effective *interoperability* has always been critical to the success of coordinated human activities. In public safety and emergency response operations, interoperability can be a matter of life and death and it requires far more than just ensuring that first responders and hospital-based first receivers are able to communicate by radio. In areas like emergency services and health care administration, effective interoperability can save lives, improve response capabilities and save billions of dollars.

This book highlights the challenges of communication and information sharing among first responders and first receivers (we will refer to them as "responders-receivers") as well as other public safety, health and emergency response stakeholders. We draw on case studies to identify the challenges to effective interoperability, then explain what today's best practices and emerging technologies offer that could dramatically improve the realities on the ground. To help clarify today's interoperability failures, we present a World Interoperability Failure model to illustrate worst-case leadership and planning deficiencies. We also establish a World Interoperability Success Plan as a plausible detailed approach to achieve holistic interoperability. This recognizes the balancing concern of preserving local control of assets and information.

Since a wide range of technologies, policies, and practices must be coordinated for receivers-responders to do their jobs most effectively, a common understanding of the term "interoperability" is vital. We begin Chapter 1 by redefining the term interoperability. We first explore the various definitions that have been ascribed to the term, starting from a detailed literature review. Then we provide one authoritative, inclusive definition that contains the elements that are key to effective and comprehensive worldwide interoperability.

Having defined interoperability, we then propose six critical planning success factors needed to achieve it in critical IT and communication systems. This clearer, big picture view will facilitate improvements to information sharing and communication among public safety and emergency response professionals for true interoperability.

1.1 Real Interoperability

The importance of *interoperability* in public safety and emergency response information technology and communication systems is well known [1].

> The *Final Report* concludes that, unless immediate measures are taken to alleviate spectrum shortfalls and promote interoperability, Public Safety agencies will not be able to adequately discharge their obligation to protect life and property in a safe, efficient, and cost effective manner.

The analysis that follows will demonstrate that the value of interoperability is entirely dependent on two key factors: inspired, effective leaders with mutual understanding and respect; and holistic planning and implementation of all elements of interoperability—*not* just technology. Thus, an adequate definition of public safety and emergency response interoperability is fundamentally important in the specification of IT and communication systems designed to achieve interoperability.

The term "interoperable" has been used to describe the ability of human-driven operations or the capability of disparate IT and communication systems to work in unison. In the day-to-day parlance of public safety and emergency response, a discussion about interoperability by newcomers to the discourse meant radio interoperability, and voice radio in particular [2].

> The ability of officials from different Public Safety agencies to communicate with each other is limited. Yet interoperability is key to success in day-to-day operations, joint task force and mutual aid operations, and many other intra- and inter-jurisdictional activities. Interoperability is hampered by the use of multiple frequency bands, incompatible radio equipment, and a lack of standardization in repeater spacing and transmission formats.

But interoperability analysts in 1996 correctly understood the importance of both data interoperability and voice interoperability [3].

> The development of individual uncoordinated networks will leave data interoperability in the same state as voice interoperability is today. Data interoperability for criminal justice users will allow investigators and field agents to travel anywhere in the country and still access not only federal and state criminal justice systems, but also the systems of her/his own agency. Data interoperability will allow agencies responding to large-scale mutual aid events such as wildland fires to transmit staffing, equipment and related logistics information to distant Emergency Operations Centers and receive specific incident briefing and assignment data potentially long before arriving at the scene, potentially saving hours of initial downtime at incident staging areas. Data interoperability provides the potential for Internet-like communications between any terminal or group of terminals, provided the addresses of the terminal(s) are known and such access is permitted by administrative authorities and policy.

Ten years later, after changes in administrations, bureaucracies, consultants, and continued government studies—as well as the added legacy of many terrorist attacks and natural tragedies—voice communications again became the focus of interoperability planning in the United States [4]. Finally in 2008, data communication also was again determined important to the understanding of interoperability [5].

The 1996 work, although accurate with respect to the consequences of failed interoperability, greatly understated the root interoperability problems from which all others are derived [6]:

> Interoperability (or the lack thereof) is often affected by non-technical factors including reluctance to adopt new approaches and funding limitations. Contending with the human factor is another critical element in achieving interoperability.

We will show that "the human factor" is *the* critical element in achieving interoperability and that all operational, planning, spectrum, and technology factors are completely dependent upon this human factor. In 2006, the human factors that had been identified in official documents ten years earlier, once again were emphasized in documents attempting to provide interoperability education [7].

1.2 Definition and Vision

This book will address interoperability on a demographic and geographic scale that ranges from local to international. We will describe *how* the various stakeholders can achieve more effective interoperability in public safety and emergency response information technology (IT) and communication systems. Therefore, our first objective is to define the term *interoperability* in a clear way for policy makers, individuals that constitute public safety and emergency responder agencies, and leaders at all levels of government. This shared understanding among stakeholders is essential for the collaboration necessary for interoperability
Several categories of collaboration, within and among public safety and emergency response agencies, have been identified as follows [8]:

4.1.3.1 Modes of Routine Intra-agency Operability

1. Normal operations within a discipline;
2. Communications that are rehearsed and practiced every day;
3. Day-to-day patrols and duties, and responses to dispatches from emergency call centers;
4. Task force operations within a discipline or agency for a specific mission.

4.1.3.2 Modes of Interagency Interoperability

1. Day-to-Day:
    - Communications that are rehearsed and practiced every day;
    - Routine operations with neighboring agencies to provide support or backup;
    - It is estimated that this form of interoperability makes up 90% of an individual first responder's multiagency activities.
2. Task Force:
    - Cooperative effort among mixed yet specific agencies and disciplines;
    - Extensive preplanning with practice;

- Operations that are planned or scheduled and are proactive;
- Operations that have a common goal, common leader, and common communications.

3. Mutual Aid:
   - Major event that causes a large number of agencies to respond and requires considerable coordination;
   - Major event that requires response from multiple jurisdictions from the local level to the state and national level;
   - Communications that operate under a state or regional mutual-aid pact;
   - Operations that are usually not planned or rehearsed, but which react to the situation.

In recent years, the United States and Europe have made significant investments to improve public safety communications interoperability, particularly with regards to radio communications. Due to this narrowed focus, several elements of our interoperability definition will ultimately be derived from this radio-focused source and subtly alter its meaning. One pervasive definition of interoperability found in the United States describes interoperability as follows [9]:

> the ability of public safety agencies to talk across disciplines and jurisdictions via radio communications systems, exchanging voice and/or data with one another on demand, in real time, when needed.

This limited definition represents efforts within the United States to address inadequate communications coordination between the operations of public safety and emergency response organizations, but it narrows the scope of the problem to "radio" and "exchanging voice and/or data." More importantly, it mistakenly focuses attention on curing the apparent lack of radio interoperability, which is only a *symptom* of the real problem, instead of concentrating on the need to identify and fix the underlying root *cause*: the human factor. Radio interoperability would never have become an issue in public safety or emergency response had its fundamental cause not been a driving factor. Accordingly, we focus on the techniques needed to eliminate the cause, in order to minimize the symptoms of that cause and achieve true interoperability in public safety and emergency response IT and communication systems.

By broadening the definition of interoperability to reflect all its essential elements, we can better understand the interoperability challenge addressed in this book.

As early as 1990, the Institute of Electrical and Electronic Engineers defined interoperability as [10]:

> The ability of two or more systems or components to exchange information and to use the information that has been exchanged.

This definition broadens the scope beyond "radio"; implying all forms of "exchange" are included. It also references another important term, namely *information*, not just "voice and/or data." Perhaps more importantly, it includes the phrase "to *use* the information that has been exchanged [italics added]."

A further refinement of the definition of "command and control," then identify the requisite communications as follows [11]:

> Public safety operations follow a command and control hierarchy that allows public safety personnel to work seamlessly on situations that may begin small, but can evolve into large incidents requiring many resources and assistance from numerous jurisdictions. As an incident grows in magnitude, the IC [Incident Commander] has to know what resources and capabilities are becoming available for use. Each of the first-responder disciplines may have its own branch commander at a large incident, and these branch commanders must be able to coordinate, communicate, and share information with the overall IC.
>
> The communications systems that support these operations must also be capable of the same command and control features, as follows:
>
> 1. *Incident Command System (ICS):* The communications systems must support the agency's incident command policies.
> 2. *System administration of users:* The communications systems must allow authorized system administrators, as well as incident and branch commanders, to establish user profiles for network access and usage, depending upon the role that the public safety user is asked to satisfy during an incident.
> 3. *User identification and location:* The communications systems must provide user identification to others during communications and, when required, must provide user geolocation information to ICs and other authorized resources. In addition, the system must allow communications and information sharing between IC or unified command operations with an emergency operations center (EOC). In some instances, the EOC requires on-demand access to GIS-based displays, video, and communications as they are occurring at the incident.

There is a priority structure to this communication framework. Priority levels for information exchange can be found in the public safety communications priorities established by the Association of Public Safety Communications Officials (APCO) in their Project 16 trunked radio specifications. These priority levels are defined as follows [12]:

> Emergency–Priority Level 1. Every subscriber unit must have an "emergency button" to enable the subscriber to gain immediate channel access for exclusive use in safety-of-life situations. The system dispatcher shall receive this highest priority emergency call within one-half second of the emergency button push.
>
> Incident–Priority Level 2. If public safety mission requirements increase safety-of-life risk, albeit not to the immediate personal emergency level, the second precedence level or below shall be available for activation. Either a dispatcher (at a console) or a mobile subscriber can activate this mission-level priority via switch activation on their portable or mobile radio. When this precedence level—or below—is activated, the system shall provide interconnection with some or all dispatchers (depending on the specific implementation) as well as others in the associated talk group. High-speed chases, exchanges of gunfire, and major fires may require activation of this mission priority. Once activated, this precedence level must remain in effect until a dispatcher restores "normal" operation, that is, regu-

> larly assigned precedence levels. Only the Emergency Access (Priority 1) takes precedence over this mission-level priority.
>
> Command–Priority Level 3. The executive (command) or supervisory personnel have this third priority level, which also may incorporate privacy features. The specifications state that subscriber devices used by command or supervisory personnel may be assigned this Priority 3 level for their normal operating mode. This "optional" capability is suggested as many command personnel may require this added precedence for higher reliability channel assignments as they provide direction for those under their authority.
>
> Operational–Priority Level 4. This priority level is intended for public safety subscribers in systems used to support nonpublic safety or administrative functions. In this way, a public safety subscriber would preempt a nonpublic safety subscriber as a matter of course during day-to-day operations, not considered Level 3 or above.
>
> Routine–Priority Level 5. This lowest priority level is intended for nonpublic safety or administrative users in public safety systems for their routine operations. For example, vehicle maintenance services at a police station may have a channel with routine priority to provide communication between motor pool subscribers and dispatchers at the garage for coordination purposes. Obviously, this function would be lower in priority to activities associated with the higher levels of priority described above. As stated in the Emergency–Priority Level 1 case, all subscribers—even those with access only to Routine Priority—can execute the Emergency–Priority Level in the event of a safety-of-life situation.

These priorities are important from equipment selection and systems integration perspectives because they determine the required communications capacity. The priorities define the number of independent communication paths (or channels) required, the necessary bandwidth (i.e., effective information or data rate) of those channels and, therefore, the size and cost of the necessary equipment. In addition, they lead to specifications for the individual wireline and wireless communication links known as needlines, the necessary built-in redundancy of those needlines, and the need for message acknowledgments. As the final report of the Public Safety Wireless Advisory Committee (PSWAC) states [13]:

> The basic command level and subsequent lower command levels must have pre-designated (and named) channels associated with those levels. Lower levels can be more flexible and dynamic. Understanding the operational characteristics does not complete the solution, but once they are defined, the correct technical solutions can then be applied.

Thus, once these communication "channels" are defined, communications capabilities must be selected to support the necessary priorities [14].

> The success of operations ... depends on the availability and redundancy of critical communications systems to support connectivity to internal organizations, other departments or jurisdictions, critical customers, and the public. When identifying communications requirements, organizations should take maximum advantage of the entire spectrum of communications media likely to be available in any emergency situation. These services may include, but are not limited to secure and/or

non-secure voice, fax, and data connectivity; internet access; and e-mail.

Interoperable communications should provide the following:

- Communications capability that adequately supports the organization's essential incident support functions and activities;
- Ability to communicate with all staff, management, and other organizational components;
- Ability to communicate with other responding organizations and with emergency personnel; and
- Access to other data and systems necessary to conduct essential incident mitigation functions.

According to the SAFECOM Statement of Requirements (SAFECOM SoR) [15]:

> The PS [Public Safety] SoR focuses on the functional needs of public safety first responders—Emergency Medical Services (EMS) personnel, firefighters, and law enforcement officers—to communicate and share information as authorized when it is needed, where it is needed, and in a mode or form that allows the practitioners to effectively use it. The communications mode may be voice, data, image, video, or multimedia that includes multiple forms of information.

For this reason, the definition of interoperability is expanded to all communications media and modes, essentially "bandwidth on demand" [16].

> Communications interoperability allows emergency management/response personnel and their affiliated organizations to communicate within and across agencies and jurisdictions via voice, data, or video in real time, when needed, and when authorized. It is essential that these communications systems be capable of interoperability, as successful emergency management and incident response operations require the continuous flow of critical information among jurisdictions, disciplines, organizations, and agencies.

In this regard, the shared situational awareness as provided by a common operational picture (COP) is of no value unless it can be shared and used by the appropriate command elements [17].

> Effective emergency management and incident response activities rely on flexible communications and information systems that provide a common operating picture to emergency management/response personnel and their affiliated organizations. ... Properly planned, established, and applied communications enable the dissemination of information among command and support elements and, as appropriate, cooperating agencies and organizations.
>
> Incident communications are facilitated through the development and use of common communications plans and interoperable communications equipment, processes, standards, and architectures. During an incident, this integrated approach links the operational and support units of the various organizations to maintain communications connectivity and situational awareness. Communications and information management planning should address the incident-related

policies, equipment, systems, standards, and training necessary to achieve integrated communications.

A recent and comprehensive survey of the various definitions of interoperability concluded [18]:

> For nearly thirty years, both government and industry have actively explored research on interoperability measurement with the goal of creating a straightforward way of measuring, reporting, then improving the interoperability of complex networks of people, equipment, processes and organizations. Researchers have created frameworks and models, proposed measures, described levels, and listed a variety of qualitative factors in support of an interoperability measure. Within extant interoperability research, the authors' research has uncovered nearly three dozen definitions of interoperability, over five dozen distinct types of interoperability, numerous interoperability attributes, and fourteen foundational interoperability measurement models and methodologies. At least eleven research groups have been the centers-of-gravity for interoperability measurement research.

After reviewing 34 definitions, these authors concluded that interoperability encompasses people, equipment, processes, and organizations and has been extensively studied [19]. Several of these definitions include a key element of the aforementioned IEEE definition, that is, the exchanged information is used to achieve some common goal. In this book, the common goal will be stated generally as one or more of the following stages of securing the homeland:

- *Prevent*—through intelligence, sensors, or other predictive means, be able to interdict manmade threats and reduce or eliminate the impact of natural events.
- *Detect*—if prevention fails, identify and assess any manmade threat or natural disaster—as well as its potential impacts—as soon as possible following its evolution or instantaneous occurrence.
- *Protect*—using in-place or executable resiliency measures, maximize defensive measures to minimize the initial impact of manmade incidents or natural events.
- *Alert*—provide instantaneous and verifiable alerts and warnings with action-oriented situational awareness, that is, the alert provides a recommended course of action.
- *Respond*—take action to minimize scope, protect life and property, mitigate (manmade or natural) cause, and maintain situational awareness.
- *Recover*—restore/improve preevent conditions and reestablish preparedness.

Shared information provides critical situational awareness and command coordination by responders-receivers using all forms of communications, so the information must be *trusted* by those who receive it. In other words, interoperability requires that *people* will *trust* the information exchanged according to established *processes*. This element of trust does not occur explicitly in any of the 34 definitions identified in the interoperability study described above. Trust is arguably assumed in the military domain, but it is also a critical factor in civilian public safety and emergency response. Trust cannot be assumed in the civilian domain. For these reasons, trust will be prominent in the comprehensive definition of interoperability we use throughout this book.

The authorized on-demand multimode and multimedia definitions of interoperability are a major improvement over the radio voice-only version, but still do not adequately define interoperability. Ultimately, interoperability must provide the same confidence of coordinated action among disparate agencies that exists, for example, among a team of firefighters. A team of firefighters act as a fully coordinated unit because when they enter the fireground, they share situational awareness and know how each team member would react to a given command. To this end, we must further broaden our definition of interoperability to include the following [20]:

> A measure of the degree to which various organizations or individuals are able to operate together to achieve a common goal. From this top-level perspective, interoperability is a good thing, with overtones of standardization, integration, cooperation, and even synergy.
>
> Interoperability specifics, however, are not well defined. They are often situation-dependent, come in various forms and degrees, and can occur at various levels—strategic, operational, and tactical as well as technological. They are also far more likely to be recognized when interoperability problems emerge and taken for granted when such problems do not.

The key words here are *measure*, the focus on organizations or *individuals*, and their ability to *operate together* to achieve a *common goal*. In addition, this definition addresses strategic, operational, tactical, and technological elements—defining the scope of planned action and capabilities that the interoperability measure must consider. This definition also points out that interoperability is apparent in its absence, but otherwise overlooked. This statement shows the difficulty of establishing a quantitative measure other than the simple binary rating of interoperable or *not* interoperable.

Rand Europe is an independent not-for-profit research institute with the mission to help improve policy and decision-making through research and analysis. Rand has refined many specific underlying elements of interoperability that must be incorporated into our efforts [21]:

> Interoperability is a "cross-cutting concern" which must be implemented pervasively if it is to be effective. Yet because it is cross-cutting, it is typically considered out of scope by individual projects or programmes, since it involves interaction with other projects or programmes that are outside the boundary of the given effort. It must therefore be motivated and governed by mechanisms that transcend normal project and programme boundaries.

A critically important term worth noting is *pervasive*, denoting that all organizations, entities, and individuals involved in achieving the common goal must be included. Here the Rand authors illuminate a key reason for failure of government and commercial programs and projects to address interoperability: it is often "out of scope" of individual projects. Often, higher (strategic) levels of government have failed to demonstrate the ability to establish the necessary overarching IT and communications interoperability framework for these programs and projects, perhaps not even recognizing that it is their responsibility to do so.

The Rand Europe authors define several of the key practical elements needed to achieve interoperability [22].

> In order to interoperate meaningfully, organizations must have compatible data definitions and interpretations, terminology, business processes, organizational cultures, and policies concerning privacy, access, transparency, accountability, etc. Achieving true interoperability therefore requires the alignment of these semantics across all interacting organizations and their systems. Such alignment requires each organization, sector, and community of interest to define and codify its own semantics and to establish appropriate correspondences with the semantics of any other such entities with which it interacts. This in turn requires cross-organizational, cross-sector, and cross-community interaction and negotiation. Because this process is labour intensive and involves significant intellectual effort, it requires considerable lead time and so should be started as early as possible.

The Rand Europe authors recognized that there must be shared understanding of exchanged information. Even this advanced characterization of interoperability arguably falls (just) short of including trust, which is a fundamental element of holistic interoperability. In other words, the individuals and entities (via formal agreement) will *trust* the shared information to the point of using it, that is, acting on it, in a coordinated way. The individuals and their entities must be willing to operate together, or collaborate, through the exchange of trusted information. Such exchange varies as follows:

- Moment-to-moment basis during tactical operations, such as a law enforcement task force sweep, a multialarm fire or wildfire, or a mass-casualty incident, in which voice, data, and video are shared.
- Day-to-day basis, such as awareness of ongoing operations in the region, local minimal incidents and responding entities, recent incidents or events, upcoming weather, community events, and so forth.
- Continuum of time operations repeated at intervals from day-to-day through weekly, monthly, yearly, and beyond (e.g., 100-year storm).

Compiling the above-mentioned elements of our interoperability definition into one phrase suggests that interoperability measures the extent of *shared trusted understanding* among public safety and emergency responders-receivers collaborating (i.e., operating together) to achieve a common goal. To clarify this definition, we must define each of the terms used within the definition. By collaboration, we mean engaging in *predictable, collaborative actions* to achieve a common goal. By *predictable*, we mean that each responder-receiver can forecast the specific actions of those responders-receivers with whom they have achieved shared, trusted understanding. The responders-receivers span many areas:

- Public safety and emergency response disciplines with the associated diverse goals, objectives, operating techniques, semantics, tools, training, and so forth;
- Levels of government, commercial, nonprofit, and volunteer organizations;
- Diverse capabilities, responsibilities, operating procedures, semantics, priorities, policies, and resources.

Finally, we arrive at the following definition for the purposes of this book:

> Interoperability is a measure of *shared trusted understanding* that drives *predictable coordinated action* among responders-receivers collaborating to achieve a common public safety and emergency response goal.

1.3 Planning Elements of Interoperable IT and Communication Systems

A report of the U.S. Public Safety Wireless Network program [23], surveyed the best planning practices of Mississippi, West Virginia, and Tennessee in the United States to extract the best practices from their planning efforts [24]:

> This report captures the key findings of the strategic approaches that were used by each of these three states. Although this report only highlights the strategic approaches of the states of Mississippi, Tennessee, and West Virginia, the actions, strategies, and best practices can be tailored for any state or region seeking to improve interoperable communications. The actionable best practices derived from the analysis of the strategy-based assistance provided to these states included:
>
> - [1a] Comprehensively develop political and stakeholder support;
> - [1b] Centralize coordination among multiple agencies through a formal committee;
> - [1c] Formalize coordination committees through executive orders, charters, or memorandums of understanding;
> - [1d] Emphasize coordination, partnership, and asset sharing;
> - [1e] Conduct detailed analysis of current needs and capabilities while planning future developments;
> - [1f] Prepare and provide a wide range of educational materials to stakeholders and decision-makers;
> - [1g] Sponsor communications and interoperability forums where officials can learn about current challenges and plans, provide input into the process, or learn how to get involved;
> - [1h] Solicit input from all interested parties or entities throughout the coordination, planning, and project processes;
> - [1i] Examine the successful strategies of similar states or regions.

The number and letter in the "[ ]" (square brackets) within the quote have been added in the quoted material to identify each bulleted item in the analysis presented at the end of this section.

A SAFECOM report [25] drew input from 18 states [26] to capture "best practices" for shared-system development and explained the approach to their effort as follows:

> Often the best place to start an effort as complex as this one is to collect insights and practices of other states and localities that have cleared their own path toward interoperability. This report profiles eighteen different states' planning efforts. The research highlights states' and localities' experiences with collaboration and participation across stakeholder groups. The information sought in the states' efforts included:
>
> - [2a] Governance,
> - [2b] Planning,
> - [2c] Unique state characteristics impacting interoperability,
> - [2d] Identified measures of success,
> - [2e] Issues and challenges faced,
> - [2f] Critical success factors.
>
> The purpose of this report is to provide examples of individual planning efforts complemented by a brief analysis of how some of these efforts have been successful and fit within SAFECOM principles. By taking this approach, one might gain a sense of the broad, strategic perspective called for when initiating interoperable communication planning efforts at the local and state levels [27].

Again, the number and letter in the "[ ]" (square brackets) has been added in the quoted material to identify each bulleted item in the analysis below.

The PSWN program report went on to summarize its findings on critical planning success factors in shared system development [28].

> Several critical success factors have been identified in the shared systems approach. A general consensus regarding the best overall approach emerged—it is one that combines effective top-down strategies (to build senior decision-maker support) with strong "grass roots" initiatives (to establish and sustain support among radio managers and users). The specific critical success factors included:
>
> - [3a] Approaching the system design from a customer/user perspective;
> - [3b] Committing to long-term interagency partnerships by opening the system up to all;
> - [3c] Developing detailed system designs and specifications by clearly defining operational needs and requirements with the vendor and participating agencies;
> - [3d] Endorsing the mission uniqueness of agencies by accommodating agency-specific requirements and ensuring the necessary autonomy through concerted network management;
> - [3e] Recognizing and pursuing cost advantages of joint service agreements or of building joint systems (achieving certain economies that are not available to smaller, single-agency, single-jurisdiction systems);
> - [3f] Training participating users to handle and use the equipment correctly, and ensuring that training actually occurs;

- [3g] Organizing and energizing core, high-stake users to engage in strong "grass roots" education and, where appropriate, lobbying efforts to increase citizen awareness and senior-level political support;
- [3h] Leveraging off of concern on the part of radio managers and users that if they did not join in the shared system they would have diminishing support and resources for maintaining their own systems;
- [3i] Developing, encouraging, and capitalizing upon strong user support for change, and for systems that provide a more capable suite of services in a manner that is more seamless than is currently the case ...;
- [3j] Establishing and maintaining strong administrative support at the highest levels of government involved (e.g., from the Governor for statewide systems, from county board chairpersons for multicounty systems) to champion the system through the legislative process and through local zoning and approval hearings;
- [3k] Building and leveraging strong support from state legislatures through the passage of enabling legislation and the establishment of sustainable sources of funding;
- [3l] Securing long-term and ongoing funding mechanisms that will endure throughout the planning, implementation, and maintenance stages of the system;
- [3m] Addressing regulatory, licensing, and coordination issues, ...;
- [3n] Developing common operational procedures among the partners ensuring that radios are programmed properly and a common "language" is used by all of the different partners on the system;
- [3o] Keeping the lines of communication open among all of the partners on the shared system;
- [3p] Defining partnership responsibilities. ... Certain shared systems arrangements require that partners give at least a one-year notice before exiting the shared system;
- [3q] Establishing a standardized fee structure for subscribers to the system;
- [3r] Developing a detailed Memorandum of Understanding (MOU) that describes the purpose and intent of the shared system, defines the users, defines the owner/operator responsibilities, creates a governing board, establishes methods of interoperability, sets the subscriber fee structure, and ensures that all partners on the system are treated equally [29].

When planning shared systems, the next step is to categorize these best practices and critical success factors from PSWN into a minimal set of interoperability planning critical success factors. From these three PSWN documents, we have defined six interrelated holistic interoperability planning success factors as follows:

- *Governance*—includes all aspects of the governance structure mandated by the leadership to empower stakeholders to use, improve, and implement the holistic interoperability defined above.
- *Resources*—all aspects of resources and resource planning for interoperability architecture, including funds for acquisition and sustainment (operation and maintenance), people, and infrastructure.

- *Approach*—all the best practices needed to design interoperable capability, such as organizational development and effectiveness, business case analysis, enterprise architecture, cost/benefit analysis, systems engineering, program and project management, funding acquisition strategy, standards development, and so forth.
- *Architecture*—this category includes the critically important integration between people and technology in the interoperability architecture, including:
  - All operational aspects (e.g., expertise, experience, standard operating procedures/guidelines and timelines, integrated training and exercises, etc.), and related operational standards, intended to "prevent, detect, protect, alert, respond, and recover."
  - All physical facilities, including structures and hardware/software systems, physical infrastructure (location, security, facilities, networks, etc.), IT/communications equipment and infrastructure and related technical standards, necessary to implement trusted shared understanding.
- *Performance*—the detailed quantitative performance requirements for interoperability to be achieved, such as speed, accuracy, reliability, sustainability, and resiliency.
- *Acceptance*—all aspects of getting stakeholder "buy in" from all levels of government leadership and the community, as well as the responder-receivers, their planning personnel and leadership, the "press," and the public—all to establish shared trust.

Tables 1.1, 1.2, and 1.3 each show matrices with the critical success planning factors defined above mapped to the corresponding (indexed) bulleted planning factors from the three referenced PSWN reports. Note that an indexed referenced factor may be mapped to more than one of the six success planning factors, which is an indication of the interrelationships between these factors. In this way, we make sure our planning factors address all best-practice recommendations derived from PSWN reviews conducted for more than 20 state shared-system programs.

From a review of these three tables, it is apparent that the six success-planning factors of governance, resources, approach, architecture, performance, and acceptance encompass all PSWN-determined best practices for interoperability planning. The presentation order of these factors has no significance, but failure to address

Table 1.1 Mapping Interoperability Planning Success Factors to Factors from the *Statewide Strategy Best Practices Report*

Success Planning Actors	*1a*	*1b*	*1c*	*1d*	*1e*	*1f*	*1g*	*1h*	*1i*
Governance	•	•	•	•	—	•	—	—	•
Resources	—	—	—	•	—	•	•	—	•
Approach	—	•	—	—	•	—	•	•	•
Architecture	—	—	—	•	•	—	—	•	•
Performance	—	—	—	—	•	—	—	—	•
Acceptance	•	•	•	•	•	•	•	•	•

Table 1.2 Mapping Interoperability Planning Success Factors to *Review of Statewide Interoperability Planning Efforts Across the Country*

Success Planning Factor	*2a*	*2b*	*2c*	*2d*	*2e*	*2f*
Governance	•	—	—	—	•	•
Resources	—	•	—	—	•	•
Approach	—	•	—	—	—	•
Architecture	—	—	•	—	•	•
Performance	—	—	—	•	—	•
Acceptance	—	—	•	—	•	•

Table 1.3 Mapping Interoperability Planning Success Factors to Factors Presented in the *PSWN Program Symposium Compilation Report: August 1997–December 1999*

Success Planning Factors	*3a*	*3b*	*3c*	*3d*	*3e*	*3f*	*3g*	*3h*	*3i*	*3j*	*3k*	*3l*	*3m*	*3n*	*3o*	*3p*	*3q*	*3r*
Governance	—	•	—	—	•	—	—	—	—	•	•	—	•	—	•	•	—	•
Resources	—	—	—	—	•	—	•	•	—	•	•	•	—	—	—	—	•	•
Approach	•	•	•	•	—	•	•	•	—	•	—	•	•	•	•	•	—	•
Architecture	•	•	•	•	—	•	•	—	•	—	—	•	—	•	—	—	—	•
Performance	—	—	—	•	—	•	—	•	•	—	—	—	—	•	—	—	—	—
Acceptance	•	•	•	•	•	•	•	•	•	•	•	•	•	•	•	•	•	•

any one of them will mean failure in the result. All of these factors are interrelated; *interoperable IT/communications systems require interoperable planning.*

In the development of the Statewide Communications Interoperability Plans (SCIPs) in the United States, required for assignment of 2007 Public Safety Interoperable Communications (PSIC) grants, the concept of an Interoperability Continuum [30] (see Figure 1.1) was put forward as a means to assess the heuristic level of interoperability.

As defined in the continuum brochure, the key elements of interoperability are:

> [4a] Governance
> Establishing a common governing structure for solving interoperability issues will improve the policies, processes, and procedures of any major project by enhancing communication, coordination, and cooperation; establishing guidelines and principles; and reducing any internal jurisdictional conflicts. ...
>
> [4b] Standard Operating Procedures
> Standard operating procedures [(SOP)]—formal written guidelines or instructions for incident response—typically have both operational and technical components. ...
>
> [4c] Technology
> Technology is a critical tool for improving interoperability, but it is not the sole driver of an optimal solution. Successful implementation of data and voice communications technology is supported by strong governance and is highly dependent on

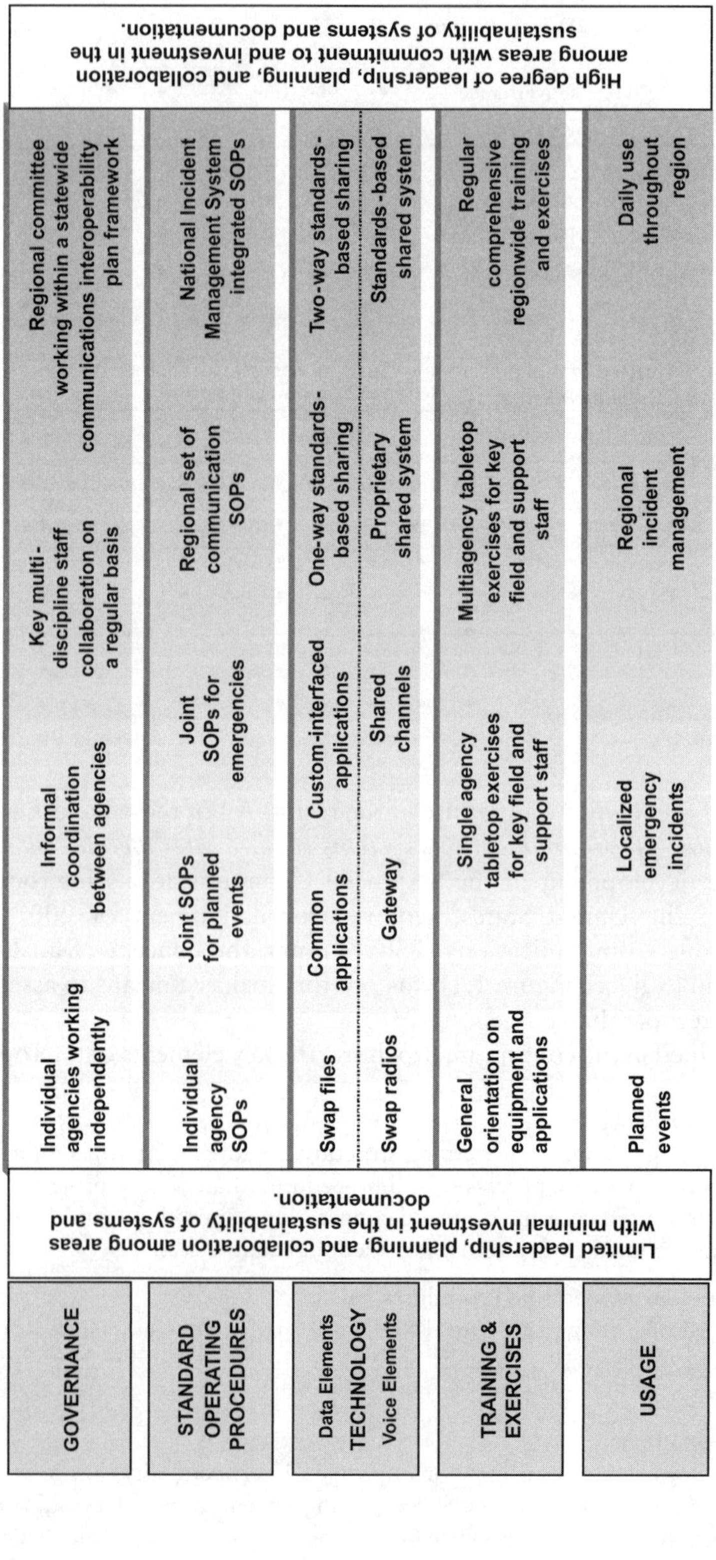

Figure 1.1 The SAFECOM interoperability continuum.

effective collaboration and training among participating agencies and jurisdictions.
...

[4d] Training and Exercises
Implementing effective training and exercise programs to practice communications interoperability is essential for ensuring that the technology works and responders are able to effectively communicate during emergencies. ...

[4e] Usage
Usage refers to how often interoperable communications technologies are used. Success in this element is contingent upon progress and interplay among the other four elements on the Interoperability Continuum. ...[31]

All of these continuum elements are mapped to the PSWN-derived planning success factors in Table 1.4.

The above table shows that both the PSWN success factors and SAFECOM continuum emphasize the critical importance of *governance* in assuring all stakeholders have a "place at the table" in developing interoperability. Governance impacts the distribution of resources, and so we have counted continuum governance ([4a]) towards this planning factor.

SOPs ([4b]) are part of the operations portion of architecture, or operational architecture, defining personnel interaction with technology and each other in the interoperability architecture. SOPs also determine *performance* of that architecture, because the SOP artifacts define how the integrated people, processes and tools, and technology will behave predictably in a collaborative environment.

The continuum *technology* ([4c]) category is limited to voice and data communications, whereas our interoperability success planning factors include not only radio communications, but all elements of IT and communications interoperability. This broader technical approach is necessary given the shrinking separation between IT and communications technologies and also since we have adopted, as part of holistic interoperability, the element of "understanding" of information in addition to the "exchange" of that information. Of course, all IT and communication systems—such as a common operational picture as a vital form of information sharing—should be included in the technology category. Thus, the continuum technology category is *included* in the PSWN-derived *architecture* success planning factor. As the table shows, the continuum technology factor also reflects *resources*, because of the cost of technology and use of best-practices *approach* (see Chapter 4). The approach factor, not identified in the continuum, is critical because

Table 1.4 Mapping Interoperability Planning Success Factors to Interoperability Elements in the SAFECOM *Interoperability Continuum*

Planning Factor	*4a*	*4b*	*4c*	*4d*	*4e*
Governance	•	—	—	—	—
Resources	•	—	•	—	•
Approach	—	—	•	—	—
Architecture	—	•	•	•	•
Performance	—	•	—	•	•
Acceptance	—	•	—	•	•

best-practice solution development processes are essential to proper development of IT and communication systems for holistic interoperability. Again, *interoperable systems require interoperable planning*.

The *training and exercise* category is critical to achieving interoperability. Ideally, the holistic interoperability capabilities to be provided are used on a moment-to-moment or day-to-day basis as part of routine public safety and emergency response work. In that way, users' familiarity with the specific technologies is continuously taught through experience and exercised during real incidents. As the scope of the incidents increase, many responders-receivers may be added, requiring training and exercise planning for these unique events. Ideally, all contributors would have a common capability and will perform predictable actions.

Finally, the *usage* category is integrated with day-to-day use of the technology and SOPs—recognizing that moment-to-moment or day-to-day use produces the best familiarity with interoperability. Resources are needed to support this frequent *usage*, while user acceptance of interoperability capabilities is fundamentally important. The continuum does not directly address user acceptance as a success-planning factor.

1.4 Holistic Interoperability

Holistic interoperability is a measure of cause and effect; the cause must be *shared trusted understanding* and the effect is *predictable collaborative action* to achieve a common public safety and emergency response goal. This cause and effect results from establishing and sustaining the proven critical interoperability planning success factors of governance, resources, approach, architecture, performance, and acceptance. We show in Chapter 2 that many leadership and planning deficiencies prevent these success factors from being established or sustained. We then postulate a worst-case World Interoperability Failure Model to illustrate the impact of these deficiencies on information sharing, and define an Interoperability Assessment Checklist to investigate whether or not potential deficiencies exist in any agency or among agencies. We also describe best practices essential to establishing and sustaining these planning success factors to achieve interoperability in critical public safety and emergency response IT and communication systems.

References

[1] Public Safety Wireless Advisory Committee, *Final Report of the Public Safety Wireless Advisory Committee, Vol. I, Executive Summary*, September 11, 1996, p. 2, (hereafter referenced as the *PSWAC Report*).

[2] Ibid.

[3] *PSWAC Report, Appendix C—Interoperability Subcommittee Final Report*, p. 144 (*PSWAC Report* p. 418).

[4] *PSWAC Report*, p. 19.

[5] SAFECOM program, *Interoperability Continuum: A Tool for Improving Emergency Response Communications and Interoperability*, U.S. Department of Homeland Security, August 2006, p. 2, http://www.safecomprogram.gov/NR/rdonlyres/65AA8ACF-5DE6-

428B-BBD2-7EA4BF44FE3A/0/Continuum080106JR.pdf, (hereafter referenced as the *SAFECOM Voice Continuum*).

[6] SAFECOM program, *Interoperability Continuum: A Tool for Improving Emergency Response Communications and Interoperability*, U.S. Department of Homeland Security, May 2008, p. 3, http://www.safecomprogram.gov/NR/rdonlyres/54F0C2De-FA70-48DD-A56E-3A72A8F35066/0/Interoperability_Continuum_Brochure_2.pdf, (hereafter referenced as the *SAFECOM Voice and Data Continuum*).

[7] Ibid.

[8] SAFECOM program, *Public Safety Statement of Requirements for Communications and Interoperability*, Vol. I, Ver. 1.2, U.S. Department of Homeland Security, October 2006, pp. 21–22, (hereafter referenced as the *SAFECOM SoR*).

[9] SAFECOM program, *Public Safety Statement of Requirements for Communications and Interoperability*, Ver. 1.0, U.S. Department of Homeland Security, March 2004, p. iv.

[10] Institute of Electrical and Electronics Engineers, *IEEE Standard Computer Dictionary: A Compilation of IEEE Standard Computer Glossaries*, New York, 1990.

[11] SAFECOM program, U.S. Department of Homeland Security, *Public Safety Statement of Requirements for Communications and Interoperability*, Vol. I, Ver. 1.2, pp. 23–24.

[12] Carlberg, K., et al., *Preferential Emergency Communications: From Telecommunications to the Internet*, Kluwer Academic Publishers: Boston, June 2003, pp. 62–63.

[13] *PSWAC Report, Appendix C—Interoperability Subcommittee Final Report*, p. 56 (*PSWAC Report*, p. 330).

[14] Federal Emergency Management Agency, *Interim Guidance on Continuity of Operations Planning for Sate and Local Governments*, Department of Homeland Security, May 2004, p. A–29.

[15] *SAFECOM SoR*, p. viii.

[16] U.S. Department of Homeland Security, "Communications and Information Management," *National Information Management System, Component II*, December 2008, p. 24.

[17] Ibid., p. 23.

[18] Ford, Thomas C., et al., "A Survey on Interoperability Measurement," *Adapting C2 to the 21st Century, International Command and Control Research and Technology Symposium*, Newport, RI, June 2007, p. 2, see http://www.dodccrp.org/events/12th_ICCRTS/CD/html/papers/096.pdf.

[19] Ibid., p.4.

[20] Hura, Myron, et. al., "A Broad Definition of Interoperability," in *Interoperability: A Continuing Challenge in Coalition Air Operations*, Rand Corporation, 2000, see http://www.rand.org/pubs/monograph_reports/MR1235/.

[21] Rothenberg, Jeff,, M. Botterman and C. van Oranje-Nassau, "Towards a Dutch Interoperability Framework: Recommendations to the Forum Standaardisatie," Prepared for the *Forum Standaardisatie*, The Netherlands, Rand Europe, 2008, p. xiv.

[22] Ibid.

[23] Derived from Science Applications International Corporation, "2.0 Methodology," *Regional Public Safety Dispatch, in the Northeast Homeland Security, Planning Region, Findings, Recommendations, and Implementation Plan*, Metropolitan Area Planning Commission, Boston, MA, December 26, 2005, pp. 19–23.

[24] Public Safety Wireless Network (PSWN) Program, *State and Local Interoperability Assistance Support—Statewide Strategy Best Practices Report*, March 2003, p. ES-1.

[25] SAFECOM program, U.S. Department of Homeland Security, *Research Report: Review of Statewide Interoperability Planning Efforts Across the Country*, April 2004.

[26] The state practices evaluated in the study were Alaska, California, Delaware, Florida, Idaho, Indiana, Maryland, Missouri, Montana, Nebraska, New York, Oregon, Pennsylvania, South Dakota, Tennessee, Washington, Virginia, and Texas.

[27] SAFECOM program, Ibid., p. 3.

[28] These recommendations were summarized after conducting symposia in Charlotte, North Carolina; Harrisburg, Pennsylvania; Sacramento, California; Boston, Massachusetts; Chicago, Illinois; Mesa, Arizona; Denver, Colorado; Lansing, Michigan; and Orlando, Florida.

[29] PSWN Program, *PSWN Program Symposium Compilation Report: August 1997—December 1999*, pp. 10–12.

[30] *SAFECOM Voice and Data Continuum*, p. 3.

[31] Ibid., pp. 2–4.

CHAPTER 2

Pearl Harbor, 9/11, and Katrina: Same Lessons Not Yet Learned

The need for true civilian public safety and emergency response interoperability can be deduced from the same reasoning that has driven major national resources into military capabilities to support the warfighter. Many countries have faced ignominious military disasters and insist on not being surprised again. "No more Pearl Harbors" is one expression that not only states this resolve, but largely reflects the investment in all aspects of military capability and sustainability worldwide. Now however, with the advent of chemical, biological, radiological, nuclear, explosive (CBRNE) weapons of mass destruction (WMD) and the increased potency of attacks by political radicals, the world's governments must focus their energy, resources, and best practices on a more holistic civilian public safety and emergency response interoperability.

The U.S. congressional investigation [1] of the surprise attack on Pearl Harbor cited 25 deficiencies exhibited in Washington, D.C. and Hawaii that answer the question "How could this have happened?" One of the greatest lessons *taught* was that [2]:

> One cannot understand the defeat which the United States suffered on December 7, 1941, by attempting to analyze it in terms of economics, sociology, technology, or any other of history's neat pigeonholes. It arose from the nature of the men involved. In our [Senators Homer Ferguson and Owen Brewster] opinion, the evidence before this committee indicates that the tragedy at Pearl Harbor was primarily a failure of men and not of laws or powers to do the necessary things, and carry out the vested responsibilities.

In this chapter, we will map these "failures of men" noted by later Congresses when trying to explain September 11th (hereafter referred to as 9/11) and the aftermath of Hurricane Katrina. Those failures also led people to wonder "How could this have happened?" This mapping will prove that the Pearl Harbor tragic lessons taught on December 7, 1941, and documented in 1946, had not yet been learned by September 11, 2001, or August 29, 2005. This book was written so that the catastrophic losses of people and property, which resulted from failed leadership and planning that continue to this day, were not in vain.

Emphasizing human failures begs the question "Can anything be done about it?" The answer is "yes." The best planning practices described in Chapter 4 and the World Interoperability Success Plan approach in Chapter 7 show that we must first fully define the problem. This chapter outlines the historical obstacles to holistic interoperability in public safety and emergency response in terms of deficiencies found by the U. S. congressional investigation of the surprise attack on Pearl Harbor. We will assess failures of leadership and planning that caused the identical failures in preparedness and response for 9/11 and Katrina. Chapter 3 organizes these deficiencies into a checklist [3] for assessing and planning holistic interoperability. Our purpose is to transform these apparent timeless *lessons taught* into a mechanism to guide the use of best practices planning tools in maximizing interoperability among our responder-receivers and other stakeholders.

As stated in Chapter 1, interoperability is only conspicuous in the consequences of its absence. Review of major interoperability failures in terms of the planning factors described in Chapter 1 will clearly demonstrate these results. To this end, we will present a comprehensive account of major interoperability failures, the deficiencies of leadership that produced these failures, and then analyze these deficiencies in terms of the interoperability planning elements of governance, resources, approach, architecture, performance, and acceptance. Such a detailed account is necessary in order to:

- Place the reader well into the shoes of those present and responsible, so that both insidiously subtle as well as blatant deficiencies can be related to one's own experiences and thought processes.
- Demonstrate how apparent tidbits of information, seemingly unimportant in their content and context, were like the inevitable stillness that seems to herald the onset of violent storms.
- Show how directly pertinent information was dismissed because of the bias of those responsible for protecting their people and assets, thereby demonstrating that jealousy and competition among leaders must be inevitably compensated for by the initiative and self-sacrifice of those being led.

As the reader considers his or her own role in public safety or emergency response systems, imagine how the deficiencies we highlight in this chapter may apply to your own staff, organizations, planners, leaders, vendors, consultants, and bureaucracies at all levels of government. While it is important to examine the actions and inactions that contributed to these overall disasters, the purpose of this study is not to assign blame, but to identify, document, and build process from these lessons learned so that we might prevent future interoperability failures.

2.1 Now and Then: Enduring Failure

The 9/11 attack began sometime after 8:14 a.m. on September 11, 2001. After conducting detailed interviews, the 9/11 investigation stated [4]:

> The most important failure was one of imagination. We do not believe leaders understood the gravity of the threat. The terrorist danger from Bin Laden and al Qaeda was not a major topic for policy debate among the public, the media, or in the Congress. Indeed, it barely came up during the 2000 presidential campaign.

The apparent complacency before 9/11/2001 was arguably founded on the belief that the broad oceans offered protection; that interorganizational collaboration would assuredly supersede personal/bureaucratic interests; and that differences between cultures and religions were interesting from an anthropological standpoint, but would have little impact on the day-to-day lives of U.S. citizens. This sort of complacency was dashed on 9/11, but was somewhat restored again by 2005, when Hurricane Katrina's aftermath again exposed the lack of planning, preparedness, initiative [5], and action in one of the most developed countries.

In both those cases, men and women with imagination saw it coming, warned others, and took action, but were continuously thwarted by bureaucratic inaction. As is so often the case, those responsible for planning for these events, which were repeatedly forecast by subject matter experts, were out of harm's way when disaster struck. Unfortunately, preparation failures resulted in forcing initiative from unsung heroes facing the disaster, and for many others, the ultimate sacrifice.

The failures of U.S. leadership and bureaucracies that allowed the 9/11 surprise attack is not a recent development—it is endemic to human civilization and those who lead. Too much complacency was the prevailing sentiment of many until this fateful moment [6]:

> At 8:46:40, American 11 crashed into the North Tower of the World Trade Center in New York City. All on board, along with an unknown number of people in the tower, were killed instantly.

The congressional report went on to detail its conclusions that "the 9/11 attacks revealed four kinds of failures: in imagination, policy, capabilities, and management" [7]. This was not the first time similar insights were collected and documented in a U.S. congressional report. In fact, 9/11 was not the first time a massive air attack, unimagined by the leadership, brought large-scale death and destruction to America.

2.1.1 Totsugeki Raigeki (TO-RA)

The moment approached when Mitsuo Fuchida knew that they had achieved complete surprise, a testament to the duration and scope of his comrades' superior imagination, policy, capabilities, and management. He could not have known how many failures of interoperability were working to his advantage. The Americans had powerful strategic evidence, timely tactical information, and advanced technology (including radio interoperability). What Americans lacked was holistic interoperability as we've defined it in this book. We show in what follows that had American leaders planned properly to achieve and verify this interoperability, the resulting disaster might have been greatly mitigated.

Near Lahilahi Point the Japanese commander ordered his radioman to signal "To, to, to" to the entire formation about to launch the attack on Pearl Harbor.

"To, to, to" is the first syllable of totsugekiseyo, which means, "to charge"[8, 9]. As he flew about 10,000 feet above Barber's Point and saw that they had achieved complete surprise, he broadcast "To-ra, to-ra, to-ra" at 0753—telling their fleet 200 miles north of Oahu that the U.S. Pacific Fleet had been surprised beyond their collective imagination. Many soldiers and civilians were soon to pay the price for failed interoperability, including those aboard targeted ships in the harbor, air force personnel at Ford Island, Hickam, Wheeler, Kaneohe, Ewa, and Barber's Point fields, and innocent Hawaiian civilians.

We selected the case of Pearl Harbor to illustrate the failure of holistic interoperability for the following reasons:

- Enough time has passed that many in-depth investigations have been performed and analyzed, so that a detailed factual account of each actor's role, conceptions, and misconceptions can be described. Too little is publicly known about 9/11 and Katrina, among others, to enable this type of accounting.
- A day-by-day and sometimes moment-to-moment tracing of interoperability failures can be illustrated to pull the reader into the lives of those proven most responsible *before* this responsibility had, sadly, become the legacy of their lives.
- The military hierarchy of the U. S. Army and Navy organizations in Washington, D.C. and on Oahu was well-structured, less susceptible to arbitrary information hording. Ostensibly, these organizations were at least as confident and well prepared as the Pearl Harbor principles until 0758, December 7th, 1941, just before 0755 when the first bomb hit Ford Island.

Below is a detailed description of the American interoperability failures that contributed directly to the strategic and tactical surprise Commander Fuchida encountered at Oahu. We will analyze this achievement, and the American leadership failures, from the perspective of governance, resources, approach, architecture, performance, and acceptance. This summary analysis of failed interoperability is based exclusively on the seminal works on the Japanese and American history leading to Pearl Harbor in:

> Gordon William Prange, Donald M. Goldstein, and Katherine V. Dillon, *At Dawn We Slept: The Untold Story of Pearl Harbor*, McGraw-Hill Companies, December 1981.

and the subsequent U.S. investigations expertly detailed in:

> Gordon William Prange, Donald M. Goldstein, and Katherine V. Dillon, *Pearl Harbor: The Verdict of History*, McGraw-Hill Companies, Penguin, December 1991 [10].

We are indebted to these authors for research so professionally and thoroughly documented, that we were able to identify the same leadership failures in the 9/11 and Katrina reports. Leaders of military or civilian personnel who require interoperability should read these texts to understand the subtle as well as "barn-door" failures that most likely await them. We recommend these books to the reader, not only to obtain a deeper understanding of failed interoperability, but more importantly, to show the initiative and self-sacrifice of heroes made necessary by failed leadership and planning.

2.1.2 The Environment

Imperialist Japan—seeking ownership of natural resources—had attacked and been at war with China since 1937. The Americans were known to be sympathetic to the Chinese cause and railed against the injustice of the destruction of many thousands of Chinese men, women, and children. Nonetheless, Americans were still isolationist in the wake of their role and sacrifice in the recent Great War in Europe, suspicious of "merchants of death" who promoted military spending, and confident that the oceans and U.S. military protected them. Also, some Americans, believed in their national—if not racial—superiority. President Franklin Delano Roosevelt had embargoed scrap metal, raw materials, and oil to fight an economic cold war against Japan and avoid an unpopular military confrontation.

With the Japanese planning acquisition of a southwest Pacific prosperity sphere, the United States moved its Pacific Fleet from San Diego to Pearl Harbor in 1940, and Japanese military leaders felt their plans were threatened. The decision was made to attack Pearl Harbor immediately.

2.1.2.1 Strategic Warning

The friction between the United States and Japan was well known by anyone reading the newspapers or listening to the radio in the early 1940s—so most Americans and their leadership knew there was a risk of war. Moreover, there were many other signs that something was going to happen.

Foreboding and Forethought Rear Admiral Richmond Kelly Turner, "a brilliant but corrosive man of enormous ego and ambition," [11] wrote a letter to be signed on November 22, 1940, by the Chief of Naval Operations (CNO) Admiral Harold R. Stark, stating that the fleet units in the Hawaiian waters were a likely target of a sudden attack and as they were vulnerable, torpedo nets should be placed within the harbor. This view was dismissed by the then Commander and Chief of the U.S. Fleet in Hawaii, Adm. James O. Richardson, who envisaged torpedoes as being launched from ships or submarines, not aircraft.

This opinion was expressed despite the fact that ten days earlier, the British had dropped specially modified aerial torpedoes from a very low altitude into Taranto Harbor, which was only 40-feet deep (a shallow depth similar to Pearl Harbor), sinking an Italian battleship. The British attack proved the feasibility of a shallow-water aerial torpedo attack and drove the Japanese to solve the shallow-water torpedo-launching problem by ensuring that torpedo aircraft would be included in Commander Minoru Genda's Pearl Harbor attack plan.

Near the end of January 1941, Peru's minister to Tokyo spoke to his friend Edward S. Crocker of the U.S. Embassy in Tokyo of a startling rumor that had been substantiated by many sources, including Japanese. As a result, the U.S. Ambassador to Japan, Joseph C. Grew, composed a dispatch to the U.S. State Department alerting them of this rumor. The dispatch described the rumor's source, and then stated that in the event of escalating hostilities, the Japanese military was planning a massive surprise attack on Pearl Harbor. Although American military officials were made aware of this story, it had no credibility with them—and no one took action to prove or disprove the rumor's credibility [12].

The U. S. Navy also understood the lessons learned from Taranto at the highest levels of government. Turner drafted a letter intended to go to the U. S. Secretary of War, Henry L. Stimson, from Secretary of the Navy, Frank Knox, approved by the CNO, with copies to Richardson and the Fourteenth Naval District at Pearl Harbor. Arriving on February 5, 1941, the letter stated that "if war eventuates with Japan, it is believed easily possible that hostilities would be initiated by a surprise attack upon the fleet or the Naval Base at Pearl Harbor...warrant[ing] taking every step, as rapidly as can be done, that will increase the joint readiness of the Army and Navy to withstand" [13] an air bombing attack and air torpedo plane attack.

Signal Intelligence Fluent in Japanese, an excellent cryptanalyst and intelligence analyst, Commander Joseph J. Rochefort was assigned to be the Officer in Charge of Station Hypo in Pearl Harbor. His responsibility was to break the Japanese Naval communication codes. His job included locating Japanese vessels, but he was unsuccessful with the carriers destined for Hawaii, because "Japan was winning the game of hide-and-seek" [14]. On December 2, Operation Hawaii, the attack plan, was already underway and maintaining radio silence when Admiral Husband E. Kimmel (commander of the U. S. Pacific fleet stationed at Pearl Harbor) heard from Lieutenant Commander Edwin T. Layton, his staff Intelligence Officer, that his staff could not locate two of the front-line Japanese carrier divisions [15]. "Thereupon Kimmel turned his ice blue glaze upon Layton. With somewhat a stern countenance and yet partially with a twinkle in his eye, he demanded 'Do you mean to say that they could be rounding Diamond Head [prominent coastal peak on Oahu] and you wouldn't know it?' To this Layton could only answer lamely, 'I *hope* they would be sighted before now [italics added].'"

Wiretaps Robert L. Shivers, Federal Bureau of Investigation (FBI) in Hawaii, was responsible for cases of subversive activity and espionage, while Captain Irving Mayfield was head of the Navy District Intelligence Office in Hawaii. As there became increasing awareness of one of their five taps, Mayfield feared his consulate taps would be uncovered—potentially leading to Japanese discontent (which the United States wanted to minimize)—so he removed all defensive information operations consulate taps (but missed one). As Prange et al. state [16]: "Here again we find a sad lack of coordination in a vital area when so much depended upon cooperation between U.S. agencies."

Counterespionage The Yugoslavian counterspy Dusko Popov (code name "Tricycle") ostensibly worked for German Military Intelligence. He presumably offered J. Edgar Hoover, the director of the U.S. Federal Bureau of Investigation, an intelligence collecting questionnaire that he surmised meant that the Japanese would attack Pearl Harbor before the end of 1941 if negotiations broke down [17]; however, the director chose not "to transmit the entire questionnaire to the White House and to military and naval intelligence agencies shows both a poverty of judgment on his part and the crippling consequences of rivalry among those government agencies charged with gathering and evaluating information essential to the defense of the United States."

A Plan for Joint Action—The Martin-Bellinger Report A standard defensive planning measure is to put yourself in your potential adversary's position and determine your most successful plan of attack. This theoretical effort was led by Rear Admiral Patrick N. L. Bellinger his Army counterpart, and Major General Frederick L. Martin, who commanded the Hawaiian Air Force. Together with excellent staff, they produced the famous Martin-Bellinger Report. Predicting the future, as Prange et al. states [18]:

> Martin and Bellinger could not have done a much better job of mind reading had they actually looked over the shoulders of Yamamoto [conceptualized and authorized Pearl harbor attack], Onishi [an attack planner], Genda [planned air assault]—and others." For in Japan [at this time] the Pearl Harbor [planning] circle was widening even as Oahu's planners labored over their report. The final document bore the date of March 31, 1941, approximately the same time that Yamamoto put his Combined Fleet staff to work on his design.

Prange et al. then conveys their report's accuracy in predicting the future. It prophesized that dawn would be the mostly likely time for an air attack and such a first strike by aircraft would draw attention away from a submarine attack [19].

The report was also appreciated in Washington and approved by Turner as it articulated the same view of the situation held in the U.S. War and Navy Departments [20]. Prange et al. also point out other intelligence about Orange (U.S. Joint Army and Navy board war plans) that completed the picture of Japan's intent. On March 31, 1941, Naval Intelligence in Washington alerted the commandants of all naval districts that the Axis Powers often initiate attacks on Saturdays, Sundays, or national holidays [21].

Plan for the Employment of Bombardment Aviation in the Defense of Oahu
Colonel William E. Farthing, commander of the Fifth Bombardment Group at Hickam Field, led a team to develop a plan for the role of American bombers in defense of Oahu. The report's authors believed that the Japanese would use six carriers in the attack (as they ultimately did). These American planners on Oahu were essentially envisaging the same high-level plan about to be implemented by the Japanese. They indicated, as Prange et al. state, that the enemy would be more concerned with delivering a successful attack than any personal losses it may incur [22]. The Farthing report was seen by the ultimate Army investigators of the Pearl Harbor attack as [23] "prophetic in its accuracy and uncanny in its analysis of the enemy's intention."

Magic By August 1940, the U.S. Army Signal Intelligence Service (SIS) deciphered the Japanese diplomatic code, called "Purple," used between the Foreign Ministry in Tokyo and its embassies and consulates around the word. Lower grade "J" codes often used from the ministry to consulates (such as Honolulu, using J-19) were also deciphered. This intelligence source was dubbed "Magic." Japanese military codes were not deciphered, in general, and the Japanese Navy changed its codes several times in 1941, so Purple was the best United States' source of strategic intelligence.

These intercepts were not provided to Admiral Kimmel or his staff stationed at Pearl Harbor by the Navy Department [24]. Neither were they available from the War Department to General Walter C. Short, Commanding General of the Hawaiian Department, responsible for protecting Hawaii from sabotage, attack, or invasion, and protecting the Pacific Fleet when it was in the harbor.

As Prange et al. state regarding the protection of the existence of Magic [25], "Indeed, those in control of Magic were so obsessively concerned with security that they failed to realize the meaning of the contents of the intercepts... As stressed through out this [Prange's] study, cumulative intelligence is basic, but to be useful it must be interpreted and disseminated." These Magic messages and the State Department's belief that negotiations with Japan were unlikely to succeed, prompted the famous "war warning" messages sent to General Short under Chief of Staff George C. Marshall's signature and to Admiral Kimmel from the Navy Department (via Stark). The message General Short received was War Department Message No. 472, signed "Marshall," which read [26]:

> Negotiations with Japan appear to be terminated to all practical purposes...hostile action possible at any moment. If hostilities cannot... be avoided the United States desires that Japan commit the first overt act. This policy should not... restrict you to a course of action that might jeopardize your defense. Prior to hostile Japanese action you are directed to undertake such reconnaissance and other measures as you deem necessary but these measures should be carried out so as not ... to alarm civil population or disclose intent. Report measures taken. Should hostilities occur you will carry out the tasks assigned in Rainbow Five [the Army's basic war plan] so far as they pertain to Japan. Limit dissemination of this highly secret information to minimum essential officers.

General Short had earlier received communication from Marshall warning him about the sabotage threat, because there were more than 100,000 Japanese in Hawaii, so he believed this to be his main concern. The fact that the message stated that he should avoid alarming the civilian population and that he also believed that the Navy—per its agreement with the Army—was effectively handling long-range reconnaissance (which it was not performing at this time) led him to the conclusion that this message had not been meant for him. He also assumed that "if the Chief of Staff [Marshall] thought there was going to be an air attack or an all-out attack on Hawaii he would have said so" [27].

Thus, General Short went on what he dubbed "Alert No. 1," against sabotage only, and not on his Alerts 2 and 3 that each included defense against air attacks [28].

> Under the Short system, Alert No. 1 was "a defense against sabotage, espionage and subversive activities without any threat from the outside"; No. 2 encompassed No. 1 plus defense against air, surface, and submarine attack; No. 3 was "a defense against an all-out attack."

As the message stated that the concerns it raised were secret, he would not have been able to tell people deployed for Alerts 2 or 3; so, he calculated Alert 1 was more appropriate. Moreover, he responded as directed to "report measures taken," and "...within an hour of receiving the War Department message, Short dispatched a brisk reply [29]: "Report department alerted to prevent sabotage. Liaison with Navy reurad [an abbreviation for "reference your radio"] for seven two twenty-seventh.'" Since General Short then received another message from Washington, D.C. stressing that negotiations had broken down and to be alert to subversive activities, he believed his Alert 1 stance was justified, understood, and supported by Washington.

The tragedy of preparation for Alert No. 1 is that American aircraft were lined up in rows in the middle of their airfields to make them easier to guard from sabotage, but inadvertently creating an ideal configuration for attacking aircraft. Additionally, Alert No. 1 meant that the many Army antiaircraft guns surrounding the harbor had their ammunition secured far from the guns. Since the U.S. Navy assumed the U. S. Army handled antiaircraft protection of their ships while in harbor, they themselves were not adequately prepared. Prange et al. summarizes that "No one, in either Washington or Hawaii, attempted to clarify these misunderstandings because no one realized that a misunderstanding existed" [30].

The "war warning" message sent to Kimmel stated [31]:

> This dispatch is to be considered a war warning. Negotiations with Japan looking toward stabilization of conditions in the Pacific have ceased and an aggressive move by Japan is expected within the next few days. The number and equipment of Japanese troops and the organizations of the naval task forces indicates [sic] an amphibious expedition against either the Philippines Thai or Kra Peninsula or possibly Borneo. Execute an appropriate defensive deployment preparatory to carrying out the tasks assigned in WPL 46 [the Navy's basic war plan].

The U.S. Navy officials in Washington, D.C., including Admiral Stark and Rear Admiral Turner, believed they had completely warned Kimmel to be prepared for attack, to "deploy" his forces, and to be ready for whatever the Japanese might do. Moreover, they thought Admiral Kimmel had access to Magic—which he didn't [32]:

> The congressional committee [investigating Pearl Harbor] jumped on both Stark and Turner for failing to ascertain the truth. It accused the Navy Department officials, among others, of being "so obsessed by an executive complex that they could not besmirch their dignitaries by 'stooping' to determine what was going on, or more especially what was going on, in their organizations." The committee pointed out specifically that "in a period over 6 months, with relations between the United States and Japan mounting in tenseness and approaching a crisis, neither of these ranking officers determined for a fact whether the fleet was receiving this [Magic] information [which it was not].

Admiral Kimmel's interpretation of this critical "war warning" message was different, perhaps due to overly careful reading of the message and the belief it superseded previous warnings. Prange et al. states that Admiral Kimmel considered the dispatch to be a pullback from the one received November 24th. Rear Admiral Turner's use of the term "war warning" had not been universally defined and did not convey the meaning to which it was intended" [33]. Because the warning had been misinterpreted, and because all available information had not been provided and direct orders not followed-up with verification, complacency continued.

2.1.2.2 Tactical Warning

We have shown that the strategic warning was not adequate to heighten American preparedness at dawn on December 7, but there was still tactical warning—immediate sensor information and operational events that if properly handled could have indicated the approaching attack.

War Warning At about 9 a.m. in Washington, D.C., on December 7 (4 a.m. Hawaiian time), Col. Rufus S. Bratton, Chief of the Far Eastern Section of the U.S. Army's Signal Intelligence Service, received the addendum to the now famous 14-part message sent to the Japanese Embassy in Washington, D.C., and captured by Magic. The 14th part of the message broke off ongoing negotiations with the United States, and the addendum stated that the Japanese Ambassador should deliver the complete message to the U.S. State Department by 1:00 p.m. Washington, D.C. time (8 a.m. in Hawaii, just prior to the planned surprise attack). This message, combined with a December 2nd intercept ordering the Embassy to destroy its coding machines (e.g., Purple machine) in the event of emergency, convinced Bratton that the United States was about to be attacked somewhere in the Pacific after the designated hour, but he never imagined that it might be Pearl Harbor [34].

> As he later explained, "Nobody in ONI [(Office of Naval Intelligence)], nobody in G-2 knew that any major element of the fleet was in Pearl Harbor on Sunday morning the 7th of December. We all thought they had gone to sea... because that was part of the war plan, and they had been given a war warning.

Fuchida woke up at about 5 a.m. December 7, Hawaiian time, 10 a.m. in Washington, D.C. He found some of his aviators in the officer's mess listening to Honolulu AM radio station KGMB; U.S. authorities had requested that KGMB broadcast all night because Army four-engine B-17 bombers were coming in from the United States and could use the signal as a navigation beacon. The signal had two other uses for the Japanese; this "soft music" [35] played by the station indicated no state of alert in Hawaii, and the Japanese could also use the signal as a navigation beacon to Honolulu (as they did).

As Fuchida's first attack wave orbited the Japanese carriers preparing to set course to Oahu, about 5,000 miles away in Washington, D.C., Bratton—after many minutes delay and much understandable angst—finally managed to alert U.S. Chief of Staff Marshall on a Sunday morning. Marshall read the entire 14-part message with critical addendum, then immediately wrote a critical alert message to transmit to all Pacific commands. Marshall had not used the telephone, as he believed it was not adequately secure. Bratton then helped Lieutenant Colonel Edward F. French, in charge of the War Department Signal Center, to decipher Marshall's difficult-to-read long-hand message so it could be typed for transmission [36, 37].

> Japanese are presenting at one pm eastern standard time today what amounts to an ultimatum also they are under orders to destroy their code machine immediately. Just what significance the hour set may have we do not know but be on alert accordingly. Inform naval authorities of this communication. Marshall.

Despite all the failed strategic warning and expert opinion pointing to the scope and timing of the Japanese surprise attack, here was a discrete message that arguably would have been compelling enough to have driven Admiral Kimmel and General Short to greater alertness and preparedness for a particular day and time. Perhaps given their respective beliefs, it would not have mattered, but even this communication was doomed with delay—atmospheric conditions had blocked off the [radio] channel to Honolulu and the subsequent choice to send it via Western Union [38] caused the message to be received by Short and Kimmel shortly after the attack was over.

The Ward Just as the Martin-Bellinger report had forecast, the Japanese used submarines as part of their attack. One of these submarines was likely sighted off Pearl Harbor by a U.S. minesweeper *Condor*, which then contacted the destroyer USS *Ward* on security patrol [39]: "Sighted submerged submarine on westerly course, speed 9 knots." USS *Ward* was unable to establish a sonar contact, so the Commander Lieutenant William W. Outerbridge who was just taking command December 5, did not report the contact. This lack of communication additionally may have caused the *Condor* to serve as an unwitting escort to the Japanese subs as it returned to the harbor [40].

At 6:30 a.m., Commander Lawrence C. Grannis, of the stores and supply ship *Antares*, spotted a submarine conning tower 500 yards from his ship and reported it to the USS *Ward* [41]. Outerbridge ordered General Quarters at 6:40 a.m. and fired two rounds upon the submarine—the first missing, the second sinking the intruder [42].

> So, strangely enough, the first shot of the Battle of Oahu came from an American ship instead of Japanese aircraft, and the first blood shed was Japanese. Just as Genda and Fuchida feared, a submarine had triggered premature action. But Japanese luck held together, pinned by misunderstanding and tied with red tape [American deficiencies persisted].

Upon hearing about the USS *Ward's* encounter, Admiral Kimmel was still not sure that they were experiencing a real attack as there had been multiple false reports of submarines to date [43]. "And so previous cries of 'Wolf!' threw Kimmel off stride at the very hour when the predators were heading for the fold" [43].

Kimmel's verification of the validity of the attack, arriving with Fuchida's first wave, was less than 60 minutes away. Prange et al. explained that had the Army been notified that a hostile submarine had been sunk in the Defensive Sea Area, the Hawaiian Department might have moved from No. 1 alert to No. 2 or 3, better preparing them for what was approaching—but no such communication occurred [44].

Aircraft Warning System The Hawaiian Department (Army) received two fixed and five transportable radio direction and ranging (radar) sets for early warning of approaching aircraft, referred to as the aircraft warning system (AWS). Use of radar had proven decisive in the Battle of Britain, as approaching enemy aircraft could be detected with enough time before their arrival to launch and direct interceptors to attack them. Despite the advantage gained by radar-directed fighters demonstrated in the Battle of Britain, the deployment of such a comprehensive warning network in Hawaii received little priority.

One of the deficiencies in supporting radar deployment was in establishing a single location where the radar operators could send their sightings. In addition, General Short was less interested in radar and was concerned more with sabotage on the islands due to the large Japanese population. The deployment of the system was riddled with issues, such as the National Park Service having restricted access to the prominent peaks that would have been the ideal location, as well as the fact that various organizations were willing to send only their "excess officers" to the Information Center (where radar sightings were interpreted), instead of their best personnel [45]. In addition, there was no communication installed between the various defensive commands that could use the sighting data (e.g., antiaircraft batteries and interceptor squadrons). Moreover, there was no automatic aircraft recognition system or standard air-traffic approach patterns to distinguish friend from foe irrespective of radar reports [46].

The leadership failed to recognize the value of AWS. Moreover, the inability to distinguish between friend or foe aircraft, a lack of information sharing between agencies, and the assignment of the least wanted staff to the Information Center, all combined to prevent an effective early warning on December 7th. There were also other operational, training, and technical problems:

- Army aircraft movements were secret, so the Information Center was not told of their movements [47].
- No trained interceptor controllers were available to respond had enemy aircraft been detected [48].
- Army interceptors were on a four-hour notice, but the radar could only see aircraft 130 miles out, meaning that even the slowest aircraft seen by the radar would be over Oahu easily two hours before American interceptors would be expected to be available [49].

Even with all these factors working against the Americans on December 7th, Funchida's 350 torpedo bombers, high-level and dive bombers, and covering fighters speeding toward Oahu were spotted by the Opana Point radar site at about 7:00 a.m. and estimated to be more than 50 planes. This sighting was communicated to the Information Center to Lieutenant Kermit Tyler (his second day as pursuit officer and assistant controller) only as the largest blip the operator had ever seen. The interceptor controller and aircraft identification officers had left for the day. Unfortunately, without any strategic warning information, it did not occur to Tyler that this could be enemy aircraft. As he had no information about scheduled arrivals, he thought it may have been planes from a Navy carrier, and as he wasn't told the number of planes estimated from the blip in Opana, this seemed a reasonable assumption. In addition, he had heard Hawaiian music on the radio that morning and remembered being told by a bomber pilot friend that the station played this music all night whenever B-17s flew from the mainland [50].

> His first reaction was one of relief that the Flying Fortresses [B-17s] were coming from the right direction. ... Of course, for security reasons, Tyler could not explain this belief to [the team at Opana], so he merely replied, "Well, don't worry about it."

They traced the flight in until it was about 74 miles distant. Tyler reported the sighting to no one else, not the antiaircraft command, not headquarters (the Hawaiian Department and General Short), and not the Navy. Thus, one of the most critical, technical early warning systems had worked well (the technology worked), but the chain of individuals, from Lockard (the original radar operator at Opana who reported only a large blip) to Tyler to General Short and Admiral Kimmel, were not notified; therefore, whatever operational plans existed from detection through interceptor response were not activated despite tactical warning. The technology worked—the people didn't.

In fact, Admiral Kimmel and the Navy were not told about the radar sightings until well after the attack. Had they known, they might have launched the subsequent naval sorties with surviving vessels to seek out the attacking ships from the direction the aircraft had approached. Instead, Navy ships and the Army aircraft sent to search for the carriers were ultimately directed south, in the opposite direction from the retiring Japanese fleet. A clear example of how a very small amount of critical information not communicated, much less understood, can have major consequences.

2.1.2.3 Attack and Aftermath

The Success of Imagination Despite strategic and tactical alerts, predictions of attack, and plans for mitigating such attacks, the U.S. Pacific Fleet and its Army defenders were unprepared for what happened next. The above deficiencies of the American leadership's planning strategy forced instant initiative—and sacrifice—from those they led.

Shortly after 7:50 a.m. when Fuchida signaled for the general attack, Rear Admiral William Rhea Furlong, commanding Battle Forces Pacific and aboard his minesweeper flagship USS *Ogala*, first heard the roar of engines, then saw a bomb drop [51]. "*What a stupid, careless pilot*, he said to himself, *not to have secured his releasing gear*. ... [then] Frulong saw the red ball of the Rising Sun and reacted instantly. 'Japanese! Man your stations!' he shouted. At his command, USS *Ogala* flashed the alarm: 'All ships in harbor sortie.'"

It had begun [52].

> Awaiting authentication of the submarine sinking report, Ramsey [Lt. Cmdr. Logan C. Ramsey of Navy Patrol Wing 2] stood near a window of the Ford Island command center, watching the color guard prepare to hoist the flag. At about 0755 he heard the scream of a plane diving over the station, turned to Ballinger, and said "Dick, get that fellow's number, for I want you to report him for about sixteen violations of the course and safety regulations." As the plane went into its dive, each man looked out separate windows to follow its course. "Dick, did you get his number?" Ramsey asked. "No, but I think it was a squadron commander's plane because I saw a band of red on it," replied Ballinger.
>
> "Check with the squadrons and find out which squadron commanders' planes

> are in the air," Ramsey ordered. That very instant Ballinger reported, "I saw something black fall out of that plane when it completed its dive."
>
> At precisely 0757 an explosion reverberated from the hangar area. Ramsey's face changed in swift comprehension. "Never mind the squadron commander, Dick," he exclaimed. "That was a Jap plane and a delayed action bomb." The words scarcely out of his mouth, Ramsey raced across the corridor to the radio room, and ordered all radioman on duty to send out in plain English: "AIR RAID, PEARL HARBOR. THIS IS NOT A DRILL!"
>
> Thus, at exactly 0758 [two minutes before Nomuru was supposed to deliver the 14-part message to the US State Department] one of the most famous radio messages ever dispatched clicked over the airwaves. ... it was Commander Logan Ramsey who first sent forth the word that woke the United States from its long sleep.

Japanese torpedo planes, bombers, and fighters attacked the U.S. Pacific Fleet in Pearl Harbor; the Hickam, Wheeler, and Bellows Army Air Force bases; the Ford Island and Kaneohe Naval Air Stations; and the Ewa Marine Corps Air Station. As Fuchida's aircraft—the last to leave the harbor area—turned for the Japanese carrier fleet, eight battleships, three light cruisers, three destroyers, and four auxiliary vessels had been sunk, capsized, or damaged. A total of 87 Navy and 77 Army aircraft had been destroyed, 128 were damaged with 80% later salvaged, and civilians from Waikiki to Pearl City had been killed—many by stray American antiaircraft fire. Because there was no effective aircraft identification system, in the midst of the melee on December 7, some American aircraft had been shot down by their own antiaircraft batteries. A total of 2,403 people (military and civilian) were killed or died of their wounds, while 1,178 of the wounded survived. Twenty-three of 37 sets of brothers were lost on the *Arizona*. The Japanese lost 29 of the 350 attacking aircraft, plus one large and five midget submarines.

Responsibility Shortly after the attack, the search began for people to blame for the fact the attack was a surprise, and the subsequent loss of life and property. The search for responsible parties began with the U.S. president on down. But many felt that this responsibility was also caste on the American people [53]:

> our blindness, our provincialism, our complacency, even our ignorance as a people [Lynchburg News, December 10, 1941]
>
> Each investigation, each development, brought a fresh outburst, culminating on August 30, 1945, when President Harry S. Truman pronounced: "...the country as a whole is basically responsible in that the people were unwilling to take adequate measures to defense [sic] until it was too late to repair the consequences of their failure to do so."

One of the more magnanimous responses understood that his leadership was a product of the society from which it came [54].

> Following publication of the report of the tragedy that Secretary of the Navy Frank Knox made shortly after the event, the blunt-penned Dorothy Thompson poured an avalanche of scorn over her countrymen:

> And I will tell you where the ultimate responsibilities lies, ... for Hawaii and for everything else. It lies with us. ...
>
> For a whole generation the American idea has been to get as much as it could for as little effort. For a whole generation the American motto has been, "*I guess its good enough*" [italics added].
>
> I accuse us. I accuse the twentieth-century American. I accuse me.

But, Americans were represented by their congressmen, and these congressmen controlled the resources needed to provide Kimmel and Short with the capabilities to defend the Pacific Fleet—the sole purpose of the Hawaiian defenses. Perhaps like the twentieth-century American, their representatives and leaders had also thought things were "good enough." Once again, acts of initiative and self-sacrifice were required by the defenders on the ground while the leaders and planners responsible were out of harm's way.

The opinion that all Americans and their representatives were partially responsible may sound compelling, but it was ultimately the American leadership in Hawaii who had been warned and those to whom they reported in Washington, D.C. that were responsible. Nevertheless, Kimmel and Short had ultimate responsibility for Pearl Harbor. As Prange et al. state [55]:

> Lawmakers cannot legislate an awareness of danger. Money cannot buy alertness. But the indifference of the Congress to the Far Eastern situation as compared to its preoccupation with the war in Europe may well have contributed to the lack of awareness, the lack of alertness that prevailed on Oahu.

Washington, D.C. and Congress had made a special effort, given what was known, to make Pearl Harbor the best defended American installation. Independent of the Navy's failures, and there were several if those, it was the American Army that had responsibility for protecting the ships in Pearl Harbor and they failed to do so. Short was the Army commander, and as Prange et al. states, he had to bear the greatest responsibility [56].

> Of all the mistakes, oversights, and blunders on the American side of the Pearl Harbor Story, none contributed more to Japan's success than did Short's. Not only did he fail to perform his mission, but he also did not truly understand it. And in the crisis of November 27 he narrowed his vision to the point where the whole military power at his command—and it was considerable—stalked a mouse while the tiger jumped through the window.

Kimmel did not escape the logic of the congressional investigators [57].

> The majority report continued relentlessly:
>
> He knew that one of the tasks before the outbreak of war was guarding against a possible surprise attack by Japan. He knew that the only effective means of detecting a surprise raiding force in adequate time to combat it was by distant reconnaissance. He knew the Japanese reputation for deceit and treachery. He knew the greatest danger to the Fleet at Pearl Harbor was the possibility of an air raid. He knew that the maintenance and protection of the Fleet while in its base constituted a fundamental element in making military dispositions at Pearl Har-

> bor. He had been categorically warned of war. He knew or must have known that the necessity of Japan's striking the first blow required of him greater vigilance consistent with his fundamental duty as commander in chief to prepare for the worst contingency. He had adequate facilities to patrol the most dangerous approaches to Pearl Harbor. The decision was not a simple one, but, failing to resolve his dilemma by seeking advice from the Navy Department, Admiral Kimmel displayed poor judgment in failing to employ every instrumentality at his command to defend his fleet.

Even the Japanese had expected such reconnaissance [58].

To assign the greatest blame to General Short, because he was the responsible authority on the scene, may be organizationally and operationally appropriate; however, this deduction does not provide a logical process to ensure that other responsible leaders don't make similar mistakes in the future. Considering the many failed efforts to prevent, detect, protect, alert, respond, and recover since Pearl Harbor, there is a reason that the "lessons" have still not been learned. The subtle, but identifiable reasons for this are directly attributable to failed interoperability as we have defined it. This failed interoperability can therefore be traced to failures in governance, resources, approach, acceptance, architecture, and performance (see Section 1.3). This book will present integrated best practices to prevent these failures and ensure interoperability that maximizes the IT and communications capability of our public safety and emergency response organizations to collaboratively prevent, detect, protect, alert, respond, and recover.

2.2 Failed Interoperability

Previous Sections 2.1.2 through 2.1.3, cause one to wonder how the U.S. Army and Navy leadership did not see the attack coming—or at least why they were not better prepared and alerted. The Army, Navy, and congressional investigators had the same questions while listening to the many witnesses, including Kimmel, Bloch, Turner, and Stark as well as Short and Marshal. As we have seen, strategic warning of the Pearl Harbor attack came from many sources, including the newspapers, diplomats, counterspies, Magic intercepts, consulate wiretaps, U.S. Army radio signal intelligence, and accurate threat analysis and prediction (the work of Martin, Bellinger, and Farthing). But this information went to different people at different times who had different levels of acceptance, interpretation, and concern which resulted in an overall lack of preparedness. The failures of tactical warnings from the December 7th Magic intercept, the USS *Ward* midget submarine sinking, and the Opana AWS sighting of Fuchida's approaching aircraft, produced the lack of alertness.

Several in-depth investigations of the Pearl Harbor attack followed that infamous day. The House Majority investigation and resulting report drafted by Assistant Counsel Edward P. Morgan, presents a detailed summary of the causes for failed American preparedness and alertness. Morgan's report identified 25 deficiencies, which ultimately traced the disaster to failed leadership. In presenting these failures, Prange et al. and his coauthors characterized Morgan's report [59].

> The final report explained that the committee posed these points "not for their novelty or profundity but for the reason that, by their very self-evident simplicity, it is difficult to believe they were ignored." The applicability of these points far transcends Pearl Harbor, so they deserve serious review.

Next, for the interoperability analysis, we will extract and summarize the deficiencies in the congressional report on the Pearl Harbor investigation. The deficiencies we will address caused a minimum of *shared, trusted understanding* at Pearl Harbor on December 7. That lack of trust and understanding resulted in too little *predicable, collaborative action* from the moment Fuchida's first wave approached Oahu. Leaders, planners, and senior-level authorities in all public safety and emergency response organizations and governance levels worldwide should review these deficiencies and avoid them in their own organizations, operational plans, and IT and communication systems. Furthermore, consultants, designers, and integrators who support the planning of these IT and communication systems should account for potential human failures in their detailed system requirements.

2.2.1 Failure of Organization

This first deficiency addressed the many intelligence and independent operational command-and-control hierarchies that prevented *shared, trusted understanding*, which resulted in inadequate *predicable collaborative action* in the event of an apparent threat [60].

> 1. Operational and intelligence work requires centralization of authority and clear-cut allocation of responsibility.
>
> Morgan warned, "The pyramiding of superstructures of organization cannot be inducive to efficiency and endangers the very function of our military and naval forces" (or of any other organization, for that matter).

Several apparent misunderstandings and communication failures between the U.S. Army and Navy defenders of Oahu, and between their respective superiors in Washington, supported the Japanese success on December 7th. These deficiencies include the following [61–65]:

- Hoover of the FBI removed portions of an intercepted espionage questionnaire about Pearl Harbor. The Army's Magic intercepts were not provided to General Short or Admiral Kimmel.
- Admiral Kimmel and Rear Admiral Bloch did not alert General Short (Army) that the Japanese were ordered to destroy their codes and code machines in an emergency, assuming he had received the same information—but he had not.
- Liaison between the Navy and Army was a young, inexperienced officer who was not provided all information about the Navy's plans.
- Rear Admiral Block arguably did not understand the annex in the Army/Navy Joint Coastal Defense Plan JCD-42 that stated that Washington or Hawaii could implement elements of the plan before war was declared if they believed

conditions warranted it, so he did not execute long-range reconnaissance (his responsibility).

- Even though Rear Admiral Bloch and Admiral Kimmel did not believe they had the numbers and types of aircraft to perform long-range reconnaissance, they had agreed with Short to execute it under JCD-42.
- There were different meanings of Short's Alert No. 1, meaning antisabotage alert, and the Navy's Alert No. 1, meaning "the highest degree of readiness." No one detected, much less corrected, this misunderstanding before December 7th.
- Short assumed Admiral Kimmel and Rear Admiral Bloch were performing long-range reconnaissance after the November 27 war-warning message (which they were not). He neither verified this assumption nor conducted his own reconnaissance.
- A fleet request to allow the Army to control antiaircraft guns on Navy vessels in the harbor coordinated by the Information Center—connected with the AWS—was rejected by Rear Admiral Bloch, debatably one reason why Navy ships shot down Navy aircraft on December 7—it was finally done on December 9.
- The AWS detection at the Opana site was not relayed to the Navy, and the USS *Ward's* attack on the midget submarine was not told to the Army.

2.2.2 Failure of Assumption

The second deficiency is too often a source of surprise in many critical situations and is a source of many "famous last words" [66].

> 2. Supervisory officials cannot safely take anything for granted in the alerting of subordinates.
>
> With such words Morgan's analysis soared far beyond the lessons taught by a military operation, however significant, and entered into a fundamental region of the human condition. How often similar phrases signal tragedies in every area of life!

There were many unfortunate assumptions in Washington, D.C. and Hawaii that characterized lack of alertness or preparedness on December 7th. Key strategic assumptions that impacted preparedness included:

- Admiral Turner assumed that Kimmel and Short had direct access to the Magic intercepts, but they didn't, and Kimmel believed he was getting whatever intelligence was available—which he was not.
- The Navy Department assumed Kimmel would deploy the Pacific Fleet after the war warning, or said they expected him to do so during the subsequent investigations—he had not.
- The Army Department (Marshall) assumed that the Hawaiian Department was performing long-range reconnaissance, but it was not doing so; Hawaii believed it had relegated this responsibility to the Navy (who also was not doing so because Bloch had not ordered it).

- On Oahu, General Short assumed Admiral Kimmel was performing long-range reconnaissance using their patrol vessels and aircraft, but, Admiral Kimmel wasn't because he believed he didn't have enough aircraft to cover all approaches to the island over 360 degrees.
- Rear Admiral Bloch was told that General Short had moved to Alert No. 1 after the November 27th war warning. As a result, Rear Admiral Bloch assumed this meant a "full alert" not just alert against sabotage, and he never verified what Army Alert No. 1 meant.
- The November 27th war-warning message left out Hawaii for an expected Japanese strike although it listed the southwestern Pacific bases. Previous alerts mentioned they might attack in any direction, so Admiral Kimmel assumed that message not mentioning Hawaii was significant—and never checked if that was the correct meaning of the war warning.
- After the November 27th war warning, General Short informed Chief of Staff Marshall he intended to conduct antisabotage defense. The recipients of this message at the War Department assumed it meant that Short's long-range reconnaissance had shown no intruders, so he was focusing instead on sabotage. Marshall and his staff never verified this key assumption.
- Short consistently put training ahead of preparedness.

Critical tactical assumptions that reduced alertness included:

- Lieutenant Tyler assumed that because he heard KMGB broadcasting in the early morning hours, that it was U.S. Aircraft—and not enemy planes—that were approaching the island when he received the Opana radar operator's message.
- The 14th Naval District (Bloch) and Admiral Kimmel thought that because there had been previous false submarine sightings, the *Ward's* attack was possibly another false alarm.
- Despite receiving a Navy Department notice to the contrary, Kimmel assumed an aerial torpedo could not be adequately launched in the harbor. For this reason no torpedo nets were needed, particularly because the nets would encumber port operations.

Just as the failure of American interoperability is grossly apparent in this analysis, the outstanding integration of intelligence and operations was apparent in the Japanese effort, particularly in their collection and use of the "bomb plot" reports via a planted operative in the Honolulu consulate [67]. In fact, "The smooth interweaving of Japanese intelligence with planning and operations forms a sharp contrast to American mishandling of military information in 1941."

In contrast, the American government had many organizations whose leaders had overlapping but parallel responsibilities (e.g., intelligence collection, analysis, and dissemination) and sometimes multiple responsibilities. For many reasons, information was not shared between these organizations, and the systems or operating procedures for such sharing were vague, undefined, unfulfilled, and unverified.

2.2.3 Failure of Omission

In the Washington and Hawaii commands, there were many situations in which information was intentionally not provided, or it was thought it would be received from another source, but not verified. In this regard, Morgan's third deficiency was [68]:

> 3. Any doubt as to whether outposts should be given information should always be resolved in favor of supplying the information.

Additional important strategic information was not shared as a matter of course, such as Kimmel not sharing the war-warning message with the Navy and Bloch not informing the fleet (Kimmel) nor the Information Center that an Axis attack would most likely be on weekends or national holidays.

Important intraservice tactical interoperability sharing failures to note:

- Marshall did not use the scrambler telephone—the fastest means—to inform the Pacific commands of the "one o'clock" message.
- Kimmel and the Fourteenth District staff failed to notify the Information Center or the Hawaiian Command about the *Ward*'s alleged attack on a submarine.
- Ellison did not tell Tyler that he saw more than 50 aircraft approach Oahu.
- Lieutenant Tyler did not inform any interceptor squadron that the Opana station had detected approaching aircraft, nor did he tell anyone else—including the Navy.

2.2.4 Failure of Verification

Assumption is the parent of failed verification. Given the uncertainty in language and meaning, much less priority and extent, it was considered incumbent on the leadership to perform inspections and verify the condition and capabilities of their subordinates [69].

> 4. The delegation of authority or the issuance of orders entails the duty of inspection to determine that the official mandate is properly exercised. The job of an administrator is only half completed upon the issuance of an order; it is discharged when he determines the order has been executed.

The failure of Stark or Marshall to follow up with their Hawaiian commanders to verify they had done what was expected of them was a critical deficiency. There had been direct intelligence (through Ambassador Grew) and from counterespionage, as well as indirectly from Magic intercepts, that a Japanese surprise attack on Hawaii was coming, and all Washington officials suspected, due to information from many sources, that this was a probability, not just a possibility.

2.2.5 Failure of Supervision

The investigation's fifth deficiency stated [70]:

> 5. The implementation of official orders must be followed with closest supervision.

Short had declared his intention to move to Alert No. 1 against sabotage after he received the November 27th war-warning message. Short had believed that with the Navy conducting long-range reconnaissance (which he had not verified) because it was in an agreement (misunderstood or ignored by Bloch), he would learn from the Navy if a fleet were approaching. Washington thought Hawaii was already conducting long-range reconnaissance, but did not verify it. Instead, they interpreted Short's response to focus on sabotage as an indication that long-range reconnaissance had not spotted anything.

2.2.6 Failure of Alertness

The argument that the many drills, exercises, and false alarms in the months leading to the Japanese attack was countered by the sixth deficiency [71]:

> 6. The maintenance of alertness to responsibility must be insured through repetition.

As Prange et al. points out [72], it is "far better to send the fire engines on nine goose chases than hesitate the tenth time and allow a major conflagration to burn out of control."

2.2.7 Failure of Complacency

In a sense, Morgan's seventh deficiency was the cause of the lapse in alertness [73]:

> 7. Complacency and procrastination are out of place where sudden and decisive actions are of the essence.

There was a general antiwar sentiment in post-World War I America during which time there was reluctance to spend public resources on "merchants of death." Americans felt protected by the broad oceans separating them from enemy forces and harbored the opinion that the next minute, the next hour, the next weekday, and the next weekend morning would be like all others.

2.2.8 Failure of Intelligence

One of the more tragic failures was the fact that so much strategic and tactical information regarding intent, capability, and timing was available to the American government and its military commanders. American experts correctly envisioned the content and timing of the Japanese attack, but none of this mattered as the information was collected by different people at different times and then "analyzed" by other people. No one person had all the information or the assumption that others had the information was not verified. As Morgan's report stated [74]:

> 8. The coordination and proper evaluation of intelligence in times of stress must be insured by continuity of service and centralization of responsibility in competent officials.

2.2.9 Failure of Attitude

Morgan points out that not once did a War or Navy Department dispatch result in a subordinate asking for clarification or question its contents, prompting the next deficiency [75]:

> 9. The unapproachable or superior attitude of officials is fatal; there should never be any hesitancy in asking for clarification of instructions or in seeking advice on matters that are in doubt.

2.2.10 Failure of Imagination

There were many situations in which failed imagination (as would be evidenced many times after December 7th to the present day) did not allow the coordination and corroboration needed to envisage the threat and its successful delivery, despite the information that was available to Washington and Hawaiian commanders [76].

> 10. There is no substitute for imagination and resourcefulness on the part of supervisory and intelligence officials.

Over sixty years after the Pearl Harbor attack, the *9/11 Report* characterized all levels of government, when it reported Richard Clarke's (special assistant to the president on counterterrorism) statement that [77]:

> he thought that warning about the possibility of a suicide hijacking would have been just one more speculative theory among many, hard to spot since the volume of warnings of al Qaeda threats and other terrorist threats, was probably hundreds of thousands. ... Yet the possibility was imaginable, and imagined. ... [yet] the FAA [Federal Aviation Administration] analysts judged such an operation unlikely ... ...A suicide hijacking is assessed to be an option of last resort.

As we have shown, the December 7th attack was also not thought possible by the leadership, despite the Martin and Bellinger report and the Farthing report, that demonstrated the possibility. Many intelligence warnings established the likelihood of the attack on Pearl Harbor, but they were summarily discounted by leadership.

2.2.11 Failure of Communications

Referring to the *language* aspects—particularly the wording of messages, Morgan stated [78]:

> 11. Communications must be characterized by clarity, forthrightness and appropriateness.

Messages from the War and Navy Department to Short and Kimmel were ambiguous and crafted with an attempt at brevity that led to further misunderstanding by omission and assumption, [resulting in] a loss of clarity [79]. The Navy Department felt sure that its November 27th message would lead to a deployment of the fleet according to the Navy's basic war plan, WPL 46. On the contrary, Kimmel's analysis of this war-warning message differed significantly from what the drafters intended and he reacted without seeking further clarification.

2.2.12 Failure of Paraphrase

Another failure of communications is reading into a message more than what was intended, and if there is a lack of confidence in the result, following up to verify [80].

> 12. There is great danger in careless paraphrase of information received and every effort should be made to insure that the paraphrased material reflects the true meaning and significance of the original.

Short's message, paraphrased by the War Department, contributed to the false assumption that he was performing long-range reconnaissance, despite that he had an established arrangement with the Navy to conduct long-range reconnaissance.

2.2.13 Failure of Adaptability

Morgan's next deficiency dealt with operating procedures, namely [81]:

> 13. Procedures must be sufficiently flexible to meet the exigencies of unusual situations.

Perhaps the failure of imagination, coupled with the desire not to disrupt day-to-day activities led to a rigidity of standard operating procedures, such as: the four-hour alert time for aircraft, the distant caching of antiaircraft ammunition, the 0400 to 0700 operation of the AWS, and the need for confirmation of the *Ward's* action. Regrettably, these poorly conceived operational plans were in place during a time when the leadership on Oahu had received strategic alerts. Contingency plans and practicing for those plans, based on the threat posed in the *Martin-Bellinger* and *Farthing Reports*, would have been advisable.

2.2.14 Failure of Disclosure

The secrecy surrounding the existence of Magic protected a valuable source of information on Japanese intentions almost to the point of eliminating any worth to its owners. As Morgan pointed out [82]:

> 14. Restriction of highly confidential information to a minimum number of officials, while often necessary, should not be carried to the point of prejudicing the work of the organization.

The top American commanders responsible for protecting the west coast of America, Kimmel and Short, did not have access to Magic intercepts, and for security reasons, they were not trusted to know that such a capability even existed.

2.2.15 Failure of Insight

Morgan pointed out that the barn-door syndrome existed, defined as the phenomenon that sometimes occurs when things are so large, so all-encompassing, and so obvious that most people fail to see them, existed. His next deficiency states [83]:

> 15. There is great danger of being blinded by the self-evident.

For example, since indications were that the Japanese were attacking toward Southeast Asia; this predisposed the military from properly interpreting signs that Hawaii could also be a target.

2.2.16 Failure of Dissemination

Related to the "failure of omission" and "failure to (properly) communicate" with one's subordinates, Morgan included this deficiency when he wrote [84]:

> 16. Officials should at all times give subordinates the benefit of significant information.

There are numerous examples of this deficiency at all levels of the Pearl Harbor story. Two notable ones are: when Washington did not inform Kimmel or Short of the "bomb plot" series of messages to the Honolulu Consulate (seeking to know which ships were at which locations in the harbor), and when Kimmel failed to give the strategic war warning to his task force commander of the carrier Lexington when they departed Oahu two days before the attack, in effect blindsiding Oahu's limited tactical warning capability.

2.2.17 Failure of Inspection

There were notable examples of those in charge not visiting, much less inspecting, critical facilities like the Information Center [85].

> 17. An official who neglects to familiarize himself in detail with his organization should forfeit his responsibility.

Given the importance of the AWS on Oahu to the protection of Kimmel's ships, it would be natural to assume he would have a thorough knowledge of the entire capability—particularly as he was not performing long-range reconnaissance after the war-warning message, nor verifying his coordination of response with Short following the November 27th war-warning message. This critical deficiency touches on many others, because failing to know one's own organizational capabilities as well as those upon which one is dependent puts everything at risk. If such large gaps in preparedness are not identified, and subordinates won't, don't, or can't effectively identify them, then they cannot be corrected.

2.2.18 Failure of Preparedness

The need to account for possibilities, not just probabilities, and to continuously review and assess "enemy" intentions and capabilities in order to be prepared for the worst was not done on Oahu by 0753, December 7th, 1941. In this regard, Morgan's next deficiency was [86]:

> 18. Failure can be avoided in the long run only by preparation for any eventuality.

This failure was at the heart of American failure and Japanese success. As Prange et al. write [87]:

> history has shown that if an enemy can launch a certain kind of attack, in all probability he will do exactly that. His intentions cannot hurt his opponent; his capabilities can, if properly utilized.

2.2.19 Failure of Consistency

Morgan expressed the 19th deficiency as [88]:

> 19. Officials, on a personal basis, should never countermand an official instruction.

In this case, Prange et al. reference Morgan's concern that Stark's official messages implied one thing, but his personal dispatches implied something else [89].

> Stark's office officially informed Kimmel... that Japan might hit both the United States and Britain. But the next day, in a personal letter to Kimmel, Stark cited his dispatch, then added, "Personally I do not believe the Japs are going to sail into us and the message I sent you merely stated the 'possibility."

2.2.20 Failure of Jealousy

Morgan expressed the next deficiency—which often manifests itself as organizational rivalry (arguably the principal cause for failed interoperability)—as simply [90]:

> 20. Personal or official jealousy will wreck any organization.

There are many such examples, such as the aforementioned request that the Information Center control Navy antiaircraft batteries and the Signal Corps wanting to maintain control of the radar sites on Oahu. Of course, this jealousy is at the heart of failed interoperability worldwide and permeates most organizations at many levels.

2.2.21 Failure of Relationships

Short and Kimmel had a cordial relationship, symbolizing good Army-Navy relations. Morgan points out the failure, which results from this personal relationship, in the next deficiency [91].

> 21. Personal friendship ... should never be accepted in lieu of liaison or confused therewith where the latter is necessary to the proper functioning of two or more agencies.

One of the more insidious causes of interoperability failures was due to the cordial relationships between the Navy (Kimmel and Bloch) and the Army (Short) [92].

> All concerned sincerely believed that their cordial relations constituted all the cooperation necessary. Unfortunately, this was not so. As the congressional committee's majority report suggested, this friendly relationship, hedged in by a disinclination to encroach upon each other's territory, exercised a baneful influence.

2.2.22 Failure of Priority

Morgan stated his next deficiency as follows [93]:

> 22. No consideration should be permitted as excuse for failure to perform a fundamental task.

Kimmel and Short explained that part of the reason they could not perform some of the undone tasks evident to investigators such as verifying the operation of the AWS, was because they had too much to do. In this regard, Kimmel was quoted as stating [94]:

> I thought I knew. You must trust somebody. I couldn't do everything.

2.2.23 Failure of Reporting

As in deficiencies No. 3 and No. 16, Morgan stressed failed information sharing mostly from subordinates to superiors [95].

> 23. Superiors must at all times keep their subordinates adequately informed and, conversely, subordinates should keep their superiors informed.
>
> As Prange et al. point out [96]:
>
> In this particular instance, he [Morgan] had in mind the failure of subordinates to get to Marshall and Stark the first thirteen parts of the final Japanese message, and Kimmel's failure to clue in Bellinger about the "war warning." Morgan also reminded his readers of Short's failure "to inform the essential and necessary officers of his command of the acute situation" upon receipt of the "war warning."

2.2.24 Failure of Improvement

The next critical deficiency suggests that organizations should continually improve themselves by seeking gaps in information sharing or trusted action and correcting them. Morgan's 24th deficiency addressed the lack of retrospection and self-improvement, particularly for organizations with critical responsibilities [97]:

> 24. The administrative organization of any establishment must be designed to locate failures and to assess responsibility.

There are many examples of this failure to identify information-sharing problems and correct them, including operational and technological gaps:

- Identification and correction of the approximately one-hour arrival time of incoming aircraft after AWS detection versus the four-hour interceptor alert status.
- Lack of standard aircraft approach patterns and friend-or-foe identification process.
- Lack of antiaircraft battery control by the Information Center.
- No direct phone lines with liaison commands from the Information Center.
- No trusted immediate voice communications between Washington and Hawaii.

Perhaps these critical capabilities were assumed "good enough."

2.2.25 Failure of Delegation

In the last documented deficiency, Morgan emphasized the importance of combining authority with responsibility when delegating tasks to subordinates [98]:

> 25. In a well-balanced organization there is close correlation of responsibility and authority.

The War Department certainly expected Short to take proactive measures, but in his view limited his course of action so as not to alarm the local population or taking any act that might indicate his objectives.

2.3 Pearl Harbor Interoperability Analysis

To summarize key information-sharing failures, Figure 2.1 depicts a summary of strategic and tactical information flows prior to the Pearl Harbor attack. As stated over half a century after Morgan's emphasis on failed "imagination," contributing to the Japanese success at Pearl Harbor, the *9/11 Report* Subsection 11.1 entitled *Imagination* stated [99].

> Understanding the Danger
>
> If the government's leaders understood the gravity of the threat they faced and understood at the same time that their policies to eliminate it were not likely to succeed any time soon, then history's judgment will be harsh. Did they understand [or were they capable of imagining] the gravity of the threat?

As Prange et al. have shown, the politically charged situation between the United States and Japan, the Magic intercepts, the wiretaps, the military radio intel-

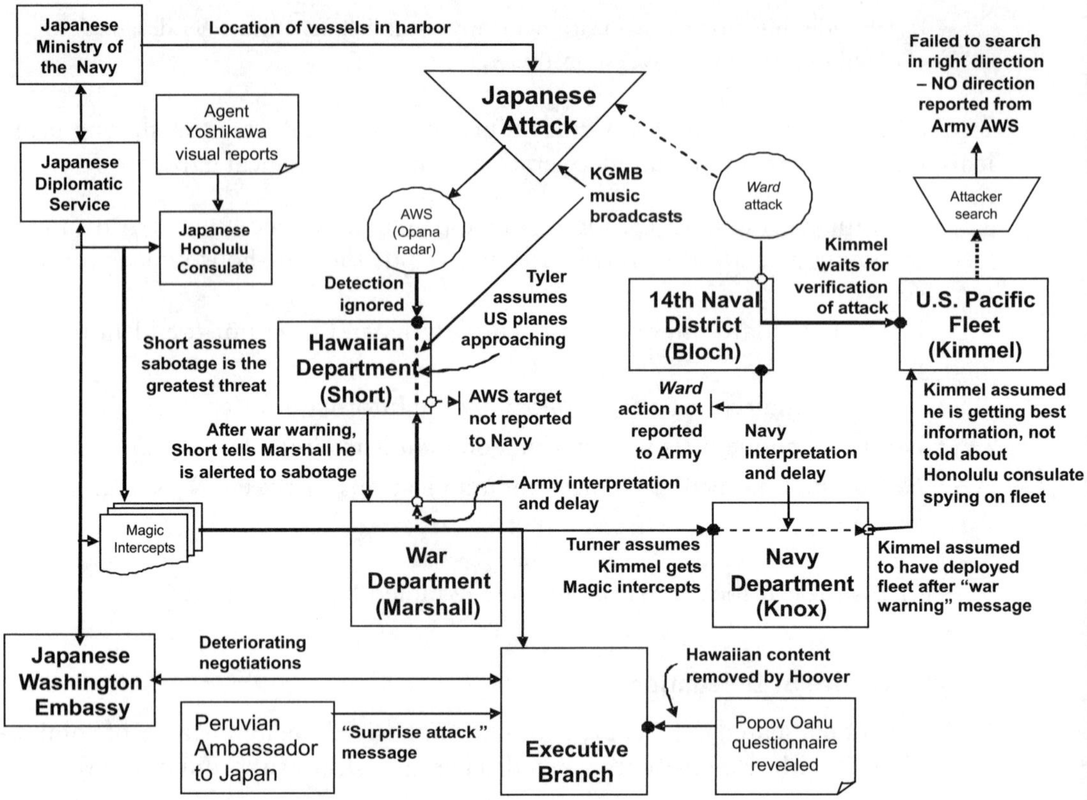

Figure 2.1 Overview of information flow before the Pearl Harbor attack.

ligence, the Popov story, and even the United State's own "Orange" planners failed to ensure preparedness on December 7. It is hard to imagine that despite all the available information, a total surprise was still not only possible, but probable—or even that much imagination would be required to at least be prepared at dawn on Saturdays and Sundays during an obviously critical period. However, the reader has seen all of this telltale evidence in one place and at one time. The strategic warning provided by the intelligence collection and analysis described above was known—if not believed—by different people at different times, and much of it was not available to the commanders at Pearl Harbor, each of whom held different assumptions and prejudices, personalities and capabilities.

The 25 deficiencies, summarized in Table 2.1, each represent a *gap* between best-practice interoperability and an example of failure, over 60 years after December 7, 1941. Unfortunately, failure to identify and correct any one gap in governance (deficiency #24), resources, approach, architecture, performance, or acceptance could lead to tragedy. Governments must empower leaders who will seek out and fill these gaps, ensuring that the system-of-systems built from people and their tools (technology) will not fail yet again because of deficiencies so apparent and publicized long ago. In summary, Prange et al. state that [100]:

> Careful consideration of Morgan's points reveals that many of them are separate facets of a single problem—the failure to communicate. The aspect of the matter may well stand as one of the basic causes of the Pearl Harbor tragedy, second only

Table 2.1 Summary of Deficiencies Identified in the Report from the Pearl Harbor Congressional Investigation

Deficiency		
No.	*Name*	*Failure of Governance—see Subsection 2.2.5.x where "x" is the "No." column*
1	Organization	Multiple parallel organizations with ambiguous authority.
2	Assumption	Information-sharing convention is not known or understood.
3	Omission	Information-sharing distribution is incomplete, people and entities excluded.
4	Verification	Commands/information sent, no follow-up to ensure understanding and action.
5	Supervision	Close supervision to verify understanding and predictable action not provided.
6	Alertness	Heightened alert is undermined by repeated training and exercises.
7	Complacency	Vigilance relaxes from the day-to-day lull of "business as usual."
8	Intelligence	Centralize intelligence services with tailored dissemination.
9	Attitude	Superiors do not engage in open dialogue with peers and subordinates.
10	Imagination	Worst-case scenarios not included in preparedness and response planning.
11	Communications	Information exchanged is ambiguous, convoluted, or contradictory.
12	Paraphrase	Messages altered according to assumption and no verification.
13	Adaptability	Alert and response thresholds are not matched to the known threat environment.
14	Disclosure	Intelligence so protected that it is inaccessible to those who urgently need it.
15	Insight	Inadequate understanding of the threat make risks poorly estimated.
16	Dissemination	Information is not provided to subordinates who need to know.
17	Inspection	Leaders do not know or understand their personnel and critical systems.
18	Preparedness	Prepare for consequences of what a threat *might* do, instead of what it *can* do.
19	Consistency	Official direction is contradicted by unofficial speculation from authorities.
20	Jealousy	Individual or organizational one-upmanship for real or perceived self-benefit.
21	Relationship	Personal friendships inhibit identification and resolution of deficiencies or gaps.
22	Priority	Failure to prioritize critical needs over day-to-day activities.
23	Reporting	Subordinates fail to report information up the command chain.
24	Improvement	Failure to identify gaps, particularly in worst-case scenarios, and correct them.
25	Delegation	Responsibility is delegated with inadequate authority to act.

> to the failure to believe in its possibility. The failure of 1941 to apply such principles as Morgan's cost the United States dearly at Pearl Harbor; hence they must be considered as lessons to be learned from the debacle. Any country, any organization, indeed any individual, can profit by pondering these points.

Recalling our definition of interoperability, a measure of *shared, trusted understanding* that ensures *predictable coordinated action* between entities in achieving a common public safety and emergency response goal, there is no doubt that it was lacking in U.S. government, army, and navy preparedness exposed at 0758, December 7, 1941, Hawaiian time.

Consider elements of interoperability, namely: governance, resources, approach, acceptance, architecture, and performance to highlight the failures of American leadership that enabled the Japanese success on that day. Table 2.2 shows an overall assignment of Pearl Harbor deficiencies to gaps in the elements of interoperability defined in Chapter 1. It shows that multiple interoperability gaps are responsible for most deficiencies, with governance, operational architecture,

Table 2.2 Summary of Pearl Harbor Deficiencies Mapped to Interoperability Gaps

Deficiency		*Deficiency Directly (●) or Indirectly (○) Caused by Interoperability Element of:*					
No.	*Name*	*Governance*	*Resources*	*Architecture*		*Performance*	*Acceptance*
				OPS	*SYS*		
1	Organization	●	●	●	—	—	○
2	Assumption	—	—	○	—	—	○
3	Omission	—	●	●	○	—	○
4	Verification	—	○	●	○	—	○
5	Supervision	○	○	●	○	—	○
6	Alertness	—	—	●	○	○	○
7	Complacency	—	—	●	—	—	○
8	Intelligence	●	●	●	—	●	○
9	Attitude	○	—	○	—	—	○
10	Imagination	—	—	○	—	—	○
11	Communications	—	—	●	—	●	○
12	Paraphrase	—	—	●	—	●	○
13	Adaptability	—	—	●	—	—	○
14	Disclosure	●	—	●	—	—	○
15	Insight	—	—	○	—	—	○
16	Dissemination	●	—	●	○	●	○
17	Inspection	●	○	○	—	—	○
18	Preparedness	—	●	●	●	●	○
19	Consistency	○	—	○	—	—	○
20	Jealousy	●	●	●	○	—	●
21	Relationship	—	—	—	—	—	○
22	Priority	—	—	○	—	—	○
23	Reporting	●	—	●	—	○	○
24	Improvement	●	●	●	—	●	●
25	Delegation	●		●	—	—	○

● - Strong (direct) impact ○ - Medium (indirect) impact OPS = Operational, SYS = System

and acceptance driving more deficiencies by far than the system architecture and performance analytic elements of interoperability. This result demonstrates that human factors were primarily responsible for the deficiencies in Washington/Pearl Harbor interoperability, not the lack or failure of technology—but how it was employed by the people trained to use it.

2.4 Lessons Taught—Not Learned

Our intent in this book is to finally convert these Pearl Harbor deficiencies *taught* into lessons *learned*—implemented using recommended best practices—through logical processes described in this book. To do so, we will expand the descriptions of selected deficiencies to include important, related nuances of the 9/11 and Katrina failures, including each deficiency's original wording in the *Congressional Pearl Harbor Report*. The quotes presented below were selected to demonstrate that the same deficiencies driving failed interoperability on December 7, 1941, drove corresponding failures on September 11, 2001, and following the natural disaster on August 29, 2005.

2.4.1 Deficiency Mapping: From Pearl Harbor to 9/11 and Katrina

A review of the *9/11 Report* and the *Katrina Report* show that the same deficiencies identified by Congress in 1946 from their Pearl Harbor investigation still existed over 60 years later on 9/11 and again for Katrina despite the significant advances in technology and planning occurring during those years [101].

> The methods for detecting and then warning of surprise attack that the U.S. government had so painstakingly developed in the decades after Pearl Harbor did not fail; instead, they were not really tried. They were not employed to analyze the enemy that, as the twentieth century closed, was most likely to launch a surprise attack directly against the United States.

This situation, however, need not remain the status quo. As we will show, the available technology is an *effect* of the fundamental problem, and not the cause. It is for this reason—as stated in Chapter 1—that lack of radio (or any technological) interoperability is a *symptom*, and not the *cause*, of failed holistic interoperability. The deficiencies do not lie in the "tools," but in the people and their processes that plan and select the tools. From the *9/11 Report* [102]:

> even the best information technology will not improve information sharing so long as the intelligence agencies' personnel and security systems reward protecting information rather than disseminating it.

This statement applies equally to all public safety and emergency response organizations worldwide, with clearly less justification for information protection than invoked by these intelligence agencies. Similarly, the *Katrina Report* states [103]:

> We reflect on the 9/11 Commission's finding that "the most important failure was one of imagination." The Select Committee believes Katrina was primarily a failure of initiative. But there is, of course, a nexus between the two.

Understanding and learning from these deficiencies is more than a theoretical exercise—it has critical tangible importance to those on the front line of any disaster. It is always our responder-receivers that must make up in their initiative and self-sacrifice for ineffective planning [104]:

> The Select Committee identified failures at all levels of government that significantly undermined and detracted from the heroic efforts of first responders, private individuals and organizations, faith-based groups, and others.

The institutional and individual failures we have identified became all the more clear when compared to the heroic efforts of those who acted decisively. Those who didn't flinch, who took matters into their own hands when bureaucratic inertia was causing death, injury, and suffering. Those whose exceptional initiative saved time and money and lives.

In order to determine how to mitigate or eliminate these prevailing deficiencies from public safety and emergency response, we first must better understand the causes of these deficiencies—as we clearly realize their effects. With this understanding, we can develop a rational checklist for assessing public safety and emergency response information-sharing architecture. These assessments will then enable interoperability to be achieved in existing and planned public safety and emergency response IT/communication systems. In the following subsections, we finalize our interoperability checklist by enhancing the Pearl Harbor deficiencies to reflect the lessons taught (again) by 9/11 and Katrina. These enhancements capture the nuances that differentiate civilian from military organizations—a distinction that impacts how information sharing is planned, the understanding of shared information, and ultimately the actions (or inactions) derived from this understanding.

2.4.2 Pearl Harbor Failures Revisited for 9/11 and Katrina

2.4.2.1 Failure of Organization Revisited

Review and Analysis The original Pearl Harbor deficiency, which we have dubbed "Organization," was stated in the congressional report as "operational and intelligence work requires centralization of authority and clear-cut allocation of responsibility" [105]. In this regard, the *9/11 Report* stated [106]:

> The U.S. government did not find a way of pooling intelligence and using it to guide the planning and assignment of responsibilities for joint operations involving entities as disparate as the CIA [Central intelligence Agency], the FBI, the State Department, the military, and the agencies involved in homeland security.

Throughout the 1990s, competing priorities, flat budgets, rivaling bureaucracies and an outmoded structure compounded this failing [107]. The tragedy of omission also affected the Federal Aviation Administration (FAA), whose 40-person intelligence unit working in conjunction with the FBI, CIA, and other agencies were, in theory, to have enough intelligence of specific and general threats to develop and deploy countermeasures. They received a large volume of data, but little to none pertaining to the presence of terrorists within the United States. In addition, they were not apprised of the FBI's investigation in 1998 into the use of flight training by terrorists [108].

Organizations and government agencies were not, as a whole, concerned with the lack of interoperability amongst them and the possible ramifications before 9/11 [109].

> the FDNY [Fire Department of New York] and NYPD [New York Police Department] each considered itself operationally autonomous. As of September 11, they

were not prepared to comprehensively coordinate their efforts in responding to a major incident. The OEM [New York Office of Emergency Management] had not overcome this problem.

Apparently, the organizational structure and resulting performance of intelligence agencies before Pearl Harbor was similar in deficiency to that realized up to—and well beyond—9/11 [110].

OEM headquarters, which could have served as a focal point for information sharing, did not play an integrating role in ensuring that information was shared among agencies on 9/11, even prior to its evacuation. There was a lack of comprehensive coordination between FDNY, NYPD, and PAPD [Port Authority Police Department] personnel climbing above the ground floors in the Twin Towers.

Although not from an intelligence perspective, multiple governance levels faced similar obstacles in communicating with each other while responding to Katrina, as "DOD, FEMA, and the state of Louisiana had difficulty coordinating with each other, which slowed the response" [111]. As Northern Command [112] was a new organization, there were coordination issues stemming from its lack of "insight into state response capabilities or adequate interface with governors, resulting in mutual *distrust* during the Katrina response [italics added]" [113]. This lack of a common understanding and trust also impacted the medical response where "Deployment confusion, uncertainty about mission assignments, and government red tape delayed medical care" [114]. Considering the significant demand for housing of the many refugees and failed interoperability, "FEMA failed to take advantage of the Department of Housing and Urban Development's expertise in large scale housing challenges" [115].

Even the National Response Plan (NRP), designed to provide a large-scale emergency response framework, includes five "levels of multi-agency centers for information sharing under the NPR [sic]. What is of concern is the number of offices information would theoretically have to travel through to get DHS in touch with a Local Center" [116].

Presumably, well documented and specific information-sharing architecture between these major government organizations was either not designed, not performed, or if performed, not developed using best practices; and if performed with best practices, it was not employed—or these results of failure would not have been reported. This deficiency was both apparent in 9/11 and Pearl Harbor from an intelligence perspective, and in Katrina from an overall failure in planning or implementation.

Bottom line: The first Pearl Harbor deficiency of organization (which we will shorten to "PHD #1"), means that organizational responsibilities overlap and ambiguous or nonexistent operating procedures produce dangerous interoperability risks and failures. Apparently, detailed and proper (best-practices) planning for efficient and effective information sharing among these organizations has not been performed.

2.4.2.2 Failure of Assumption Revisited

Review and Analysis The PHD #2 Assumption deficiency was defined as "Supervisory officials cannot safely take anything for granted in the alerting of subordinates" [117]. We have expanded this definition to include a leader or subordinate's assumption of what other individuals or organizations intend to do, are doing, or have done. This assumption should be assessed in terms of the consequences of being wrong. For example, in their interview with key law enforcement officials after 9/11, the congressional committee members were told that [118]:

> Attorney General Ashcroft was briefed by the CIA in May [2001] and by Pickard [acting FBI Director, summer 2001] in early July about the danger.
>
> The Attorney General told us he asked Pickard whether there was intelligence about attacks in the United States and that Pickard said no. ...[and while] he could not assure Ashcroft that there would be no attacks in the United States, the reports of threats were related to overseas targets. Ashcroft said he therefore *assumed* [italics added] the FBI was doing what it needed to do. He acknowledged that in retrospect, this was a dangerous assumption. He did not ask the FBI what it was doing in response to the threats and did not task it to take any specific action. He also did not direct the INS [Immigration and Naturalization Service], then still part of the Department of Justice, to take any specific action.

This "dangerous assumption" was comparable to Rear Admiral Stark assuming that Kimmel was deploying the fleet after the November 27, 1941 war warning message—and failing to verify his assumption. In the case of Katrina, these assumptions were built into the NRP [119]:

> This tension was reflected in the National Response Plan [NRP] before Katrina. The Catastrophic Incident Annex assumes that local response capabilities may be "insufficient," as they will be "quickly overwhelmed." But the NRP plan states federal resources will only be integrated into the response effort *upon a request by state and local authorities* [italics in original] and *assumes* [italics added] state and local officials will be able to do the integrating themselves.

These assumptions plagued follow-up on support expectations as well, when [120]:

> according to Louisiana and FEMA officials, state and local officials verbally requested specific assets or commodities during conference calls that were never fulfilled. In these cases no immediate action was taken because FEMA officials *assumed* [italics added] the state would follow up the verbal requests with official written requests.

It is therefore essential to design operational and system architecture based on established, agreed, and verified evidence, not expectations or assumption alone—especially when lives and property are at stake—holistic interoperability was needed, but absent.

Bottom line The PHD #2 Assumption has been expanded to include planners as well as leaders who make assumptions when important consequences are in the balance, and fail to verify these assumptions. Well-planned IT/communication systems that eliminate the need for assumptions, and well-trained leadership who avoid acting on assumptions in critical scenarios would mitigate or eliminate this deficiency.

2.4.2.3 Failure of Omission Revisited

Review and Analysis The original PHD #3 Omission was stated as "any doubt as to whether outposts should be given information should always be resolved in favor of supplying the information" [121]. In other words, if people or organizations would benefit from having information in relevant public safety and emergency response scenarios, then the appropriate information should be provided. In the case of the 9/11 attack within the United States, it was understandable that many subordinate individuals or organizations would be "out of the loop," but even [122]:

> At more senior levels, communication was poor. Senior military and FAA leaders had no effective communication with each other. The chain of command did not function well. The President could not reach some senior officials. The Secretary of Defense did not enter the chain of command until the morning's key events were over. Air National Guard units with different rules of engagement were scrambled without the knowledge of the President, NORAD [North American Air Defense Command], or the National Military Command Center [the principal Pentagon command center].

Omission of key organizations cost the lives of those who were able to reach the emergency (911) operators from the World Trade Center after the first plane struck [123].

> The NYPD's 911 operators and FDNY dispatch were not adequately integrated into the emergency response. In several ways, the 911 system was not ready to cope with a major disaster. These operators and dispatchers were one of the only sources of information for individuals at and above the impact zone of the towers. The FDNY ordered both towers fully evacuated by 8:57, but this guidance was not conveyed to 911 operators and FDNY dispatchers, who for the next hour often continued to advise civilians not to self-evacuate, regardless of whether they were above or below the impact zones. Nor were 911 operators or FDNY dispatchers advised that rooftop rescues had been ruled out. This failure may have been harmful to civilians on the upper floors of the South Tower who called 911 and were not told that their only evacuation hope was to attempt to descend, not to ascend.

Again, proper planning of an information-sharing architecture would have accounted for critical command elements and ensured their inclusion in the information-sharing architecture for situational awareness and command coordination [124].

> The preparation for and response to Hurricane Katrina show we are still an analog government in a digital age. We must recognize that we are woefully incapable of storing, moving, and accessing information—especially in times of crisis.

Many of the problems we have identified can be categorized as "information gaps"—or at least problems with information-related implications, or failures to act decisively because information was sketchy at best. Better information would have been an optimal weapon against Katrina. Information sent to the right people at the right place at the right time. Information moved within agencies, across departments, and between jurisdictions of government as well. Seamlessly. Securely. Efficiently.

This book will describe how to achieve this vision for public safety and emergency response, beyond the rudimentary planning performed today.

After Pearl Harbor, 9/11, and Katrina, the proper planning for detailed information sharing had not occurred; key organizations and individuals were omitted from what little information-sharing architecture existed. Again, either it was poorly planned—or not planned—at the required detailed operational level, or detailed plans existed but were not used. Either way, interoperability failed independent of the myriad of telecommunications or radio technologies implemented at the time by civilian authorities.

Bottom Line No organization which could potentially play a role in the prevention or mitigation of an incident or disaster should be omitted from detailed planning of the relevant information-sharing architecture. Excluding them is a deficiency in planning, which will inevitably result in preventable disaster or unmitigated consequences from interoperability failure. Thus, we have expanded the PHD #3 Omission to include not only subordinates in the chain of command, but all individuals and organizations (stakeholders) with a potential role in prevention, detection, alert, protection, response, and recovery.

2.4.2.4 Failure of Verification Revisited

Review and Analysis The original PHD #4 Verification was defined as "the delegation of authority or the issuance of orders entails the duty of inspection to determine that the official mandate is properly exercised" [125]. This verification was of vital importance when expected actions expected from orders sent to Hawaii were interpreted and performed quite differently than Washington had imagined. The resulting misinterpretation was not perceived since Washington never verified the orders were being carried out as intended. Thus, automatic verification—or incessant human prompting for verification is a critical element of an information-sharing architecture.

During the 9/11 attack, the Northeast Air Defense Sector (NEADS) had been given the necessary order to shoot down commercial airliners not obeying fighter direction. In responding to the congressional inquiry leading to the *9/11 Report* [126],

> The NEADS [Northeast Air Defense Sector] commander told us he did not pass along the order because he was unaware of its ramifications. Both the mission commander and the senior weapons director indicated they did not pass the order to the fighters circling Washington and New York because they were unsure how the pilots would, or should, proceed with this guidance. In short, while leaders in Washington believed that the fighters above them had been instructed to "take out"

> hostile aircraft, the only orders actually conveyed to the pilots were to "ID type and tail."

The leaders assumed their order had been executed, but it had not—and they did not follow-up by verifying that it had been done—they merely assumed it. This PHD #4 Verification also rung true again during Hurricane Katrina, when critical information regarding the status of the levees was not confirmed [127].

> Perhaps the single most important piece of information during Katrina was confirmation of the levee breaches in New Orleans. Beyond the importance of the information itself, the implications of the information determined whether or not Katrina would be just another bad storm in New Orleans or the nation's worst-case hurricane disaster. Because DHS failed to anticipate the likely consequences of the storm and procure the buses, boats, and aircraft that were ultimately necessary to evacuate the flooded city prior to Katrina's landfall, the next critical decision point of the federal response became confirmation of the levee breaches. If the levees breached and flooded a large portion of the city, then the flooded city would have to be completely evacuated. Any delay in confirming the breaches would result in a delay in the post-landfall evacuation of the city.
>
> On Monday evening the HSOC [Homeland Security Operations Center] failed to conclude that levees had breached in New Orleans despite a FEMA eyewitness report and the presence of numerous Coast Guard air assets over New Orleans, which had the ability to communicate to almost anywhere in the country.

Lack of verification for such critical situational awareness must be included in our definition of this deficiency, particularly when—just like with Pearl Harbor—there were both clear intelligence warnings and experts had predicted the very high-level attack plan used by the Japanese on December 7th. As the *Katrina Report* states, President Bush "apparently forgetting what he had been told at the August 28th briefing, stated: 'I don't think anybody anticipated the breach of the levees'." [128]. As with the Farthing and Martin-Bellinger reports before Pearl Harbor, many had predicted the levee breeches well before Katrina—but there was apparently no information-sharing architecture developed for just such a scenario, which would have included early warning sensors (such as those "$5.00" detectors employed to alert homeowners of basement drain backup). The Hurricane Pam exercise reflected recognition by all levels of government of the dangers of a category 4 or 5 hurricane striking New Orleans [129] but "implementation of lessons learned from Hurricane Pam was incomplete." Thus, we have expanded the PHD #4 Verification definition to include verification of critical information under the given circumstances from any source, not only the need to follow-up on the issuance of critical issued orders.

Bottom Line The failure of verification deficiency (PHD #4) is the assumption of critical information in crises is an important deficiency, and IT/communication system architecture must be designed to verify such information automatically, or to prompt human operators to do so. This PHD #4 therefore includes failure to include verification capability within the information-sharing architecture (operational and system), particularly for predicted scenarios like the Pearl Harbor attack and the impact of a category (Cat) 4 or 5 hurricane on the New Orleans levees.

2.4.2.5 Failure of Supervision Revisited

Review and Analysis In the original PHD #5, the Failure of Supervision meant that "the implementation of official orders must be followed with closest supervision" [130]. This supervision is particularly vital in tactical public safety and emergency response scenarios, such as was found on 9/11 in New York City, when [131]:

> Effective decision making in New York was hampered by problems in command and control and in internal communications. Within the Fire Department of New York, this was true for several reasons: ... more units were actually dispatched than were ordered by the chiefs; some units self-dispatched; and once units arrived at the World Trade Center, they were neither comprehensively accounted for nor coordinated.

Leadership failure to properly supervise activities—which in the case of prelandfall Katrina meant understanding the risks and taking timely action to mitigate those risks—was again evident despite expert warning when [132]:

> The day before Katrina hit, the NWS [National Weather Service] office in Slidell, Louisiana issued a warning saying, "MOST OF THE AREA WILL BE UNINHABITABLE FOR WEEKS ... PERHAPS LONGER ... HUMAN SUFFERING INCREDIBLE BY MODERN STANDARDS"

The warning—like many others—was clear and concise, and existing response plans, though dependent on human initiative for activation, provided the necessary direction [133]:

> In addition to the Louisiana state plan, local governments have emergency evacuation plans. The City of New Orleans Comprehensive Emergency Management Plan ("New Orleans Plan") provides: "The authority to order the evacuation of residents threatened by an approaching hurricane is conferred to the Governor by Louisiana statute." But this power "is also delegated to each political subdivision of the State by Executive Order."

The New Orleans Plan further explains: "This authority empowers the chief elected official of New Orleans, the Mayor of New Orleans, to order the evacuation of the parish residents threatened by an approaching hurricane." Under this authority, the Mayor of New Orleans is responsible for giving the order for a mandatory evacuation and supervising the actual evacuation of his population. The Mayor's Office of Emergency Preparedness must "coordinate with the State . . . on elements of evacuation" and "assist in directing the transportation of evacuees to staging areas."

The plan provides the direction, but no means of supervision or automated verification to ensure this required supervision is being performed, but [134]

- Despite adequate warning 56 hours before landfall, Governor Blanco and Mayor Nagin delayed ordering a mandatory evacuation in New Orleans until 19 hours before landfall.
- The failure to order timely mandatory evacuations, Mayor Nagin's decision to shelter but not evacuate the remaining population, and decisions of individuals led to an incomplete evacuation.
- The incomplete pre-landfall evacuation led to deaths, thousands of dangerous rescues, and horrible conditions for those who remained.

The failure to supervise activities essential for success in well-known scenarios—particularly when existing plans foresee such scenarios—is a major deficiency. The information sharing that failed in this case was that resources from all governance levels were not engaged in time to compensate for the ill-advised "shelter in place" order, particularly for people living behind inadequate Cat 4 levees below water level and well-below storm-surge level. For these reasons, we will enhance PHD #5 Supervision to include not only supervision of one's own subordinates, but direction and orders provided by other critical authorities and plans when safety-of-life crisis looms. As has been demonstrated again, failing to do so risks inadequate situational awareness, command confusion, and misdirection of responders-receivers and other stakeholders when most needed.

Bottom Line Failed supervision during critical events, much less day-to-day activities that could become critical, is a particularly insidious deficiency to detect and mitigate because (1) many responders-receivers depend upon the best-informed decision to be made regarding their activities, and (2) those who select and back government leaders and planners have the right to expect best practices are employed in developing emergency plans and following through with them, particularly in proven crisis conditions with forewarning.

2.4.2.6 Failure of Alertness Revisited

Review and Analysis The original PHD #6 for Alertness applies universally when it states "The maintenance of alertness to responsibility must be insured through repetition" [135]. The pre-2000 Y2K or Millennium threat apparently served to trigger significant antiterrorism concerns and coordinated action, but then this alertness subsided [136]:

> In the period between December 1999 and early January 2000, information about terrorism flowed widely and abundantly. The flow from the FBI was particularly remarkable because the FBI at other times shared almost no information. That from the intelligence community was also remarkable, because some of it reached officials—local airport managers and local police departments—who had not seen such information before and would not see it again before 9/11, if then. And the terrorist threat, in the United States even more than abroad, engaged the frequent attention of high officials in the executive branch and leaders in both houses of Congress. ...
>
> ...*After the millennium alert, the government relaxed* [italics added]. Counterterrorism went back to being a secret preserve for segments of the FBI, the Counterterrorist Center, and the Counterterrorism Security Group. But the experi-

> ence showed that the government was capable of mobilizing itself for an alert against terrorism. While one factor was the preexistence of widespread concern about Y2K, another, at least equally important, was simply shared information.

The leaders and planners were well alerted to Katrina by accurate predictions form the National Weather Service (NWS).

> federal officials fully informed Blanco [Louisiana Governor] and New Orleans Mayor C. Ray Nagin about the threat to New Orleans. On the evening of August 27, National Hurricane Center Director Max Mayfield called Blanco and later spoke to Nagin about the power of Hurricane Katrina. Also on Sunday, President Bush called Blanco to express his concern for the people of New Orleans and the dangers they faced and urged a mandatory evacuation. On Sunday, the Slidell Office of the National Weather Service issued a very strongly worded warning at approximately 10:00 a.m. [137]:
>
>> Devastating damage expected . . . Hurricane Katrina . . . a most powerful hurricane with unprecedented strength . . . rivaling the intensity of Hurricane Camille of 1969... Most of the area will be uninhabitable for weeks . . . perhaps longer. At least half of well constructed homes will have roof and wall failure. All gabled roofs will fail . . . leaving those homes severely damaged or destroyed... Water shortages will make human suffering incredible by modern standards.

This alertness failure was prevalent in the case of Katrina, in which predictions and exercises had shown the consequences of exactly the type of storm Katrina had become—and was known to be—before landfall, yet [138]...

> despite express authority to commandeer resources and enforce or facilitate the evacuation of the City of New Orleans and despite recognition of the probability that Hurricane Katrina would cause breaches of the levees and flooding of the city, Blanco and Nagin did not exercise those authorities by declaring a mandatory evacuation and enforcing it or using state and city resources to facilitate the evacuation of those who could not or would not, absent extraordinary measures and assistance, evacuate. This extraordinary storm required extraordinary measures, which the Governor and Mayor did not take.

Whether Pearl Harbor, 9/11, or Katrina, the operational information-sharing architectures sounded the alert, but it was dependence on human initiative—particularly that of the leadership—which served as the "weak link" in prevention and response.

Bottom Line Alertness is too often hampered by the deficiencies of those who are alerted, particularly when they are not personally involved in the actual activities of prevention and response. Thus, a deficiency exists in any public safety and emergency response system when the decision to take action, much less the awareness of the alert, is limited to few individuals who neither have the insight, imagination nor training to respond optimally in a crisis. Should public safety and emergency response information-sharing architecture fail to provide appropriate alerting—pushing awareness ahead of orders to act—then failed initiative will invariably again lead to otherwise avoidable lives lost and property destroyed.

2.4.2.7 Failure of Complacency Revisited

Review and Analysis Failed alertness can be considered the offspring of the Failure of Complacency (PHD #7). This complacency was apparent in the activities of the leadership in the days leading to 9/11 [139]:

> In sum, the domestic agencies never mobilized in response to the threat. They did not have direction, and did not have a plan to institute. The borders were not hardened. Transportation systems were not fortified. Electronic surveillance was not targeted against a domestic threat. State and local law enforcement were not marshaled to augment the FBI's efforts. The public was not warned.

This same complacency was apparent by those top leaders removed from the direct impact of the disaster before, during, and immediately after Katrina's landfall [140].

> Indeed, Mr. Chairman, so nonchalant were top officials in the aftermath of Katrina that they did not even break from their vacations to attend to the disaster. While [President] Bush, Brown [FEMA Director] and Chertoff [DHS Secretary] had all been briefed on August 28th of the possibility of a levee failure, Secretary Chertoff made a trip to Atlanta to visit the CDC to discuss avian flu on the 29th, the day of landfall, and did not become fully aware of the levee failures until mid-day on August 30th, some 30 hours after the initial breach. Chertoff did not declare Katrina an Incident of National Significance until August 30th, 36 hours after landfall, when he also designated Brown principal federal officer. It must be asked why this was not called before the storm struck and whether it contributed to the delay in the response.
>
> Secretary of Defense Rumsfeld was busy taking in a baseball game in San Diego on the night of the 29th, in anticipation of Bush's visit to deliver a speech on the war in Iraq the next day.

Perhaps these officials considered they had little to do with engaging the necessary preparedness or response, and assumed (PHD #2) that others would handle whatever was needed, though failed to verify this assumption (PHD #4).

Bottom Line Complacency is a state of mind that leaders impacting public safety and emergency response apparently cannot afford to enjoy—particularly given the lack of moment-to-moment situational awareness possible regarding the levees, which was known to be lacking. Again, the role of information-sharing architecture here would have been to provide the required awareness and verified, as well as coordinated, command and supervision for evacuation. Even in the presence of complacency, a properly designed information-sharing architecture would not be reliant on the imagination and initiative of single individuals, no matter their governance role. There is too much at stake to continue time and again to make this mistake, and technology has been available for many years to mitigate misplaced complacency.

2.4.2.8 Failure of Intelligence Revisited

Review and Analysis The proper information-sharing architecture for intelligence was apparent in the Pearl Harbor disaster. As it was shown, "the coordination and proper evaluation of intelligence in times of stress must be insured by continuity of service and centralization of responsibility in competent officials" [141] (PHD #8). This failure of intelligence was one of the primary causes for the loss of life and property on 9/11 [142]:

> Those working counterterrorism matters did so despite limited intelligence collection and strategic analysis capabilities, a limited capacity to share information both internally and externally, insufficient training, perceived legal barriers to sharing information, and inadequate resources.

In addition [143],

> The U.S. government did not find a way of pooling intelligence and using it to guide the planning and assignment of responsibilities for joint operations involving entities as disparate as the CIA, the FBI, the State Department, the military, and the agencies involved in homeland security.

Uncertainty over the rules governing information sharing and who would interpret intelligence left the best analysts out of the loop—just like intelligence organizations before the Pearl Harbor attack. As a result of this confusion, the criminal investigators who were knowledgeable about al Qaeda and experienced with criminal investigative techniques, including finding suspects and possible criminal charges, were thus excluded from the search [144].

As the bulwark against terrorism in the United States [145],

> the FBI did not have an effective intelligence collection effort. Collection of intelligence from human sources was limited, and agents were inadequately trained. Only three days of a 16-week agents' course were devoted to counterintelligence and counterterrorism, and most subsequent training was received on the job. The FBI did not have an adequate mechanism for validating source reporting, nor did it have a system for adequately tracking and sharing source reporting, either internally or externally. The FBI did not dedicate sufficient resources to the surveillance and translation needs of counterterrorism agents. It lacked sufficient translators proficient in Arabic and other key languages, resulting in a significant backlog of un-translated intercepts.

This failed operational architecture drove, and similarly impacted, the technologies employed for sharing information, because [146]

> the FBI's information systems were woefully inadequate. The FBI lacked the ability to know what it knew: there was no effective mechanism for capturing or sharing its institutional knowledge. FBI agents did create records of interviews and other investigative efforts, but there were no reports officers to condense the information into meaningful intelligence that could be retrieved and disseminated.

In the case of Katrina, we expand this PHD #8 to include the ability to forecast the impact of a disaster and predict its consequences. As we have shown, predictions by the NWS were timely, accurate, and well communicated to all concerned, including the public. The deficiency did not lie within the forecasting, but in the generation of actionable information, much less action, on the part of the leaders who received these forecasts. Therefore, the information-sharing architecture must account for failed leadership initiative by awaiting its timely response, forecasting the results of failed initiative, and alerting other leadership to take specific action before its too late (i.e, circumvent the command chain to avoid ultimate human disaster).

Bottom Line The Intelligence PHD #8 has been expanded to include forecasting and consequences assessment as well as classic intelligence collection and analysis. We also include the actionable information derived from this intelligence—and how it is being used—as it must be simultaneously addressed in any architecture deficiency assessment. In other words, not only should the consequences of the incident/event itself be forecast, but also the impact of leadership decisions from these forecasts on those consequences—a "snowball-rolling-down-the-mountain" effect. Thus, intelligence and forecasting (consequence prediction) must regressively account for the actions of those who are provided with this information in order to ensure complete situational awareness and command coordination.

2.4.2.9 Failure of Attitude Revisited

Review and Analysis One of the more well-known deficiencies apparent at Pearl Harbor is a common characteristic in hierarchical organizations, where those in leadership positions feel a certain entitlement associated with their clout. This entitlement can lead to a breakdown in understanding, because "the unapproachable or superior attitude of officials is fatal; there should never be any hesitancy in asking for clarification of instructions or in seeking advice on matters that are in doubt" [147]. This behavior produces a gap in information sharing that negatively impacts both operability and interoperability. We will expand this Attitude PHD #9 to include difficulty in accepting others' authority or their participation in a common goal—which is particularly damaging to the development of multiorganizational, or joint, operational architecture. Before 9/11, this move to accept and thrive in a joint operational architecture had been adopted by the U.S. military [148]:

> in organizing national defense, the Goldwater-Nichols legislation of 1986 created joint commands for operations in the field, the Unified Command Plan. The services—the Army, Navy, Air Force, and Marine Corps—organize, train, and equip their people and units to perform their missions. Then they assign personnel and units to the joint combatant commander, like the commanding general of the Central Command (CENTCOM). The Goldwater-Nichols Act required officers to serve tours outside their service in order to win promotion. The culture of the Defense Department was transformed, its collective mind-set moved from service-specific to "joint," and its operations became more integrated.

Unfortunately, this modern understanding of the benefit of joint or combined operations has not pervaded the civilian public safety and emergency management organizations in general [149].

> The ESU [Emergency Service Unit] team on the 31st floor conferred with the FDNY personnel there to ensure that they, too, knew that they had to evacuate, then proceeded down stairwell B. During the descent, they reported seeing many firefighters who were resting and did not seem to be in the process of evacuating. They further reported advising these firefighters to evacuate, but said that at times they were not acknowledged. In the opinion of one of the ESU officers, some of these firefighters essentially refused to take orders from cops. At least one firefighter who was in the North Tower has supported that assessment, stating that he was not going to take an evacuation instruction from a cop that morning. However, another firefighter reports that ESU officers ran past him without advising him to evacuate.

Similar attempts have failed on the civilian side, to the detriment (again) of public safety [150].

> The chief vehicle for INS and for state and local participation in law enforcement was the Joint Terrorism Task Force (JTTF), first tried out in New York City in 1980 in response to a spate of incidents involving domestic terrorist organizations. This task force was managed by the New York Field Office of the FBI, and its existence provided an opportunity to exchange information and, as happened after the first World Trade Center bombing, to enlist local officers, as well as other agency representatives, as partners in the FBI investigation. The FBI expanded the number of JTTFs throughout the 1990s, and by 9/11 there were 34. While useful, the JTTFs had limitations. They set priorities in accordance with regional and field office concerns, and most were not fully staffed. Many state and local entities believed they had little to gain from having a full-time representative on a JTTF.

Once intelligence (PHD #8) had failed to alert (PHD #6) the Federal Aviation Administration (FAA) and others of the imminent attack on the United States, prevention failed and coordinated response was required [151].

> To some degree, the Mayor's directive for incident command was followed on 9/11. It was clear that the lead response agency was the FDNY, and that the other responding local, federal, bistate, and state agencies acted in a supporting role. There was a tacit understanding that FDNY personnel would have primary responsibility for evacuating civilians who were above the ground floors of the Twin Towers, while NYPD and PAPD [Port Authority Police Department] personnel would be in charge of evacuating civilians from the WTC complex once they reached ground level.

The NYPD also greatly assisted responding FDNY units by clearing emergency lanes to the WTC. In addition, coordination occurred at high levels of command. For example, the Mayor and Police Commissioner consulted with the Chief of the Department of the FDNY at approximately 9:20. There were other instances of coordination at operational levels, and information was shared on an ad hoc basis. For example, an NYPD ESU team passed the news of their evacuation order to firefighters in the North Tower.

It is also clear, however, that the response operations lacked the kind of integrated communications and unified command contemplated in the directive. These problems existed both within and among individual responding agencies.

As was shown above, multiple federal, state, and local emergency response organizations did not coordinate adequately in the Katrina aftermath. Typical of the difficulties were issues that arose between Disaster Medical Assistance Teams (DMATs) and the FEMA National Disaster Medical System (NDMS) [152]:

> Beall [NDMS Chief Jack Beall] agreed there was tension and speculated DMAT [Disaster Medical Assistance Team] members are accustomed to being in control of their environments and are not used to taking orders from federal officials. He also said most of the FEMA NDMS officials deployed during Katrina and giving orders to DMATs were unseasoned and their inexperience contributed to the friction.

Bottom Line We are expanding the Failure of Attitude (PHD #9) to include a cultural stovepipe effect that resides largely among civilian organizations at different governance levels, even if involved in the same public safety and emergency response discipline. This cultural stovepipe is comparable to the deficient behavior that arises among different public safety and emergency response disciplines and their organizations at the same governance level.

2.4.2.10 Failure of Imagination Revisited

Review and Analysis PHD #10, the failure of imagination, was a common deficiency in all three tragedies. As stated in the Pearl Harbor Report, "There is no substitute for imagination and resourcefulness on the part of supervisory and intelligence officials" [153]. The likelihood of aircraft as weapons had been imagined before 9/11, but this imagination was not shared [154].

> The most important failure was one of imagination. We do not believe leaders understood the gravity of the threat.

But such a possibility as occurred on 9/11 had been foreseen [155].

> In early August 1999, the FAA's Civil Aviation Security intelligence office summarized the Bin Ladin hijacking threat. After a solid recitation of all the information available on this topic, the paper identified a few principal scenarios, one of which was a "suicide hijacking operation." The FAA analysts judged such an operation unlikely, because "it does not offer an opportunity for dialogue to achieve the key goal of obtaining Rahman and other key captive extremists."

In addition to failed imagination, the FAA analysts lacked insight (PHD #15) as to the terrorists' mindset—as plain as it had been in the Middle East for decades. The announcement to the occupants of the South Tower to remain in place after the North Tower had been hit, suggests that the idea of this airplane strike was unimaginable to those with the gravest responsibility [156].

> At about this time, an announcement over the public-address system in the South Tower stated that the incident had occurred in the other building and advised tenants, generally, that their building was safe and that they should remain on or return to their offices or floors. ... We do not know the reason for the announcement, as both the deputy fire safety director believed to have made it and the director of fire

> safety for the WTC complex perished in the South Tower's collapse. Clearly, however, the prospect of another plane hitting the second building was beyond the contemplation of anyone giving advice. According to one of the first fire chiefs to arrive, such a scenario was unimaginable, "beyond our consciousness." As a result of the announcement, many civilians remained on their floors. Others reversed their evacuation and went back up.

Since the forecast of Katrina's outcome had both been made and publicized, so it in itself had been clearly imagined. Unfortunately, the forecast of experts only fell within the imagination of the leadership after it was too late. Anticipation, imagination, and foresight are all necessary in the absence of hard facts in the form of validated predictions, which were available even to the public at large before Katrina made landfall.

Bottom Line Imagination is the seed of initiative; it is at least the imagination to envision the world in the next moment if one does not demonstrate initiative "now." We see that the failure to imagine the possible, and prepare actively for it, is a critical leadership deficiency. It is more so when experts predict these events in advance and their warnings are not heeded.

2.4.2.11 Failure of Communications Revisited

Review and Analysis In PHD #11, the failure of communications, the original deficiency stated only that "communications must be characterized by clarity, forthrightness and appropriateness" [157]. We will expand that definition to include all elements of communications from transmission to reception, including intelligibility and understanding of what is sent to the listener (the original PHD #11). Thus, not only verbal communications and its understandability, but all forms of communications media—including data and video communications—must be included in this deficiency. This deficiency is difficult at times to differentiate from PHD #2 Omission, in which organizations are simply not planned to be "in the loop." But, even if these agencies are in the loop, this is no guarantee the information will be properly exchanged or understood or trusted at its intended destination.

Many of the failures of communications that occurred on 9/11 were due to a lack of operational knowledge and procedures, even those as simple as the depression of a single button. These failures are not tied to the performance of the technologies themselves (much as the radar at Pearl Harbor did what it was designed to do) [158].

> Earlier, an FDNY chief had asked building personnel to activate the repeater channel, which would enable greatly-enhanced FDNY portable radio communications in the high-rises. One button on the repeater system activation console in the North Tower was pressed at 8:54, though it is unclear by whom. As a result of this activation, communication became possible between FDNY portable radios on the repeater channel. In addition, the repeater's master handset at the fire safety desk could hear communications made by FDNY portable radios on the repeater channel. The activation of transmission on the master handset required, however, that a second button be pressed. That second button was never activated on the morning of September 11.

Although some firefighters in the North Tower, the second tower to collapse, received the evacuation order after the South Tower collapsed, many did not [159].

> Other firefighters did not receive the evacuation transmissions, for one of four reasons: First, some FDNY radios did not pick up the transmission because of the difficulties of radio communications in high-rises. Second, the numbers trying to use tactical 1 after the South Tower collapsed may have drowned out some evacuation instructions. According to one FDNY lieutenant who was on the 31st floor of the North Tower at the time, "[Tactical] channel 1 just might have been so bogged down that it may have been impossible to get that order through." Third, some firefighters in the North Tower were off-duty and did not have radios. Fourth, some firefighters in the North Tower had been dispatched to the South Tower and likely were on the different tactical channel assigned to that tower.

Terrestrial physical communications—largely dependent upon elevated structures in the flat terrain of the southern United States—was largely devastated by Katrina, but this was not the only kind of communication failure that occurred [160].

> Poor situational awareness and its resulting effect on command and control contributed to the negative effects of inaccurate media reports because public officials lacked access to the facts to address media reports. Throughout the early days of the response, media reports from New Orleans featured rampant looting, gunfire, crime, and lawlessness, including murders and alleged sexual assaults at the Superdome and Convention Center. Few of these reports were substantiated, and those that were—such as the gunfire—were later understood to be actually coming from individuals trapped and trying to attract the attention of rescuers in helicopters.

Other Katrina communication failures were endemic to failed planning [161].

> Blum also noted that "one critical area where we lack integration is in interoperable communications. National Guard units do not have the equipment necessary to effectively share information with Title 10 forces. This caused significant challenges on the ground that then bubbled up the chains."

Bottom Line Operational and system architectures for public safety and emergency response must specifically address assured communications independent of the scenario. Information-sharing architectures must agree beforehand on the semantics—such as natural speech with common terminology—to be used, common data communications, and so forth. The resulting IT/communication systems must be designed to minimize end-to-end ambiguity and confusion, so assessments of these systems must include the full meaning of the term "communications" in the deficiency checklist.

2.4.2.12 Failure of Paraphrase Revisited

Review and Analysis The Failure of Paraphrase (PHD #12) is the interpretation and retransmission, or relay, of modified information either (1) to fill in communication gaps of understanding with assumption, or (2) to somehow filter or distort the information conveyed, resulting in misunderstanding and erroneous assumptions. As Morgan wrote in the congressional report on Pearl Harbor [162], "There is great danger in careless paraphrase of information received and every effort should be made to insure that the paraphrased material reflects the true meaning and significance of the original."

We will expand PHD #12 to include paraphrasing in the interpretation, and misinterpretation, of laws and guidelines for sharing information. In the months leading up to 9/11, such paraphrasing of laws reduced the chance of capturing some of the terrorist's intent on doing harm before 9/11 [163, 164].

> The FBI had different tools for law enforcement and intelligence. For criminal matters, it could apply for and use traditional criminal warrants. For intelligence matters involving international terrorism, however, the rules were different. ...
> ...Agents in the field began to believe—incorrectly—that no FISA [Foreign Intelligence Surveillance Act (1978)] information could be shared with agents working on criminal investigations.
>
> This perception evolved into the still more exaggerated belief that the FBI could not share any intelligence information with criminal investigators, even if no FISA procedures had been used. Thus, relevant information from the National Security Agency and the CIA often failed to make its way to criminal investigators.

Paraphrasing also affected evacuation orders before Hurricane Katrina [165]:

> A wide variety of terms were used to describe the levels of evacuation orders, indicating a lack of clarity and a potential point of confusion for the resident population. For example, the different parishes used a wide variety of terms to describe the level of evacuation imposed before declaring a mandatory evacuation. These terms included a "precautionary" evacuation, a "voluntary" evacuation, a "recommended" evacuation, a "highly recommended" evacuation, and a "highly suggested" evacuation. It appeared many of these officials were bending over backward to avoid using the term mandatory.

More importantly, serious disruption to public safety and emergency response can occur when unsubstantiated reports are paraphrased into significant, but false, situational awareness for responders-receivers [166].

> Many stories of rape, murder, and general lawlessness were at best unsubstantiated, at worst simply false. And that's too bad, because this storm needed no exaggeration. As discussed in our report, widely-distributed uncorroborated rumors caused resources to be deployed, and important time and energy wasted, chasing down the imaginary. Already traumatized people in the Superdome and elsewhere, listening to their transistor radios, were further panicked.

Bottom Line The paraphrase deficiency is critical to understand in public safety and emergency response systems in which messages from callers are often relayed through the dispatcher to the responders-receivers. Moreover, misinterpretation of the intent of laws or regulations protecting how information is collected and for what purpose, as well as other laws and regulations, can risk decisive action when the public safety is at risk. For these reasons, information sharing must convey precise meaning and provide inherent means for validation and assurance.

2.4.2.13 Failure of Adaptability Revisited

Review and Analysis The Failure of Adaptability (PHD #13) is the deficiency associated with planning prevention and response to a fixed set of preconditions and events, rather than maintaining adaptability to the form taken by the real—and not predicted—threat. PHD #13 was stated as "Procedures must be sufficiently flexible to meet the exigencies of unusual situations" [167].

The Adaptability PHD #13 and Organization PHD #1 were prevalent shortly before 9/11 [168].

> The September 11 attacks fell into the void between the foreign and domestic threats. The foreign intelligence agencies were watching overseas, alert to foreign threats to U.S. interests there. The domestic agencies were waiting for evidence of a domestic threat from sleeper cells within the United States. No one was looking for a foreign threat to domestic targets. The threat that was coming was not from sleeper cells. It was foreign—but from foreigners who had infiltrated into the United States.

So, the intelligence and criminal justice systems did not adapt to account for the new threat. Similarly, Hurricane Katrina did not fit the plans of emergency planners [169].

> The failure of initiative was also a failure of agility. Response plans at all levels of government lacked flexibility and adaptability. Inflexible procedures often delayed the response. Officials at all levels seemed to be waiting for the disaster that fit their plans, rather than planning and building scalable capacities to meet whatever Mother Nature threw at them.

Responders-receivers rely on well-practiced tasks and activities as if they were drawn from a toolset, with different actions required as the scenario evolves. They do not and cannot follow a single game plan throughout the incident or event, but necessarily must adapt practiced techniques to the actual situation confronting them.

Bottom Line The failure to adapt is a major deficiency in public safety and emergency response and, as will be shown, such adaptation is part of best practice for all responders-receivers. Thus, operational and system architecture must be designed to accommodate this flexibility, because no two incidents are identical, even the most mundane. If not, the response will be fundamentally flawed.

2.4.2.14 Failure of Disclosure Revisited

Review and Analysis "Restriction of highly confidential information to a minimum number of officials, while often necessary, should not be carried to the point of prejudicing the work of the organization" [170]. This Failure of Disclosure (PHD #14) reflects the policy to minimize the distribution of information gained from intelligence sources to protect not only the information and the fact it is known, but also avoid implication as to how it was collected. In the Pearl Harbor tragedy, failure to provide such information to trusted officials aided the Japanese success on December 7th. The identical situation aided surprise in the 9/11 attacks [171].

> Information Sharing
> We have already stressed the importance of intelligence analysis that can draw on all relevant sources of information. The biggest impediment to all-source analysis—to a greater likelihood of connecting the dots—is the human or systemic resistance to sharing information.
>
> The U.S. government has access to a vast amount of information...but has a weak system for processing and using what it has. In interviews around the government, official after official urged us to call attention to frustrations with the unglamorous "back office" side of government operations.

The Disclosure PHD #14 implies information is intentionally not provided to individuals or organizations for security reasons, which is supportable if it is justifiable under clear policy or doctrine. This PHD #14 differs from other deficiencies such as Omission (PHD #3) where planners do not (for whatever reason) include all necessary organizations in their information-sharing architecture; and it differs from PHD #9 Attitude in which planners do not consider certain individuals or organizations in this architecture because of institutional rivalry or Jealousy (PHD #20).

There was no apparent disclosure issue in the case of Hurricane Katrina. Predictions about landfall and likely damage were provided publicly, as well as to government leaders, well ahead of landfall. In fact, actionable information about evacuation was provided to the public, and there is no evidence that a Disclosure (PHD #14) deficiency occurred.

Bottom Line Information-sharing architecture must include filters to ensure that intelligence is converted into actionable information for responders/receivers as necessary, particularly if the event is major, while protecting sources and methods as much as possible. Failure to do so produces a critical deficiency in a moment-to-moment crisis situation.

2.4.2.15 Failure of Insight Revisited

Review and Analysis The Failure of Insight (PHD #15) is basically a lack of understanding of the threat and the environment or consequences of the threat [172], namely "There is a great danger of being blinded by the self-evident." This blindness was also evident in the hindsight of the events leading to 9/11 [173].

> We learned that the institutions charged with protecting our borders, civil aviation, and national security did not understand how grave this threat could be, and did not adjust their policies, plans, and practices to deter or defeat it. We learned of fault lines within our government—between foreign and domestic intelligence, and between and within agencies. We learned of the pervasive problems of managing and sharing information across a large and unwieldy government that had been built in a different era to confront different dangers.

Just as specific predictions were made regarding the form and timing of the Japanese attack on Pearl Harbor, a detailed and—it would turn out—accurate forecast of the effects of a major hurricane on New Orleans had recently been performed [174].

> We also examined why FEMA seemed unable to implement lessons that should have been learned well in advance of Katrina. There were the lessons of previous hurricanes. Further, FEMA officials participated in the now-widely-known exercise called Hurricane Pam in July 2004, an exercise that predicted with eerie similarity Katrina's impact on New Orleans, including an evacuation of a million people, overflowing levees, and the destruction of hundreds of thousands of buildings.

So despite experts predicting the consequences of both events for the leadership, the results were effectively ignored. In part, this result occurs because leaders at many governance levels may have no background in public safety or emergency response or the threats and their precursors or environment, or assume others will handle it, and that their preparations are "good enough." Nevertheless, these leaders can have significant impact on planning and sustainment of responder-receiver information-sharing architecture as well as strategic and tactical decision-making affecting prevention and response.

Bottom Line The Deficiency of Insight (PHD #15) can impact the improvement or development of information-sharing capabilities, which requires significant time and resources to achieve. This failed insight generally affects understanding of the threat and the corresponding scenarios for which information-sharing architecture is developed. Thus, it leads to built-in flaws in the IT/communication systems built in at-risk jurisdictions, in other words, failed leadership and deficient planning effects the technical systems put in place for public safety and emergency response.

2.4.2.16 Failure of Dissemination Revisited

Review and Analysis The Failure of Dissemination (PHD #16) was originally defined as "Officials should at all times give subordinates the benefit of significant information" [175]. This definition will be expanded to include dissemination outside one's own organization under other deficiencies, such as PHD #1 Organization, PHD #3 Omission, or PHD #14 Disclosure, in which external agencies are left "out of the loop." This failure, much like Omission or PHD #1 Organization, is based on original intent not to share the information with certain individuals or organizations, not on a lack of capability to do so, which we identify as PHD #11 Communications. In the 9/11 event, PHD #16 occurred within the FBI as FISA prevented dissemination of intelligence information to FBI agents who may have located one or more of the future 9/11 terrorists.

Unquestionably, the preparation and response to Hurricane Katrina demonstrated a failure to disseminate information [176].

> The Department of Defense has significant assets for the collection of intelligence, as well as communications and satellite equipment needed in all military operations ... [that] were not optimally used during the Katrina response.

This lack of dissemination reduced situational awareness available to DoD units responding to the hurricane. As a result, one of the Katrina Report's main findings was that " DOD lacked an information sharing protocol that would have enhanced joint situational awareness and communications between all military components" [177].

Bottom Line Information-sharing architectures must disseminate information with clear communications—no paraphrasing—to all those within their organizations, and verify this dissemination. This dissemination is critical in all phases of homeland security, from prevention and alert to response and recovery.

2.4.2.17 Failure of Inspection Revisited

Review and Analysis The Failure of Inspection (PHD #17) is founded on the Pearl Harbor experience, in which Congress determined that "an official who neglects to familiarize himself in detail with his organization should forfeit his responsibility" [177]. In other words, a leader or planner should be directly familiar with the people, process, and tools he or she directs or commands in public safety or emergency response—failing to do so suggests the individual be discontinued in that role. Such familiarity is highly likely for leaders and planners of responder-receiver organizations who have "come up through the ranks" to lead their organizations, but it is far less often true of the civilian leaders and planners who directly impact the capabilities of these organizations. As in Pearl Harbor, the failure of leaders, planners, and responder-receiver command elements to inspect their critical facilities (e.g., dispatch center, operations center, deployable facilities, etc.) is essentially dereliction of duty. This PHD #17 Inspection deficiency caused failed interoperability between NEADS and the FAA in the case of 9/11, and failed interoperability among DoD, FEMA, and state and local public safety and emergency response organizations in the case of Katrina.

Bottom Line Leader and planner familiarity with critical capabilities within their planning and operational domains is fundamentally important to operational and system architecture. As we will observe, it does not guarantee the use of best practices in their improvement or design, but it does offer at least the consideration that resources may be needed to do so. These leaders and planners need to verify operation of their architecture through each of it components; inspecting the people and tools that compose this architecture is an essential duty of their positions. Such inspection by experienced officials can be instrumental to uncovering other deficiencies, and if inexperienced, provides an ideal format to meet about their subordinates and understand their abilities and needs.

2.4.2.18 Failure of Preparedness Revisited

Review and Analysis One of the greatest deficiencies spanning Pearl Harbor, 9/11, and Katrina is the overarching state of preparedness, which is analogous to the performance of the integrated capabilities to prevent, detect, protect, alert, respond, and recover. As the *Pearl Harbor Report* stated, "Failure can be avoided in the long run only by preparation for any eventuality" [178]. Certainly this preparedness was lacking in the case of 9/11 [179]:

> Al Qaeda's new brand of terrorism presented challenges to U.S. governmental institutions that they were not well-designed to meet. Though top officials all told us that they understood the danger, we believe there was uncertainty among them as to whether this was just a new and especially venomous version of the ordinary terrorist threat the United States had lived with for decades, or it was indeed radically new, posing a threat beyond any yet experienced.

As late as September 4, 2001, Richard Clarke, the White House staffer long responsible for counterterrorism policy coordination, asserted that the government had not yet made up its mind how to answer the question: "Is al Qida [sic] a big deal?"

A week later came the answer.

And there was this [180]:

Effective decision making in New York was hampered by problems in command and control and in internal communications. Within the Fire Department of New York, this was true for several reasons: the magnitude of the incident was unforeseen; ... The Port Authority's response was hampered by the lack both of standard operating procedures and of radios capable of enabling multiple commands to respond to an incident in unified fashion.

Lack of preparedness in New York was due to the unprecedented scope of the event, outstripping the NYFD's scenario planning to date. The scope of the Katrina aftermath also exposed similar coordination issues in FEMA [181].

According to Carwile [a leading FEMA Mississippi federal coordinating officer (FCO)], over the past four years, there has been no operational doctrine developed by FEMA. He said, as a consequence

"[t]here is no clear understanding of the responsibilities of each level (Washington, the Regions, and deployed Emergency Response Teams) and how they are to interact. This lack of operational doctrine results in unacceptable levels of overlap, double and triple ordering of resources, and long video teleconferences and conference calls [which can] disrupt field operations."

Carwile believes "well-understood and defined operational methodologies based on doctrine would minimize the need for lengthy conferences and would achieve other efficiencies."

Again, it is apparent that a scenario-dependent information-sharing architecture was required.

Bottom Line Preparedness from an information-sharing perspective requires that a resilient, integrated IT/communications architecture is free of deficiency, particularly Organization (PHD #1), Omission (PHD #3), Intelligence (PHD #8), Adaptability (PHD #13), and Dissemination (PHD #16), among others. From Pearl Harbor to 9/11 and Katrina, it is apparent that such a scenario-dependent capability would have provided a much more stable foundation for responder-receiver tasks and activities before, during, and after the disaster.

2.4.2.19 Failure of Consistency Revisited

Review and Analysis In the Failure of Consistency (PHD #19), there are both formal (official) and informal (relationship-based) information-sharing paths between leaders and their command elements. The associated PHD #19 Consistency was defined as "Officials, on a personal basis, should never countermand an official instruction," [182] and arose because Rear Admiral Stark indicated to Admiral

Kimmel that despite the war warning, he did not expect Pearl Harbor to be attacked.

The *9/11* and *Katrina* reports do not provide insight into the personal communications between leaders and their principal officials with which they have unofficial relationships, so there is no evidence of a leader providing conflicting direction to official statements. However, they do demonstrate that an expanded scope for this Consistency PHD #19 is in order. In particular, we believe that deviating from standard operating procedures and *not* communicating these changes to those dependent upon these standard operating procedures is a deficiency. In this case, such events occurred on 9/11 [183].

> Military Notification and Response. Boston Center did not follow the protocol in seeking military assistance through the prescribed chain of command. ... The center also tried to contact a former alert site in Atlantic City, unaware it had been phased out.

And there was also this [184]:

> Evacuation orders did not follow the protocol for giving instructions when a building's collapse may be imminent—a protocol that includes constantly repeating "Mayday, Mayday, Mayday"—during the 29 minutes between the fall of the South Tower and that of the North Tower. In addition, most of the evacuation instructions did not mention that the South Tower had collapsed.

As an example of this expanded consistency deficiency in the Katrina aftermath, consider [185]:

> For example, DOD by-passed the NRP mandated unified command, taking requests from the states directly, absent the necessary input and coordination by FEMA. This was apparent in the evacuation of the Superdome. Parr [Deputy Federal Coordinating Officer, FEMA] completed a plan to evacuate the Superdome Wednesday morning with the support of the Louisiana National Guard. Shortly before implementation of the plan, Parr was informed of the decision by General Honoré of Northern Command to proceed with a different evacuation plan. Unknown to Parr, Blanco had requested DOD's involvement in the evacuation the day before. The Governor's request was made outside the unified command and without the knowledge of FEMA officials, resulting in a duplication of efforts and a delay in the evacuation. Additionally, Parr stated that the actual evacuation under Honoré's plan resulted in an additional 24 hour delay to evacuees.

The deviation from standard protocol had occurred much earlier when Chertoff appointed Brown as the principal federal official [186].

> According to DHS officials, Brown had not taken the required PFO [principal federal official] training program and was not on the approved PFO roster. It is unclear why Chertoff deviated from the requirements of the NRP and designated an untrained individual to serve as PFO for such a catastrophic disaster.

Bottom Line Deviation from standard protocol, SOPs, and best practices, standard practice, or established rules is necessary in special situations, but such deviation should be communicated to all concerned, and not performed unilaterally. This expanded PHD #19 Consistency includes redirection from official edicts, whether implemented by personal relationships or other arrangements.

2.4.2.20 Failure of Jealousy Revisited

Review and Analysis As described in above, PHD #20 Jealousy is one of the greatest impediments to interoperability and is often caused by PHD #9 Attitude, and it can be a driver for other deficiencies, such as PHD #1 Organization and PHD #3 Omission. It is fundamentally important to identify this deficiency as it invariably prevents information sharing among responders-receivers and other stakeholders who would otherwise collaborate to achieve a common public safety and emergency response goal. In the case of Pearl Harbor, the congressional report stated that "Personal or official jealousy will wreck any organization" [187]. We interpret this deficiency as individual or organizational protectiveness of information, much less maximizing control of available personnel, equipment and other resources (e.g., spectrum, network bandwidth, etc.).

In the case of the *9/11 Report*, the authors summarized the conditions leading to PHD #8 Intelligence as "The combination of an overwhelming number of priorities, flat budgets, an outmoded structure, and *bureaucratic rivalries* [italics added] resulted in an insufficient response to this new challenge" [188]. As shown above in PHD #9 Attitude, these rivalries also existed between NYPD and the FDNY. Even during a disaster such as Katrina [189]:

- Information was not being effectively communicated to the DMATs from either of the MSTs [management support teams].
- There was considerable friction between DMATs and the MSTs. An "us and them" attitude was prevalent.
- Only one fulltime FEMA/NDMS employee was present at the airport MST (arriving after operations had started). All other Airport MST staff were taken from onsite DMATs, reducing the number of team personnel for patient treatment and operations support.

Bottom Line Although not as evident in major disasters because of the public gaze, failure to share information on a moment-to-moment basis because of PHD #20 is one of the primary causes of failed interoperability. For this reason, documented reasons for not sharing information with each specific relevant organization in each driving scenario must accompany the checklist results.

2.4.2.21 Failure of Relationships Revisited

Review and Analysis PHD #21 was documented by the congressional investigation as "Personal friendship ... should never be accepted in lieu of liaison or confused therewith where the latter is necessary to the proper functioning of two or more agencies" [190]. We will expand this definition slightly to include both friendly and unfriendly relationships—particularly if they impact what should otherwise be a professional liaison. From the 9/11 Report [191]:

> Responsibility for domestic intelligence gathering on terrorism was vested solely in the FBI, yet during almost all of the Clinton administration the relationship between the FBI Director and the President was nearly nonexistent. The FBI Director would not communicate directly with the President. His key personnel shared very little information with the National Security Council and the rest of the national security community. As a consequence, one of the critical working relationships in the counterterrorism effort was broken.

In addition there was the apparent example of patronage in bringing Joe Allbaugh in as FEMA director despite his total lack of emergency response experience [192]. However, although our checklist searches for negative impacts of such friendships—because they may prevent proper information sharing—benefits of such relationships must not be forgotten [193].

> We salute the exceptions to the rule, or, more accurately, the exceptions that proved the rule. People like Mike Ford, the owner of three nursing homes who wisely chose to evacuate his patients in Plaquemines Parish before Katrina hit, due in large part to his close and long-standing working relationship with Jesse St. Amant, Director of the Plaquemines Office of Emergency Preparedness.

Bottom Line Positive and negative personal relationships can have detrimental as well as beneficial impacts on public safety and emergency response interoperability. If detrimental, they must be captured in the checklist and these negative impacts eradicated from the information-sharing architecture. Leaders, planners, and command elements must put emotion aside and concentrate instead on institutionalizing information sharing independent of their personal feelings, which have no place in professional public safety and emergency response organizations.

2.4.2.22 Failure of Priority Revisited

Review and Analysis The Deficiency of Priority (PHD #22) describes a situation where governance leaders are too focused on other issues to allot appropriate time, attention, and resources on priority public safety and emergency response needs, particularly when there is an evident community or jurisdictional threat. As stated in the Pearl Harbor case, "no consideration should be permitted as excuse for failure to perform a fundamental task" [194].

Such a priority deficiency was apparent in building U.S. central intelligence collection and analysis capability as described in the *9/11 Report* [195].

> DCI [Director of Central Intelligence] Tenet and his deputy director for operations told us they did have a management strategy for a war on terrorism. It was to rebuild the CIA. ...
>
> As some officials pointed out to us, there is a tradeoff in this management approach. In an attempt to rebuild everything at once, the highest priority efforts might not get the maximum support that they need. Furthermore, this approach attempted to channel relatively strong outside support for combating terrorism into backing for across-the-board funding increases. Proponents of the counterterrorism agenda might respond by being less inclined to loosen the purse strings than they would have been if offered a convincing counterterrorism budget strategy. The DCI's management strategy was also focused mainly on the CIA.

FEMA had selected a major hurricane hitting New Orleans as its number one disaster scenario for information-sharing response planning [196].

> FEMA and Louisiana officials accelerated the planning process because of the overwhelming consensus that a category five hurricane hitting New Orleans was one of the most likely and devastating disaster scenarios our nation faced.

Yet for Hurricane Katrina [197]:

> Under the Homeland Security Act, the Secretary of Homeland Security reports to the President and is the department's top disaster official; yet emergency management is just one of the Secretary's many responsibilities. According to Chertoff's testimony before the Select Committee, "he is not a hurricane expert, nor does he have much experience with disasters."

Reviewing the Katrina Report, it becomes apparent through PHD #9 Attitude that the hurricane's landfall was not a priority for the nation's principal leadership—or perhaps their information-sharing roles were unimportant in the prevailing (and undocumented) information-sharing architecture for the high-level National Response Plan.

Bottom Line PHD #21 Priority, by definition, is interrelated with all other deficiencies. If the public safety and emergency response activities of the leaders and planners are consistently ranked below other responsibilities, then all other 24 deficiencies will be evident in the resulting interoperability failures. Critical information-sharing needs and the resulting architecture must be an utmost priority.

2.4.2.23 Failure of Reporting Revisited

Review and Analysis The Deficiency of Reporting (PHD #23) considers reporting up and down the command chain, that is, "Superiors must at all times keep their subordinates adequately informed and, conversely, subordinates should keep their superiors informed" [198]. This reporting may be for situational awareness or command coordination, but its absence was apparent on 9/11. We expand this deficiency to include ambiguity in reporting "up the chain" as well as failure to report, that is, no clear chain of command—so reporting is ambiguous.

On 9/11, parallel command conferences ensued as [199]:

> senior officials from FAA headquarters participated in this video teleconference at various times. We do not know who from Defense participated, but we know that in the first hour none of the personnel involved in managing the crisis did. And none of the information conveyed in the White House video teleconference, at least in the first hour, was being passed to the NMCC. As one witness recalled, "[It] was almost like there were parallel decisionmaking [sic] processes going on; one was a voice conference orchestrated by the NMCC ... and then there was the [White House video teleconference[... [I]n my mind they were competing venues for command and control and decisionmaking."

Perhaps, more importantly, there was not common situational awareness through automatic reporting within a key agency on that day [200].

> FAA Control Centers often receive information and make operational decisions independently of one another. On 9/11, the four hijacked aircraft were monitored mainly by the centers in Boston, New York, Cleveland, and Indianapolis. Each center thus had part of the knowledge of what was going on across the system. What Boston knew was not necessarily known by centers in New York, Cleveland, or Indianapolis, or for that matter by the Command Center in Herndon or by FAA headquarters in Washington.

Parallel command structure also existed when the DoD and National Guard both supported the relief effort for Katrina [201].

> This dual chain of command structure, lengthy federal troop activation system, and, in the case of Katrina, devastated local authorities, contributed to a poorly coordinated federal response to Katrina.

Bottom Line Whether command elements and their subordinates or governance leaders fail to report situational awareness or consolidated command up or down the command chain, an interoperability failure results. This PHD has been expanded to include situations where information-sharing is divided among two or more parallel paths, confusing situational awareness and coordinated command for those engaged because of ambiguous command chains.

2.4.2.24 Failure of Improvement Revisited

Review and Analysis From the perspective of this book, PHD #24, Failure of Improvement, is one of the most important deficiencies preventing interoperability form being developed as part of other system improvements. In the Pearl Harbor investigation [202], this deficiency was defined as requiring that "the administrative organization of any establishment must be designed to locate failures and to assess responsibility." This book's mission is to describe how to identify these deficiencies and make the necessary improvements. As shown in Chapter 1, interoperability improvement involves:

1. Establishing governance and providing adequate resources;
2. Employing a best-practices approach (see Chapter 4) to develop operational and system architecture (see Chapters 5 and 6, respectively);
3. Defining important performance metrics achieved;
4. Gaining acceptance for the developed architecture (see Chapter 4).

Critical improvements that still need to be established in civilian bureaucracies could have already been implemented that would have negatively impacted the terrorists' success on 9/11 [203, 204].

> Before 9/11, the United States tried to solve the al Qaeda problem with the capabilities it had used in the last stages of the Cold War and its immediate aftermath. These capabilities were insufficient. Little was done to expand or reform them ...

> Those working counterterrorism matters did so despite limited intelligence collection and strategic analysis capabilities, a limited capacity to share information both internally and externally, insufficient training, perceived legal barriers to sharing information, and inadequate resources.

In addition, the Congress was itself not without inadvertent complicity [205].

> The Congress, like the executive branch, responded slowly to the rise of transnational terrorism as a threat to national security. Its attention to terrorism was episodic and splintered across several committees. It did not systematically perform robust oversight to identify, address, and attempt to resolve the many problems in national security and domestic agencies that became apparent in the aftermath of 9/11.

Major improvements in disaster response had been underway for several years before 9/11, but were accelerated by those events. By the time of Katrina, progress had been made in defining high-level plans such as the NRP. If not evoked in a timely manner, however, the important early response in the case of Katrina is thwarted [206].

> Smith [Deputy Director, Louisiana Office of Homeland Security and Emergency Preparedness] ... believed, "the biggest single failure of the federal response was the Department of Homeland Security's failure to recognize that Katrina was a catastrophic event and implement the catastrophic incident annex to the National Response Plan"...Had DHS implemented the catastrophic incident annex to the NRP, Louisiana should have had a significant number of federal troops and federal assets, days prior to their actual arrival. . . Instead federal troops did not arrive in number until Saturday, after the evacuations of the Superdome, Convention Center and cloverleaf were complete.

The need for improvement in these intelligence and disaster management capabilities requires the proper development of documented information-sharing architecture based on a range of *possible* scenarios. This information-sharing would target specific situational awareness and consolidated command with decision-support functions so such damaging personnel and deployment decisions are handled properly in the future.

Bottom Line Best practices must be used to improve existing information-sharing gaps in our public safety and emergency response architectures. PHD #24 in the interoperability checklist will identify some other types of deficiencies to be addressed, while PHD #24 will be "checked" when improvements are planned and implemented.

2.4.2.25 Failure of Delegation Revisited

Review and Analysis PHD #25, Failure of Delegation, is a classic situation in which individuals and organizations have the responsibility for success, but have no authority to control or achieve it. This deficiency is related to PHD #24 Improvement, and thus it is connected to all other deficiencies. In the Pearl Harbor congressional report, PHD #24 [207], the authors wrote that "in a well-balanced organization there is close correlation of responsibility and authority."

In leading up to the 9/11 scenario, one particular federal position falls into this deficiency of responsibility without authority [208]:

> The DCI is confirmed by the Senate but is not technically a member of the president's cabinet. The director's power under federal law over the loose, confederated "intelligence community" is limited. He or she states the community's priorities and coordinates development of intelligence agency budget requests for submission to Congress.

This responsibility gives many the false impression that the DCI has line authority over the heads of these agencies and has the power to shift resources within these budgets as the need arises. Neither is true. In fact, the DCI's real authority has been directly proportional to his personal closeness to the president, which has waxed and waned over the years, and to others in government, especially the secretary of defense.

A similar role is defined for the PFO in an NRP-activated disaster [209].

> The PFO is the primary point of contact and source of situational awareness for the Secretary of DHS for incidents of national significance. The PFO is expected to facilitate federal support to the unified command structure that is set up in conjunction with state and local officials. Also, PFOs coordinate the overall federal incident management and assistance activities throughout all of the phases of emergency management—prevention, preparedness, response, and recovery. In carrying out this coordination role, the PFO does not have direct authority over the FCO or other federal and state officials.

This "responsibility without authority" condition is replicated at all levels of governance as well as command levels in public safety and emergency response organizations, where proactive leaders and planners are given improvement responsibility for improving interoperability (see PHD #24), but neither the resources nor authority to achieve it.

Bottom Line The Delegation deficiency (PHD #25) is an important category of interoperability failure, because it has the effect of identifying the driven individual who desires interoperability improvement, but is thwarted in doing so. Giving this person the responsibility but no authority (e.g., over resources) is a classic leadership deficiency—perhaps in part due to PHD #20 (Jealousy), PHD #9 Attitude, and others.

2.5 Leadership Failures and the Need for Holistic Interoperability

In this book, we have employed the interpretation of "communication" used by Prange et al. to convey *understanding*, not just "exchange information" as some interoperability definitions have put forward (see Chapter 1). The naïve view that interoperability is just about "fixing the radio interoperability problem" or only "exchanging information" will continue to produce the tragedies of loss in the future. This book presents (in Chapter 4) a methodology for leaders, planners, organizations, governments, and cooperating nations to identify these gaps of understanding–thus failing to ensure predictable collaborative action–and take steps to make certain they are not the cause of oft-repeated interoperability failures. It is a methodology to be tailored by leaders and their planners to minimize the risk that they and their organizations will have to suffer the tragedy and lifelong "what if's" from their own Pearl Harbor. These leaders are only to blame if, once exposed to these "lessons learned," they do not intensely scrutinize themselves, their subordinates, their interagency relationships, and all other elements of interoperability needed to avert preventable disaster.

As authors of *The Verdict of History* state at the end of that volume [210]:

> On the twenty-fifth anniversary of Pearl Harbor, *The New York Times* declared that the day "must live in memory not only of things past and men long dead, but as a perpetual reminder to future generations that ambition must be tempered with restraint and courage with wisdom.
>
> ... Application of that principle could halt many an ill-considered action in its tracks. It implies an attitude of mutual understanding and respect has become more and more essential. As early as December 8, 1945, the *Saturday Review of Literature* warned, "unless the December 7, 1941, episode is placed in its long-range historical perspective, the real lesson of Pearl Harbor will have been overlooked. That lesson, fundamentally, is that American security is inseparable from world security."
>
> Substitute the name of any other nationality for "American" and this statement remains valid. Humanity cannot afford to forget the lessons of Pearl Harbor. The world is much too small; the risk is much too great; the time is much too late."

The many preventable manmade and natural tragedies, and the associated failed alertness and preparedness, show the lessons are not learned—and it is an arguable deficiency of leadership to assume they will be learned without specific verifiable and provable processes or that correcting this deficiency is only a matter of procuring the right technology. In this regard, the outcome at Pearl Harbor, which generated similar shock and confusion to other disasters—Titanic (1912), Lusitania (1915), Poland (1939), Coventry (1940), Russia (1941), Bastogne (1944), Hiroshima (1945), Beirut (1983), 9-11 (2001), Madrid (2004), Tsunami (2004), Katrina (2005), and Mumbai (2008)—remains a classic example of failed interoperability as we have defined it.

The Pearl Harbor failure of interoperability emphasizes the major point of this book, that the lack of public safety and emergency response IT/communications interoperability and its repeated consequences is entirely a human failure. Again, these leadership failures result in the necessity for rapid initiative and selfless heroism on the part of responders and receivers when "the storm breaks"—and it is about to break again.

Next, Chapter 3 will use these and other deficiencies to define a worst-case World Interoperability Failure model of public safety and emergency response entities given the leaders, planners, bureaucracies, consultants, and vendors at all governance levels worldwide. The Failure model and its bureaucratic layers will provide a basis for the application of best practices to establish governance, resources, approach, architecture, performance, and acceptance to identify gaps, correct them, and finally achieve interoperability for this fundamental government function.

References

[1] *Hearings Before the Joint Committee on the Investigation of the Pearl Harbor Attack*, Congress of the United States, Seventy-ninth Congress, Part 39 (Washington, D.C., 1946).

[2] Prange, Gordon William,, Donald M. Goldstein, and Katherine V. Dillon, *Pearl Harbor: The Verdict of History*, McGraw-Hill Companies, Penguin, 1991, pp. 267–68 (hereafter cited as *The Verdict of History*).

[3] Suggested by Dr. Donald M. Goldstein, coauthor of Pearl Harbor: *The Verdict of History* and *At Dawn We Slept: the Untold Story of Pearl Harbor*.

[4] National Commission on Terrorist Attacks Upon the United States, "Executive Summary," *The 9/11 Commission Report: Final Report of the National Commission on Terrorist Attacks Upon the United States*, July 2004, p. 9, (hereafter referenced as *9/11 Executive Summary*).

[5] Select Bipartisan Committee to Investigate the Preparation for and Response to Hurricane Katrina, *Failure of Initiative, Final Report of the Select Bipartisan Committee to Investigate the Preparation for and Response to Hurricane Katrina*, U.S. Government Printing Office (Washington, D.C., 2006), available at http://www.gpoacess.gov/congress/index.html, (hereafter cited as the *Katrina Report*).

[6] National Commission on Terrorist Attacks Upon the United States, The *9/11 Commission Report: Final Report of the National Commission on Terrorist Attacks Upon the United States*, July 2004, p. 7, (hereafter cited as the *9/11 Report*).

[7] Ibid., p. 339.

[8] Concatenation of the first syllables of "Totsugeki Raigeki," the concatenation "To-ra" also formed the word "tiger" leading to the well-known phrase signaling the start of the Pearl Harbor attack.

[9] Prange, Gordon William, Donald M. Goldstein, and Katherine V. Dillon, *At Dawn We Slept: The Untold Story of Pearl Harbor*, McGraw-Hill Companies, 1981, p. 503, (hereafter cited as *At Dawn We Slept*).

[10] For the detailed perspective on Japanese planning, see *The Pearl Harbor Papers: Inside the Japanese Plans*, Donald M. Goldstein and Katherine V. Dillon, ed., Brassey's, 2000.

[11] *At Dawn We Slept*, p. 40.

[12] Ibid., p. 31.

[13] Ibid., p. 45.

[14] Ibid., p. 362.

[15] Ibid., p. 440.

[16] Ibid., p. 443.

[17] *The Verdict of History*, pp. 309–310

[18] *At Dawn We Slept*, p. 95.

[19] Ibid., p. 94.

[20] Ibid., p. 95.

[21] Ibid., p. 96.

[22] Ibid., p. 186.

[23] Ibid., p. 188.

[24] *The Verdict of History*, p. 210.

[25] Ibid., p. 558.

[26] *At Dawn We Slept*, p. 402.

[27] Ibid., p. 403.
[28] *The Verdict of History*, p. 350.
[29] *At Dawn We Slept*, p. 404.
[30] Ibid., p. 405.
[31] Ibid., p. 406.
[32] *The Verdict of History*, p. 215.
[33] At *Dawn We Slept*, p. 406.
[34] Ibid., p. 486.
[35] Ibid., p. 487.
[36] Ibid., pp. 494–95.
[37] Ibid.
[38] Ibid., p. 495.
[39] Ibid., p. 484.
[40] Ibid., pp. 484–85.
[41] Ibid., p. 495.
[42] Ibid., p. 496.
[43] Ibid., p. 497.
[44] Ibid., pp. 497–98.
[45] *The Verdict of History*, pp. 367–368.
[46] Ibid., pp. 368–369.
[47] Ibid., p. 369.
[48] Ibid., p. 370.
[49] Ibid., p. 371.
[50] *At Dawn We Slept*, pp. 500–501.
[51] Ibid., p. 506.
[52] Ibid., p. 517.
[53] *The Verdict of History*, p. 3.
[54] Ibid., p. 4.
[55] Ibid., p. 33.
[56] *At Dawn We Slept*, p. 731.
[57] *The Verdict of History*, p. 442.
[58] *At Dawn We Slept*, p. 219.
[59] *The Verdict of History*, p. 552.
[60] Ibid.
[61] Ibid., p. 378.
[62] Ibid., p. 391.
[63] Ibid., pp. 403–04.
[64] Ibid., p. 395.
[65] Ibid., p. 393.
[66] Ibid., p. 552.
[67] Ibid., p. 277.
[68] Ibid., p. 553.
[69] Ibid.
[70] Ibid., p. 554.
[71] Ibid.
[72] Ibid.
[73] Ibid.

[74] Ibid.
[75] Ibid., pp. 555–556.
[76] Ibid., p. 556.
[77] *9/11 Report*, p. 345.
[78] *The Verdict of History*, p. 556.
[79] Ibid., p. 557.
[80] Ibid.
[81] Ibid.
[82] Ibid., pp. 557–58.
[83] Ibid., p. 558.
[84] Ibid.
[85] Ibid.
[86] Ibid., p. 559.
[87] Ibid.
[88] Ibid., p. 560.
[89] Ibid., p. 219.
[90] Ibid., p. 560.
[91] Ibid.
[92] Ibid., p. 396.
[93] Ibid., p. 560.
[94] Ibid., p. 448.
[95] Ibid., p. 561.
[96] Ibid.
[97] Ibid.
[98] Ibid.
[99] *9/11 Report*, p. 340.
[100] *The Verdict of History*, 562.
[101] *9/11 Report*, pp. 347–48.
[102] *9/11 Report*, p. 88.
[103] *Katrina Report*, Executive Summary of Findings, p. 1.
[104] Ibid.
[105] *The Verdict of History*, p. 552.
[106] *9/11 Executive Summary*, p. 11.
[107] Ibid., p. 12.
[108] *9/11 Report*, p. 83.
[109] Ibid., p. 285.
[110] *9/11 Report*, p. 321.
[111] *Katrina Report*, Executive Summary of Findings, p. 3, (hereafter referenced as *Katrina Executive Summary*).
[112] Northern Command is a unified combatant command of the United States military created in October of 2002 to provide homeland defense in support local, state, and federal authorities.
[113] Ibid., p. 4.
[114] Ibid., p. 5.
[115] Ibid., p. 5.
[116] *Katrina Report*, Appendix 8, *Supplementary Report to the Findings of the Select Bipartisan Committee to Investigate the Preparation for and Response to Hurricane Katrina*, pre-

sented by the Select Committee on Behalf of Rep. Cynthia A. McKinney, (Washington, D.C., February 6, 2006), p. 26, (hereafter referenced as *Katrina Appendix 8*).
[117] *The Verdict of History*, p. 552.
[118] *9/11 Report*, p. 265.
[119] *Katrina Executive Summary*, p. 15.
[120] Ibid., p. 139.
[121] *The Verdict of History*, p. 553.
[122] *9/11 Executive Summary*, p. 15.
[123] *9/11 Report*, p. 318.
[124] *Katrina Executive Summary*, p. 1.
[125] *The Verdict of History*, p. 553.
[126] *9/11 Report*, p. 43.
[127] *Katrina Report*, pp. 139–140.
[128] *Katrina Appendix 8*, p. 29.
[129] *Katrina Report*, p. 2.
[130] *The Verdict of History*, p. 554.
[131] *9/11 Executive Summary*, p. 15.
[132] *Katrina Report*, p. 12.
[133] *Katrina Appendix 8*, p. 5.
[134] *Katrina Executive Summary*, p. 2.
[135] *The Verdict of History*, p. 554.
[136] *9/11 Report*, p. 359.
[137] *Katrina Report*, p. 108.
[138] Ibid., p. 109.
[139] *9/11 Report*, p. 265.
[140] *Katrina Appendix 8*, p. 29.
[141] *The Verdict of History*, p. 555.
[142] *9/11 Report*, p. 13.
[143] Ibid., p. 11.
[144] *9/11 Report*, p. 271.
[145] Ibid., p. 77.
[146] Ibid., p. 77.
[147] *The Verdict of History*, pp. 555–56.
[148] *9/11 Report*, pp. 408–09.
[149] Ibid., pp. 309–10.
[150] Ibid., p. 81–82.
[151] Ibid., p. 319.
[152] *Katrina Report*, p. 298.
[153] *The Verdict of History*, p. 556.
[154] *9/11 Executive Summary*, p. 9.
[155] Ibid., p. 345.
[156] Ibid., pp. 287–289.
[157] *The Verdict of History*, p. 556.
[158] *9/11 Report*, p. 297.
[159] Ibid., p. 307.
[160] *Katrina Report*, p. 169.
[161] Ibid., p. 226.

[162] *The Verdict of History*, p. 557.
[163] *9/11 Report*, p. 78.
[164] Ibid., p. 79.
[165] Ibid., p. 108.
[166] *Katrina Report*, pp. 360–61.
[167] *The Verdict of History*, p. 557.
[168] *9/11 Report*, p. 263.
[169] *Katrina Report*, pp. 1–2.
[170] *The Verdict of History*, pp. 557–58.
[171] *9/11 Report*, pp. 416–17.
[172] *The Verdict of History*, pp. 558.
[173] *9/11 Report*, p. xvi.
[174] *Katrina Report*, p. 13.
[175] *The Verdict of History*, pp. 558.
[176] *Katrina Report*, p. 223.
[177] Ibid., p. 224.
[178] *The Verdict of History*, pp. 559.
[179] *9/11 Executive Summary*, p. 9.
[180] Ibid., p. 15.
[181] *Katrina Report*, p. 326.
[182] *The Verdict of History*, p. 560.
[183] *9/11 Report*, p. 20.
[184] Ibid., p. 307.
[185] *Katrina Report*, p. 144.
[186] Ibid., p. 136.
[187] *The Verdict of History*, p. 560.
[188] *9/11 Executive Summary*, p. 12.
[189] *Katrina Report*, p. 298.
[190] *The Verdict of History*, p. 560.
[191] *9/11 Report*, p. 358.
[192] *Katrina Appendix 8*, p. 24.
[193] *Katrina Executive Summary*, p. 1.
[194] *The Verdict of History*, p. 560.
[195] *9/11 Report*, p. 358.
[196] *Katrina Report*, p. 81.
[197] Ibid., p. 132.
[198] *The Verdict of History*, p. 561.
[199] *9/11 Report*, p. 36.
[200] Ibid., p. 16.
[201] *Katrina Report*, p. 201.
[202] *The Verdict of History*, p. 561.
[203] *9-11 Report*, pp. 350–351.
[204] *9-11 Executive Summary*, p. 13.
[205] Ibid., p. 15.
[206] *Katrina Report*, p. 137.
[207] *The Verdict of History*, p. 561.

[208] *9/11 Report*, p. 86.

[209] *Katrina Report*, p. 39.

[210] *The Verdict of History*, pp. 566–67.

Selected Bibliography

Flynn, Stephen, *The Edge of Disaster: Rebuilding a Resilient Nation*, Random House, 2007.

National Commission on Terrorist Attacks Upon the United States, "Executive Summary," *The 9/11 Commission Report: Final Report of the National Commission on Terrorist Attacks Upon the United States*, July 2004.

Prange, Gordon William, Donald M. Goldstein, and Katherine V. Dillon, *At Dawn We Slept: The Untold Story of Pearl Harbor*, McGraw-Hill Companies, 1981.

Prange, Gordon William, Donald M. Goldstein, and Katherine V. Dillon, *Pearl Harbor: The Verdict of History*, McGraw-Hill Companies, Penguin, 1991.

Select Bipartisan Committee to Investigate the Preparation for and Response to Hurricane Katrina, *A Failure of Initiative, Final Report of the Select Bipartisan Committee to Investigate the Preparation for and Response to Hurricane Katrina*, 109th Congress, 2nd Session, U.S. Government Printing Office, Washington, D.C., February 2006.

CHAPTER 3

World Interoperability Failure Model

The *Katrina Report* asks a simple question regarding the performance of the bureaucracy that had been funded by the American people [1].

> After so many tens of billions of dollars have been spent in establishing and maintaining a new super-agency, the Department of Homeland Security, attentive Americans must ask not only how such a stunning failure of government response to a natural disaster is possible, but also how it is possible that in the process of expending vast sums on new bureaucratic infrastructure charged with enhancing our safety we now learn that lessons of the past were ignored in a way that led directly to the scenario of the present failure. The Committee's Report offers at least a large part of the answer when it points repeatedly to a failure to take initiative, which can easily be interpreted as a failure of leadership. As we read in the *Report*.
>
> ...a dispassionate review made it clear that even an extraordinary lack of situational awareness could not excuse the feckless, flailing, and organizational paralysis evident in the documents and communications the Committee reviewed. Leadership requires decisions to be made even when based on flawed and incomplete information. Too often during the immediate response to Katrina, sparse or conflicting information was used as an excuse for inaction rather than an imperative to step in and fill an obvious vacuum.

3.1 The Cause

It is in part to answer this question that we introduce a worst-case World Interoperability Failure Model. The purpose of this model is to represent the worst-case environment in which holistic interoperability must be created and sustained for our responders-receivers and other stakeholders. To do so, our World Interoperability Failure model is assumed to be composed of multiple governance levels. These governance layers include the leaders and their planners, bureaucracies, consultants, and vendors that determine the IT and communication capabilities of our responder-receivers and other public safety and emergency response stakeholders.

These levels are depicted in Figure 3.1 as follows:

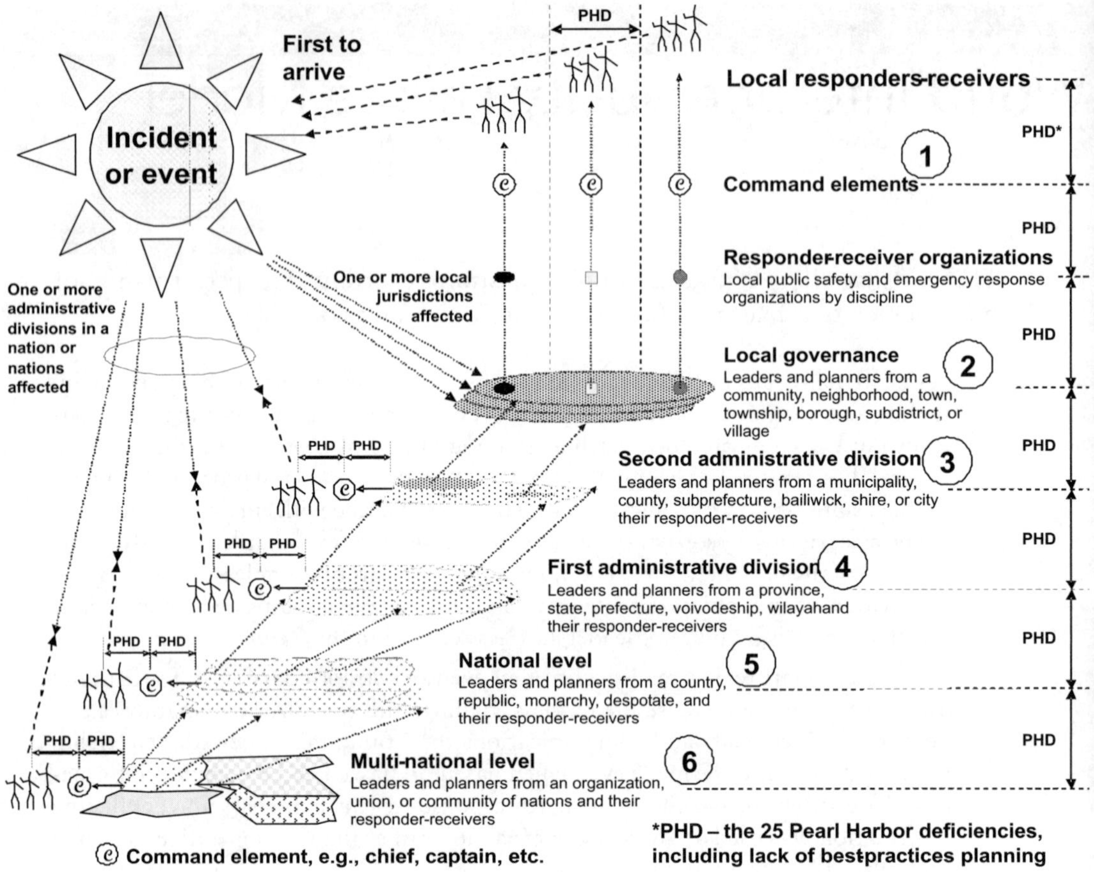

Figure 3.1 Governance layers that determine holistic interoperability.

1. Command elements for the individual public safety and emergency response organizations at all governance levels.
2. Local governance composed of the leaders and planners from their town, community, township, village, subdistrict, and so forth.
3. Second administrative division of their nation, such as their municipality, city, county, shire, and so forth.
4. First administrative division of their nation, such as their state, province, prefecture, and so forth.
5. National level, including, republics, monarchies, oligarchies, or dictatorships.
6. Multinational collectives, such as their organization, union, or community of nations.

In this figure, we show the 25 Pearl Harbor deficiencies residing between governance layers, as well as between command or supervisory elements and their corresponding responder-receiver organizations or stakeholder organizations, respectively. In the figure, the local responder-receivers are closest to the would-be incident and first to become involved.

The figure also shows the 25 Pearl Harbor deficiencies impacting the intraoperability between command elements and their responders-receivers organizations, and laterally both among responders-receivers from different disciplines or local jurisdictions. We stress that this World Interoperability Failure Model is intended as a conceptual worst-case model only, and not meant to represent the capabilities or deficiencies of any specific individuals, organizations, agencies, or departments of any nation, region within a nation, commercial or nonprofit enterprise. We note that although there are likely individuals who seek to correct these deficiencies. We also suggest, however, that the 25 Pearl Harbor deficiencies are supplemented by internal political, social, economic, and many other pressures that prevent such people from being heard or regarded, much less supported to improve the status quo, often stating (as suggested before Pearl Harbor) that "things are 'good enough.'"

3.2 Responder-Receivers Organizations in the Failure Model

3.2.1 Public Safety, Emergency Response, and Other Stakeholder Disciplines

In our hypothetical World Interoperability Failure model, public safety and emergency response organizations at all governance levels include the equivalent disciplines of emergency medical services, fire service, law enforcement, emergency management, public works, hospitals, and local boards of health [2]. In this day of chemical, biological, radiological, nuclear, and explosive (CBRNE) weapons of mass destruction (WMDs), 1972 Olympics Mumbai-style terrorists, and massive tsunamis, each of our public safety and emergency response responder-receiver organizations and stakeholders could possess information or capabilities vital to a population's survival. Thus, one or more of the responder-receiver disciplines or stakeholders in Table 3.1 could possess—or need—critical information. At these times, a simple yes/no response to a question asked can result in tragically missed opportunities to prevent or minimize the impact of disaster, such as, "are the lower wards evacuated," or "are the hijacked aircrafts headed south"? The individuals or organizations possessing this information may play the most critical role in the events about to unfold.

Imagine, for example, that when George Elliot at the Opana Point radar station told the Information Center on Oahu that a large fleet of aircraft was approaching, Lieutenant Tyler had described this event to the Navy, who'd just attacked a submarine at the harbor entrance. Instead of telling Elliot "Well, don't worry about it," perhaps American interceptors would have launched to break up the approaching Japanese attack formation, perhaps the *Arizona* would not have received its singular fatal blow, perhaps fewer families would have had to have been notified about the loss of their loved ones. Imagine if the FBI had not become excessively reluctant to share intelligence information, or Chertoff named a trained PFO days earlier. In other words, it is proven folly to artificially prioritize (PHD #22 Priority) one individual, organization, or discipline over another with regards to public safety and emergency response interoperability.

In our worst-case failure model, the leading supervisory elements and their staff are subject to all 25 Pearl Harbor deficiencies applied to intraoperability, also called

Table 3.1 Stakeholder Organizations in Disaster

Agriculture, food, and water utilities and companies.	Banking, finance, and insurance companies.	Chemical industry and hazardous material clean-up/disposal companies.
Criminal justice, including local, state, and federal law enforcement, particularly the investigative and intelligence elements.	Education, including private as well as public schools, as shelter	Emergency services, including government emergency management (nonlaw enforcement) and public/private forecasting services (e.g., weather).
Energy, including utilities, and environmental services—both public and private.	Government (local, state, federal), including legal and policy elements.	Health and public health services (including hospitals).
Hospitality and lodging, including camp grounds and National Parks.	Information and telecommunications, including Internet and intranet providers, carriers, and others.	Military facilities (including National Guard and militia) and the defense industrial base.
Postal and shipping, including private carriers.	Public and private security services.	Public works (municipal departments, utilities) and waste disposal services.
Real estate, for available land, vacant buildings, and so forth.	Retail companies (e.g., large discount retail chains, particularly for building supplies, food, etc.).	Social services, including housing, welfare, protective services, and others.
Transportation agencies and companies, including ground, air, water transport services.	Venues, such as large sports and entertainment facilities.	Red Cross and volunteer organizations active in disaster.

operability, *within* their organizations. These command or supervisory elements may be elected or selected, but they have "come up through the ranks" and have in-depth familiarity with the discipline—much like Kimmel and Short had regarding their corresponding commands at Pearl Harbor, or Marshall and Stark in Washington. However, we will nevertheless assume that these leaders and their planners within responder-receiver and stakeholder organizations are:

- Thoroughly expert in the tasks and activities, SOPs, modern technologies, and pseudo-military hierarchical command structure available for their discipline, but are nevertheless susceptible to the same 25 expanded deficiencies.
- Not professionally trained and experienced in the application of best practices needed to identify and correct these deficiencies (PHD #24 Improvement), nor can they therefore accurately judge the expertise of those proposing to do so.
- Intensely protective and prevent dissemination of their operational and situational information—even if filtered and well beyond the protection of criminal or health care records, based on vague and undocumented operational security requirements (PHD #3 Disclosure).
- Garnering and protecting available resources, and masking any personal or organizational lack of ability or capability to perform optimally when necessary (PHD #20 Jealousy), or may even be unaware of such discrepancies (PHD #15 Insight).
- Assume their own personal relationships with leadership and planners of other public safety and emergency response organizations to achieve

interoperability, producing a fear to delve into the operational activities of their sister agencies (PHD #21 Relationships) lest these relationships be strained.

3.2.2 IT and Communication Systems

As has been emphasized, technology in and of itself in the form of IT and communications system architecture plays a role—albeit minimal—in these interoperability failures. This role is minimal because the IT and communication systems used by our responders-receivers and the way it performs is the *effect*, and not the *cause*, of failed interoperability. Differently stated, these systems reflect the inherent failings represented by the 25 deficiencies. For this reason, and as the *Katrina Report* states, these deficiencies have produced information-sharing stovepipes similar to those in the intelligence services [3].

> The problem is nearly intractable because of the way the government is currently structured. Lines of operational authority run to the expanding executive departments, and they are guarded for understandable reasons: the DCI [Director of Central Intelligence] commands the CIA's personnel overseas; the secretary of defense will not yield to others in conveying commands to military forces; the Justice Department will not give up the responsibility of deciding whether to seek arrest warrants. But the result is that each agency or department needs its own intelligence apparatus to support the performance of its duties. It is hard to "break down stovepipes" when there are so many stoves that are legally and politically entitled to have cast-iron pipes of their own.
>
> Recalling the Goldwater-Nichols legislation of 1986, Secretary Rumsfeld reminded us that to achieve better joint capability, each of the armed services had to "give up some of their turf and authorities and prerogatives." Today, he said, the executive branch is "stove-piped much like the four services were nearly 20 years ago." He wondered if it might be appropriate to ask agencies to "give up some of their existing turf and authority in exchange for a stronger, faster, more efficient government wide joint effort." Privately, other key officials have made the same point to us.

These stovepipes characterize failed information sharing at all governance levels shown in Figure 3.2, where:

- Stovepipe height depicts the amount of economic and other resources committed to acquire and sustain the IT and communication system;
- Width indicates breadth of capabilities achieved by the stovepiped system;
- Spacing between stovepipes indicates information sharing; the greater the separation, the less information sharing occurs.

Again, the technology "does what you design it to do," and if those design plans are flawed, that is, they did not use best practices in development, implementation, and sustainment, then the technology will be flawed in its performance—but will work as designed. The radar worked fine at Opana Point on December 7 and on 9/11, and the satellite imagery of Katrina didn't miss a thing—but consider what happened anyway despite people employing this technology.

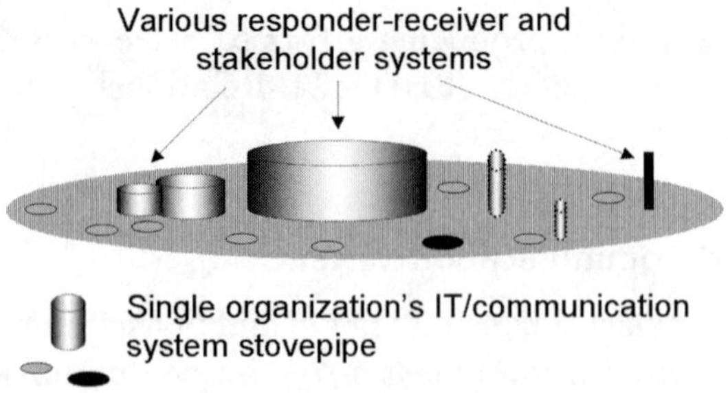

Figure 3.2 Information stovepipes at the local governance level.

3.3 Governance Layers in the Failure Model

The governance layers in Figure 3.1 include overlapping responsibilities, and conflicts with and among disciplines and jurisdictions that will affect information sharing for split-second preparedness and response. These primarily civilian organizations are less regimented, organized, or centrally directed than the military organizations in Washington, D.C. and Pearl Harbor. Each governance level has its own responder-receiver and stakeholder organizations. The local governance level is shown in Figure 3.2, the second governance level in Figure 3.3, the third in Figure 3.4, the national level in Figure 3.5, and the multinational level in Figure 3.6. These figures show the complexity of developing an information-sharing architecture that accounts for each organization depending on the chosen scenario.

In our World Interoperability Failure Model, the 25 expanded deficiencies described in Chapter 2 apply to every responder-receiver organization. In addition, elected and appointed leaders overseeing one or more agencies at all governance levels:

- Believe that nothing significant would happen in their jurisdiction (PHD #7 Complacency and #15 Insight) and cannot envision what could happen (PHD #10 Imagination), or instead, focus on what *might* happen versus what *could* happen (PHD #18 Preparedness)—so their underestimate of the likelihood of the event occurring lowers their preparedness for that event (PHD #22 Priority).

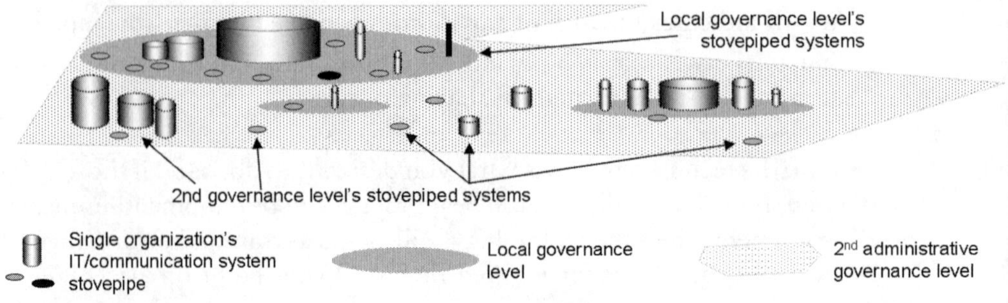

Figure 3.3 Cumulative second governance level stovepipes.

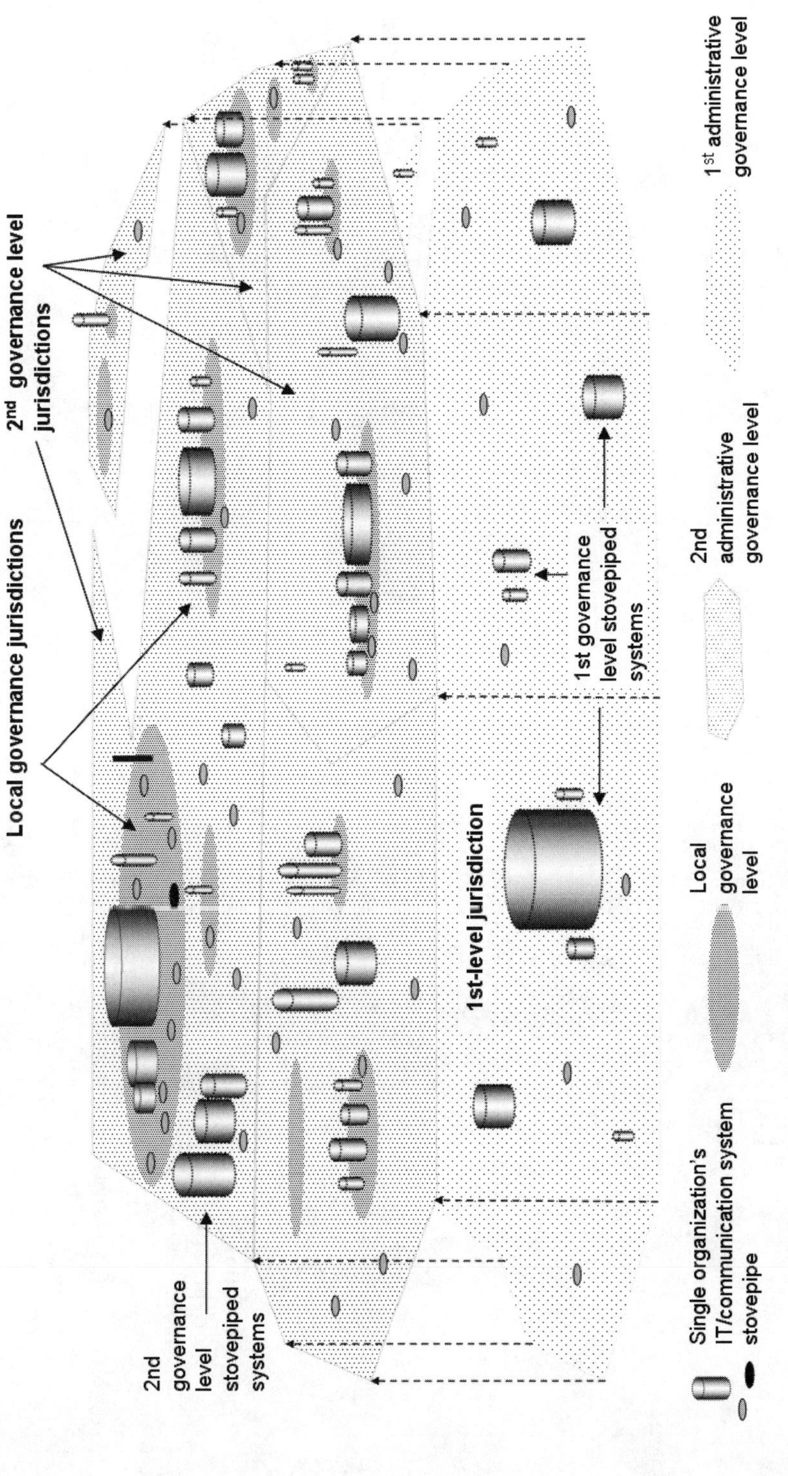

Figure 3.4 Cumulative first-level stovepipes.

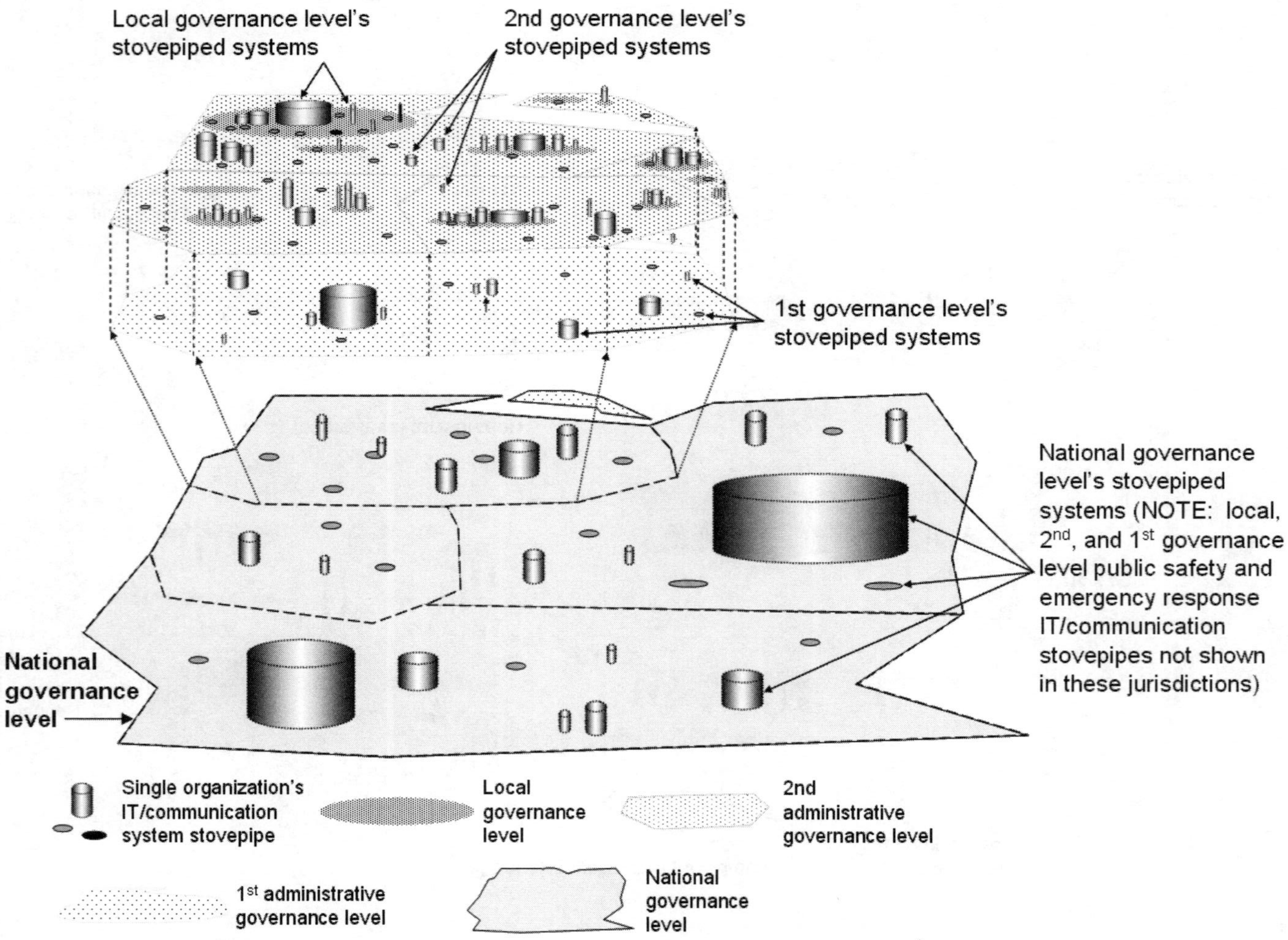

Figure 3.5 National-level stovepipes.

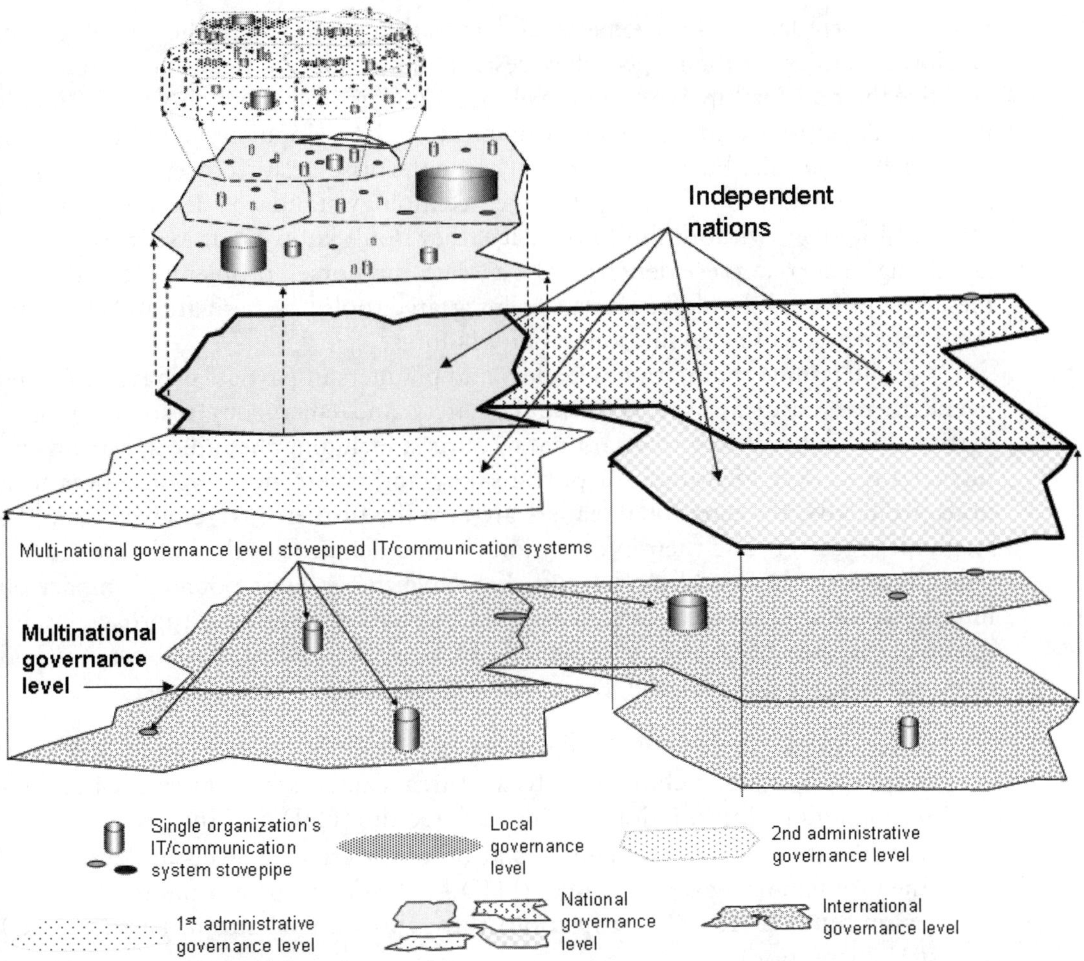

Figure 3.6 Multinational stovepipes.

- Are too overextended (PHD #22 Priority) to verify adequate information sharing (PHD #17 Inspection) among the organizations within their jurisdiction.
- Don't know how to determine the necessary information sharing (PHD #15 Insight) among other agencies or what processes to follow to improve any apparent information-sharing needs (PHD #24 Improvement), and so assume (PHD #2 Assumption) others are addressing any needs—but never verify their assumption (PHD #4 Verification).
- Assume (PHD #2 Assumption) that the leaders of their public safety and emergency response organizations are aware of these deficiencies and will address them.

Most importantly, in our theoretical World Interoperability Failure Model, the leaders and planners of these systems simply have not had the professional exposure, much less the training or practical experience, in the application of best practices for identifying deficiencies and systematically eliminating them (PHD #24

Improvement). This lack of experience has nothing to do with their intelligence, patriotism, ethics, or the respect they deserve in any specific field. No matter how effective their leadership, we assume they are not trained as practicing professionals in the fields of project management, organizational development and effectiveness, cost/benefit analysis, business case analysis, enterprise architecture, systems engineering and integration, funding strategy, technology standards development, and others. Moreover, without this knowledge, they don't know what expertise to seek in helping to correct these deficiencies—a "cart and horse" problem. Nevertheless, these best practices should be used as an integrated toolset to mitigate and eliminate the deficiencies leading to interoperability failures.

We will assume that national leaders and planners in particular have sweeping oversight over critical aspects of public safety and emergency response policy. National-level leaders control national resources, including the radio spectrum and financial resources identified for public safety and emergency response organizations. Ironically, these national leaders are also the furthermost removed from the responder-receivers and their information-sharing deficiencies than all other governance layers of the world model, but they have the greatest potential impact on interoperability success planning factors of governance, resources, approach, architecture, performance, and acceptance. In our theoretical worst-case World Interoperability Failure Model, these national leaders and their planners:

- Have no direct sustained experience in the public safety and emergency response domain, including analyzing threats and assessing their own organization and its information-sharing architecture (PHD #15 Insight).
- Assume that leaders and planners at other governance levels are addressing the problem using best practices (PHD #2 Assumption), yet fail to verify this assumption (PHD #4 Verification) through direct on-site inspection (PHD #17 Inspection).
- Lack firsthand understanding of the depth of information-sharing deficiencies within their national governance level, much less other levels (PHDs #15 Insight, #2 Assumption, #4 Verification, and #17 Inspection), or if they know these deficiencies, either do not know how to improve them or have no authority to do so (PHD #25 Delegation).
- Have no knowledge of the best practices needed to improve the status quo (PHD #24 Improvement) on a nationwide level, or if they have such knowledge, they have no authority or resources to do so (PHD #25 Delegation).
- Create barriers to development of positive relationship with subordinates and contemporaries in other organizations, prohibiting the collaboration necessary to address holistic interoperability issues (PHD #20 Jealousy and #9 Attitude).
- Provide resources (e.g., grants) to stakeholder organizations without the necessary specificity in grant application requirements, thus ensuring deficient development of information-sharing architecture and the continuation of noninteroperable stovepipe architectures.
- Believe that their preparations are "good enough"—why "fix it if it is not broken" (PHDs #2 Assumption, #4 Verification, #7 Complacency, #9 Attitude, #15 Insight, #18 Preparedness, and #22 Priority).

In our worst-case world model, leaders responsible for the work of ineffective interoperability planners cannot—or will not—accurately assess their employees' performance (PHD #20 Jealousy and #24 Improvement), because they have no experience or metrics upon which to base such an assessment (PHD #15 Insight), or because they believe there are many more important problems to address (PHD #22 Priority). In our hypothetical world model, public safety remains government's highest priority, but is not systematically supported or improved at the national level in any country. Thus all leaders, planners, and command elements in our worst-case World Interoperability Failure model are assumed to suffer from the Pearl Harbor deficiencies. They either do not want holistic interoperability (PHD #20 Jealousy or PHD #9 Attitude), or they are unaware of any deficiencies in their organizations (PHD #7 Complacency); or they are not permitted—due to political or budget constraints—to use the appropriate best practices to achieve holistic interoperability (PHD #25 Delegation and PHD #24 Improvement).

3.4 Other Deterrents to Interoperability

Governance leaders and planners are not alone in the worst-case interoperability failure model. In the worst case, bureaucracies, consultants, and vendors can also be deterrents to holistic interoperability.

3.4.1 Bureaucracies in the Failure Model

Although elected or appointed leaders are inherently transient, many government levels maintain administrative bureaucracies with personnel that remain in place for many years. In our world model, we will assume that these bureaucrats maintain their positions for life—or until some form of retirement—irrespective of the success they achieve in bringing about true interoperability. Job titles may change, organizations may be merged or reorganized, but the same individuals remain in place, so the fundamental deficiencies continue and the public unknowingly remains at risk. Thus, in our World Interoperability Failure Model, the in-place bureaucracies are not capable of changing the status quo—and there is no accountability to the public other than Pearl Harbor, 9/11, and Katrina.

3.4.2 Consultants in the Failure Model

In our worst-case world model, leaders and planners hire consultants to effectively outsource the study of their interoperability problems. These consultants are assumed to have little or no direct experience in public safety and emergency response interoperability, much like their sponsors. They provide high-level plans or observations already well-known by responders-receivers—translating the status quo into terms understandable to the sponsoring organization or, most frequently, lending support to decisions already taken by their sponsoring organization. In addition, we assume that the scope of these efforts is limited to radio interoperability, because the information-sharing deficiencies at the heart of holistic interoperability failures are not addressed or resolved in our worst-case model. As with the bureaucrats, these planning consultants do not follow systematic best practices. In our worst-case world model, consultants are used extensively by agencies that need to show progress towards interoperability and other objectives. For many reasons, we assume these consultants know nothing of the Pearl Harbor deficiencies or their analogies, and are not fundamentally capable of systematically planning their mitigation or elimination. Even if the consultants are competent, the in-place bureaucrats will tie their continued employment to consultant recommendations that retain the existing set of policies.

3.4.3 Vendors in the Failure Model

In our worst-case world model (see Figure 3.7), the IT and communication systems planned and implemented by independent agencies achieve what they were designed to do. Holistic interoperability, as defined in this book, is not a design consideration, much less a source of detailed requirements. In addition, each vendor contributing products to IT and communication systems at any governance level seeks to maximize their sales. In this regard, it is in the natural best interest of any vendor—in their stovepipe solution to the interoperability problem—to have all public safety and emergency response organizations buy products from them alone, and this motivation is a natural deterrent to interoperability.

We also assume that national leaders are not capable or, if capable, are prevented from mandating such critically important standards for civilian public safety and emergency response for fear of adverse commercial impact. Our worst-case World Interoperability Failure Model therefore assumes the public remains at risk

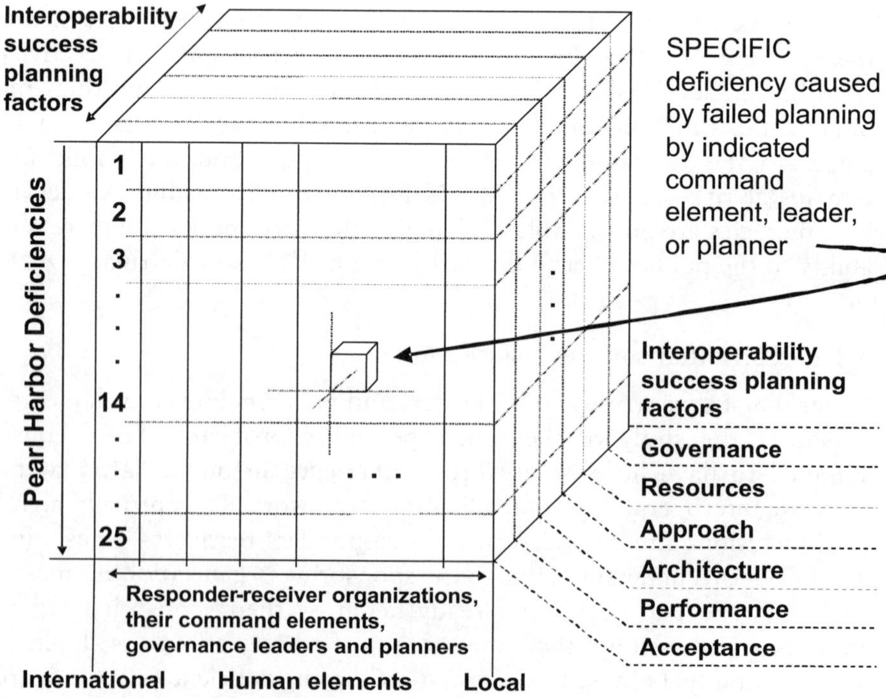

Figure 3.7 Multidimensional World Interoperability Failure model.

for perceived commercial benefit. In fact, the majority of vendors would benefit from such standards—and there are precedents for mandatory IT/communication standards and compliant vendors in the military as well as civilian domains (see Chapter 6).

There are examples of national and international initiatives to standardize IT and communications technology and mandate its use for the world's war fighters. This same treatment has not been applied with the same speed, resources, and commitment for information sharing in the civilian public safety and emergency response community. In our theoretical World Interoperability Failure Model, we add the ranks of our responders-receivers to the war fighter and seek equal priority for national commitment.

3.5 The Interoperability Assessment Checklist

Table 3.2 provides the Interoperability Assessment Checklist for public safety and emergency response organizations at different governance levels to determine—through answering a set of deficiency-derived questions—if they suffer from one or more deficiencies. This checklist produces a set of likely deficiencies to be corrected in order to achieve holistic interoperability. The best practices designed to address these deficiencies are described in Chapter 4. An example of their application to the worst-case World Interoperability Failure Model is provided in Chapter 7 for a catastrophic evacuation scenario. Chapter 7 describes the World Interoperability Success Plan, which describes an integrated best-practices approach to solve the complex multidimensional holistic interoperability problem defined by our World Interoperability Failure model.

3.6 Summary

Failed interoperability is the result of poor planning and not failed technology. Unfortunately, one "slipped bit" can mean tragedy, so our public safety and emergency response capabilities *must never* be considered good enough. In the worst case, as depicted in our World Interoperability Failure Model, our responders-receivers and their command elements exhibit the same deficiencies as their counterparts on Oahu and in Washington, D.C.; through December 7, 1941, in New York and in Washington, D.C., in 9/11/2001; and on the Gulf Coast of the United States and Washington, D.C., following August 29th 2005. These deficiencies are tragic enough given the split-second initiative and self-sacrifice these responder-receivers are expected to display.

In our failure model, however, the many associated governance levels and their associated leaders, planners, organizational bureaucracies, their paid consultants, and vendors/suppliers of noninteroperable products have largely ensured this endemic lack of interoperability. After all the failures over the years, the current leaders and planners addressed by this failure model must either not know of these deficiencies and therefore do not seek to correct them; or if known, they either:

Table 3.2 Interoperability Assessment Checklist

PHD	*Failure of ...*	*PHD definition expanded to include nuances of related 9/11 and Katrina failures*
1	Organization	Do multiple organizations or departments within an organization have overlapping or conflicting responsibilities for day-to-day operations for different incidents or events, compete for available resources, and develop independent IT/communication systems to meet stovepiped needs? Is there a detailed information-sharing architecture that documents the sources and receivers of specific information as a function of time in collaborative public safety and emergency response scenarios?
2	Assumption	Do leaders and planners in an organization assume that information sharing within and among their responders-receivers and other public safety and emergency response stakeholders is operating properly or is "good enough?" Do they believe that any significant information-sharing gaps would have been found and eradicated in the past, and if there are gaps, that the involved organizations have or will work to fill them? Do command and operational elements assume they will provide or receive certain information in the absence of documented information-sharing architecture, SOPs, and memoranda of understanding (MOUs)?
3	Omission	Have information-sharing architectures spanning all relevant organizations been developed for critical scenarios for the jurisdiction or region? Have all relevant organizations been included in operational planning for critical scenarios? Are key individuals or organizations not included in the documented information-sharing architecture for these critical scenarios?
4	Verification	Have leaders and planners verified that their organization's information-sharing needs are being met for critical scenarios? Does the documented information-sharing architecture have provisions to verify all internal and external organizational information?
5	Supervision	Do leaders and their planners have the awareness, background knowledge, expertise, or authority, to request that their public safety and emergency response organizations achieve holistic interoperability in their jurisdiction or beyond? If they do request this achievement, have they developed the information-sharing architecture within which this sharing is defined? Have an organization's leaders and planners put in place best practices or SOPs so that they can verify that critical tasks and activities, such as evacuation, are being executed according to existing plans or common sense?
6	Alertness	Do leaders maintain situational awareness of man-made or environmental threats? Do these leaders assume (PHD #2) without verification (PHD #4) that not only is it being done by someone else (PHD #2) but, if detected, will be rapidly and accurately reported to them by the responsible individuals or organizations? Do they pay proper respect to expert alerts and warnings?
7	Complacency	Do leaders and planners—largely due to a much broader range of issues than public safety and emergency response alone, and the assumption that other command elements or the media will inform them of approaching incidents or events, expect "business as usual" on a moment-to-moment and day-to-day basis? Are leaders relaxed because they believe the information-sharing and preparedness architecture is "good enough"—no improvements required?
8	Intelligence	Does one organization not, in general, collect, properly interpret, or share information regarding potential threats? Do those who collect such information have the proper training or background to properly assess the consequences or develop actionable information for their public safety organizations? Is there a capability to forecast these consequences? Do the leaders understand the importance of timely action based on expert forecast? Do they understand the consequences each moment that they fail to act?
9	Attitude	Are leaders of critical public safety and emergency response organizations who should participate in an information-sharing architecture for their jurisdiction failing to do so because of personal attitude? Are these leaders openly aloof and consider the needs of the pubic safety and best-practice information sharing to be subordinate to their positions and responsibilities?

Table 3.2 (continued)

PHD	*Failure of ...*	*PHD definition expanded to include nuances of related 9/11 and Katrina failures*
10	Imagination	Do the leaders and planners have the background, experience, or place priority on the need to evaluate and respond to *possible* threats? Do they assume it is not their role to do so—and that their public safety and emergency response organizations *are* doing so for them? If they assume it is not their job to so imagine, have they verified that someone is doing so?
11	Communications	Are there information-sharing architectures designed to achieve interoperability among the necessary organizations in critical scenarios? Do leaders and their planners in different organizations "speak the same language" of public safety and emergency response, information-sharing architecture, best practices, and so forth, so misinterpretations and misunderstandings do not occur?
12	Paraphrase	Largely due to (11) above, does communications among command elements and governance leaders produce misunderstanding? Are the orders from leadership conveyed to responders-receivers based on these misunderstandings? Is situational awareness and command coordination conveyed to others based on assumed intent or meaning? Are information gaps filled in based on assumption (PHD #2)?
13	Adaptability	Have the planners in each organization developed well-documented information-sharing architecture to address a series of possible public safety and emergency response scenarios relevant for their jurisdiction and discipline? Have training and exercises been performed to demonstrate and teach adaptation of organization practice and techniques to tailor prevention, alert, and response to the unexpected?
14	Disclosure	Have the planners established which critical information will be shared and by what method with all essential responder-receivers and stakeholders, based on operational need to know? Are steps being taken to vet responders-receivers and other stakeholders for trustworthiness relative to different levels of sensitive information disclosure? Have planners developed an information-sharing architecture with multilevel security?
15	Insight	Do leaders and planners have in-depth knowledge or consequence assessment tools for natural and man-made threats that *could* threaten their jurisdiction? Are they aware of information needed by their public safety and emergency response organizations to meet these threats, and do they know how to define a framework for identifying or implementing these needs? Are they aware of the deficiencies in their own organizations?
16	Dissemination	Do the leaders and planners have moment-to-moment understanding of situations or events, and ensure information is adequately distributed as needed among all those with a need to know in a timely fashion? Do these leaders and planners all take action to share information or *verify* that such sharing has occurred?
17	Inspection	Do leaders and their planners (1) perform on-scene review and assessment of the fixed facilities, mobile capabilities, and personnel abilities within their subordinate organizations; (2) seek external, unbiased assessment of these capabilities, and (3) assume that their subordinates are doing so (removing their need to do so)—despite public safety and emergency response being their primary governance responsibility? Do leaders, planners, and command elements inspect the people, processes, and tools as well as their scenario-based information-sharing architecture?
18	Preparedness	Have leaders and planners prepared for what they think *might* happen, versus what *could* happen? Have planners considered all rational possibilities and their unforeseen consequences in their information-sharing architectures?
19	Consistency	Do leaders, planners, or command elements have a "backchannel" to other command elements or personnel that have been used to distort official announcements, alerts, plans, and so forth? Have they themselves communicated one view officially, but then provided an unofficial view to others—whether or not in the same organization? Do leaders, planners, or command elements deviate from established protocols and procedures, without communicating that they are doing so, to all concerned?

Table 3.2 (continued)

PHD	*Failure of ...*	*PHD definition expanded to include nuances of related 9/11 and Katrina failures*
20	Jealousy	Do individual public safety or emergency response organizations, or their leaders and planners, seek to maximize their share of the available local budget, and withhold—or fail to provide the means—of sharing this information with other discipline agencies in their jurisdiction? Are governance leaders at this level unaware of this jealousy, and if aware, do they attempt to determine the impact of this gap on their jurisdiction, or know how to fill this gap?
21	Relationships	Are there personal friendships or other relationships among leaders and planners in different organizations that pass for "interoperability," but actually prevent an in-depth mutual understanding of each organization's operations and information sharing in an emergency or disaster? Have key leadership, planning or tactical command positions been given to individuals because of patronage or other relationships independent of their qualifications, including extensive public safety and emergency response domain knowledge and experience? Is an unfriendly relationship between leaders, planners, or command elements preventing essential information sharing and architecture planning and implementation?
22	Priority	Do leaders, planners, and command elements place too low a priority on information sharing? Do they fail to make holistic interoperability within and among their public safety and emergency response services an important objective in their leadership role?
23	Reporting	Do command elements report the real status of their organization to their associated governance leaders and planners? On the contrary, have they attempted to preserve the appearance that their respective organization is performing optimally, particularly in the realm of interoperability, even if close scrutiny would prove otherwise? Are there ambiguous command chains (e.g., multiple command centers that form when emergencies or disasters strike)?
24	Improvement	Do leaders or command elements and their planners: (1) employ best practices to avoid the interagency information-sharing deficiencies described in this chapter; or (2) use proven approaches for specifying, procuring, testing, and sustaining the tools needed to achieve (1) above, as well as optimally perform, their operational activities? Do governance leaders and their planners assume that interoperability (and other) deficiencies can be repaired by "just buying something" (i.e., technology). Do leaders who are *not* technologists often look solely to technology for interoperability solutions, which tends to be limited to "radio" alone? Do leaders, their planners and command elements understand that a formal information-sharing architecture is essential and that it must be developed for scenarios spanning many type of incidents or events and the activities expected of their organizations?
25	Delegation	Do leaders push responsibility for critical decisions to their planners and their command elements, but do not provide the authority to develop the information-sharing architecture across all stakeholder organizations?

- Don't want them corrected (to maintain one's own stovepipe);
- Have not been allowed (sanctioned or funded) to correct them;
- Don't know how to correct them.

This book cannot convince those leaders and planners who preserve their own stovepipes to open their information to others as appropriate (unless seeing how failure to do so has marked those responsible in history), or to allow their subordinates or coworkers to do so. It can, however, describe the best practices (Chapter 4) used in a World Interoperability Success Plan (Chapter 7), which shows how to correct these deficiencies and achieve interoperability.

References

[1] *Katrina Report*, Appendix 8, *Supplementary Report to the Findings of the Select Bipartisan Committee to Investigate the Preparation for and Response to Hurricane Katrina*, presented by the Select Committee on Behalf of Rep. Cynthia A. McKinney, (Washington, D.C., February 6, 2006), p. 5, (hereafter referenced as *Katrina Appendix 8*).

[2] George W. Bush, *Homeland Security Presidential Directive/HSPD-8*, Office of the Press Secretary, December 17, 2003.

[3] National Commission on Terrorist Attacks Upon the United States, *The 9/11 Commission Report: Final Report of the National Commission on Terrorist Attacks Upon the United States*, July 2004, p. 403.

CHAPTER 4

Best Practices for Achieving Interoperability

Organized and repeatable planning and implementation following established best practices is crucial to achieving a robust information-sharing architecture for true interoperability. Human factors, along with the lack of repeatable managerial and technical processes, have remained the primary cause of the 25 deficiencies found in the Pearl Harbor investigation. Chapter 2 showed that these same deficiencies were evident from the 9/11 and Katrina investigations as well. In Chapter 3, these deficiencies were used to define a worst-case World Interoperability Failure model for public safety and emergency response organizations and the leaders, planners, bureaucracies, consultants, and vendors associated with all governance levels. We deduced that in the worst case, best-practice planning techniques were not used or properly (universally) applied—or if used, the results were incomplete or not implemented. For this reason, ensuring that planners at all governance levels employ best practices to avoid these deficiencies is critical. If they are not employed, true interoperability will remain elusive.

The use of these best practices described in this chapter defines the *approach* interoperability planning success factor identified in Chapter 1. Hence, "achieving interoperability requires interoperable planning," that is, *integrated*—not stovepiped—best practices. In other words, these best practices are the fundamental planning tools needed to achieve holistic interoperability. Use of integrated best practices to establish holistic interoperability is an innovation in the field—but it should have been standard practice for many years. These best practices form the key elements of the World Interoperability Success Plan outlined in Chapter 7 to establish and sustain information-sharing architecture that spans all governance layers, jurisdictions, responder-receiver disciplines, and other stakeholders relative to selected threat and disaster scenarios.

4.1 Overview of Best Practices

Perhaps the many public safety and emergency response IT/communication systems worldwide were developed through ad hoc and uncoordinated attempts to achieve interoperability, but the time and effort required to "do it right" were not applied.

In our World Interoperability Failure model, we postulated that these practices were not applied either because the responsible leaders and planners did not know such practices existed, or if known, did not want or were otherwise unable to apply them. Perhaps there was the pre-Pearl harbor deficiency that things were "good enough." As was stated in the *Katrina Report* [1]:

> The greater failure was the government's delayed and confused response. In general, poor coordination, poor planning and execution, and inadequate efforts at communication are cited in the Select Committee Report as causes of this failure. On its own terms, the Report also does a fair job of balancing, on the one hand, the fact that Hurricane Katrina simply overwhelmed existing capacity of the Federal Emergency Management Agency (FEMA), as well as relief organizations such as the Red Cross to cope with the scale of the catastrophe, with the overall inadequacy of preparation and planning.

As a result, the old adage continues to apply, "there's never time to do it right, but there's always time to do it over." This chapter describes the integrated best practices needed to "do it right."

To mitigate the risks inherent in the World Interoperability Failure model, leaders and planners at all governance levels responsible for public safety and emergency response interoperability (see Figure 3.1) must use these best practices to effectively plan, implement, and sustain information-sharing architecture. These practices must be integrated at all governance levels—selected, tailored, and documented as a detailed architecture framework at the multinational and national levels. Such best practices are drawn from processes associated with a wide range of disciplines, including:

- Program/project management,
- Business case analysis,
- Organizational development,
- Funding strategy,
- Strategic planning,
- Systems engineering,
- Enterprise architecture,
- Standards development.

Figure 4.1 provides a diagram of these integrated best practices with annotated arrows defining the information flows between them. The arrows and their descriptions will be better understood after reading the following subsections. For this reason, the information flows between best practices shown in the figure will be described in detail at the end of this chapter.

In the past, these best practices were employed individually and often on an ad hoc basis by those who understood their value. In this chapter, we describe these best practices and show how they can be integrated to mitigate or eliminate the Pearl Harbor, 9/11, and Katrina deficiencies. Instead of their ad hoc implementation, the resulting integrated practices described at the end of this chapter cover the breadth of activities that provide leaders and planners with proven and repeatable

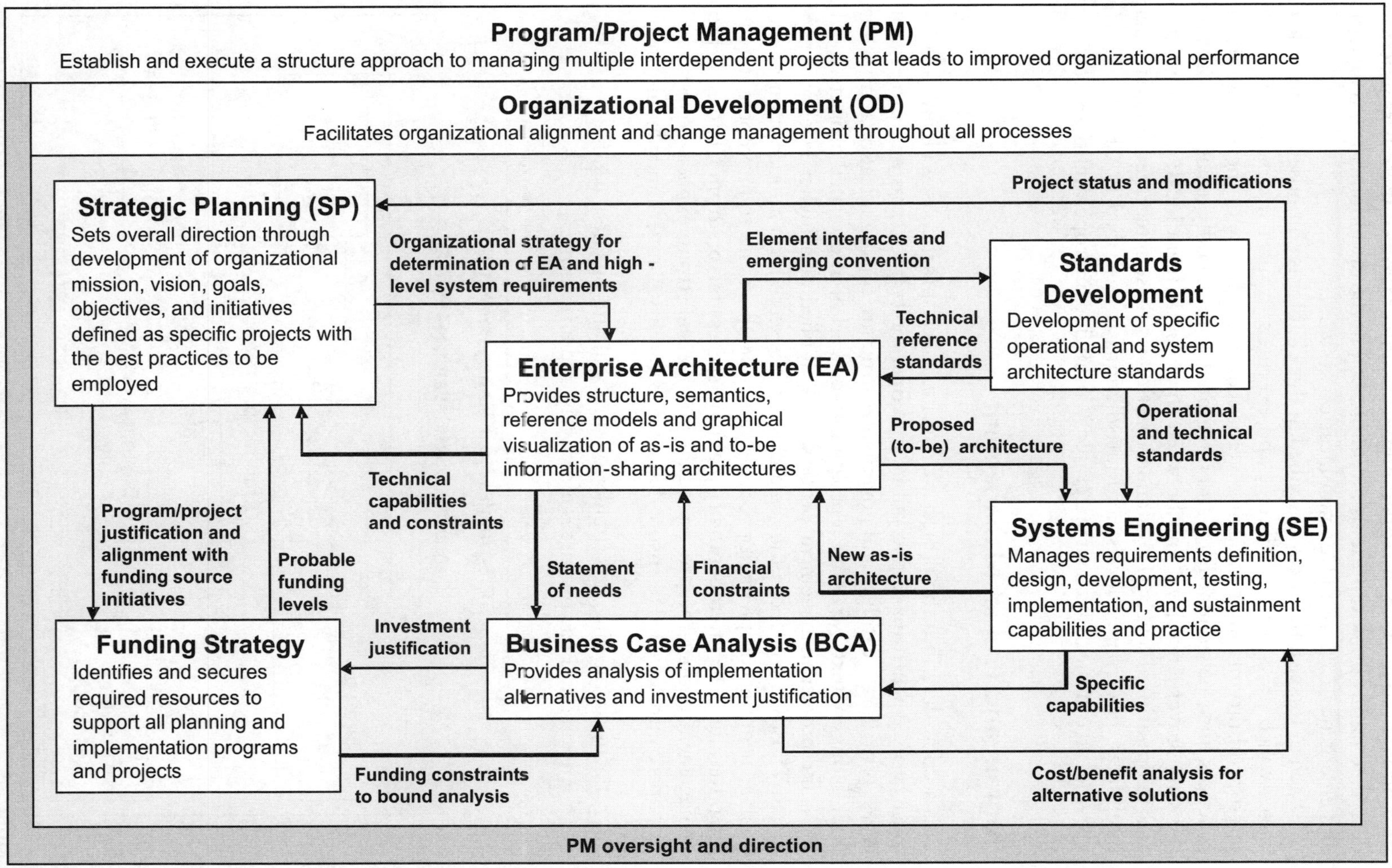

Figure 4.1 Best practices for interoperability success planning.

methodology to achieve interoperability. The integration of these best practices as described in this book represents an *innovation* in the development of public safety and emergency response IT/communication systems.

Each of the best practices described in this chapter is composed of multiple processes which are selected, tailored, and executed to meet the particular needs of interoperability planning at each governance level. The use of these integrated best practices must be standardized and integrated both laterally across jurisdictions at the same governance level and between these levels from local; through national and multinational levels. The standardization of these integrated practices across disciplines, governance levels, and jurisdictions is described in Chapter 7, which provides an outline of such an integrated Worldwide Interoperability Success Plan.

4.2 Program and Project Management

This section describes the project management best practice (see Figure 4.2) which must be used when managing public interoperability programs and projects of all types, providing a catalyst and governance structure for organizational transformation. In this context, we define a program to be a set of interrelated projects, in which the same project management processes are applied at a higher level of reporting detail. Thus, we will focus the remainder of this section on a description of project management, its complement processes, and its importance to interoperability planning. Recalling the interoperability planning success factors from Chapter 1, project management defines the processes that provide the strong explicit *governance* and *resource* control needed to ensure that interoperability planners and planning teams:

- Focus on meeting stakeholder operational and business needs;
- Allocate and use resources effectively and efficiently;

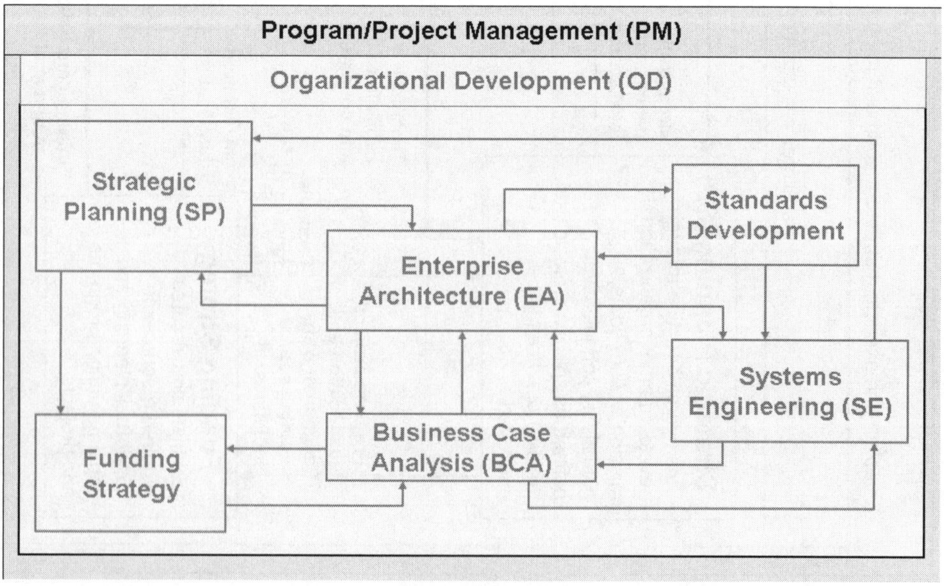

Figure 4.2 Program and project management best practice.

- Get project work done according to budgeted schedule (time) and resources (money);
- Ensure with feedback that work outcomes meet project requirements;
- Maintain program or project alignment with the strategic mission, vision and goals of an organization.

The project management best practice provides its interoperability planners with a standard, yet flexible and highly adaptable methodology that enables successful execution of their projects as well as increasing reuse of their adopted project management processes for the multiple projects that such planning will identify. In addition, high-level management practices are described below that create a solid foundation for a successful interoperability program at each governance level, thus supporting and complementing overall project management activities for other government priorities. Within the project management best practice, technical activities are planned; and standards and procedures are established to ensure technical quality, manage data, manage configurations, manage technical risk, and report status.

The detailed project management best practice is defined by the Project Management Institute (PMI) in *A Guide to the Project Management Body of Knowledge* [2], simply known as the *PMBOK Guide*. Given the importance of the relationship between the project management processes and all other best practices required for interoperability projects (e.g., strategic planning, enterprise architecture, business case management, systems engineering, etc.), understanding project management processes is essential to effectively employing all of these processes as well as effectively managing the integration of their results. For this reason, project management was shown as the overarching best practice in Figure 4.1.

The project management processes apply to all projects in public safety sectors, including civilian government and commercial enterprises as well as military and intelligence organizations at all governance levels. While certain aspects of the project management work will vary from sector to sector and influence how the best practice is adapted or tailored to the specific planning environment, the processes themselves are universally applicable. The following description is intended to be a high-level description of project management fundamentals for readers with little or no knowledge of the associated processes. We strongly recommended that program and project managers at all levels refer to the Project Management Institute's *PMBOK Guide* [3] and the associated training courses and accompanying tutorial materials [4] that are offered. With this caveat, this section will provide an introduction to project management artifacts with key definitions, an overview of the PMI approach, tailoring guidance, and guidelines for implementation.

4.2.1 Definitions

PMI defines a project in the simplest terms as a unique and temporary endeavor or effort, with established beginning and end times and conditions. Such projects are continuously performed in every public safety and emergency response organization and at all governance levels, whether or not PMI best practices are employed. Project managers—those professionals who lead projects consistently with time-tested processes—are vital to effective and efficient performance of any project. Project managers contribute to quality, efficiency, and operational or business results across the enterprise. Formally, the PMI defines project management as [5] "the application of knowledge, skills, tools and techniques to project activities to meet the project requirements."

4.2.2 Overview of the Project Management Institute Approach

The *PMBOK Guide* defines the core project management standard and provides a set of processes that are performed to ensure both that disciplined project management practices are in place and that quality results are delivered. These baseline practices drive rationally managed projects for a large number of industries as well as local, regional, national or multinational government organizations. The *PMBOK Guide* defines five process groups, nine knowledge areas, and 44 basic project management processes, with each process described by key inputs, tools and techniques, and outputs.

The five *PMBOK* process groups shown in Figure 4.3 are:

1. *Initiating*—authorizing the project or a project phase;
2. *Planning*—selecting the course of action required for attaining the project objectives;
3. *Executing*—integrating all project resources to carry out the project plan;
4. *Monitoring and Controlling*—ensuring that project objectives are met by regularly monitoring progress against plan to identify variances and take corrective actions when necessary;
5. *Closing*—formally accepting results of project or project phase.

The nine knowledge areas described in the *PMBOK Guide* [6] must be performed to ensure a project delivers the objectives for which it was undertaken, namely:

1. *Integration management*—defines the processes that need to be performed to ensure the elements of the project are integrated, including the use of integrated project teams [7].

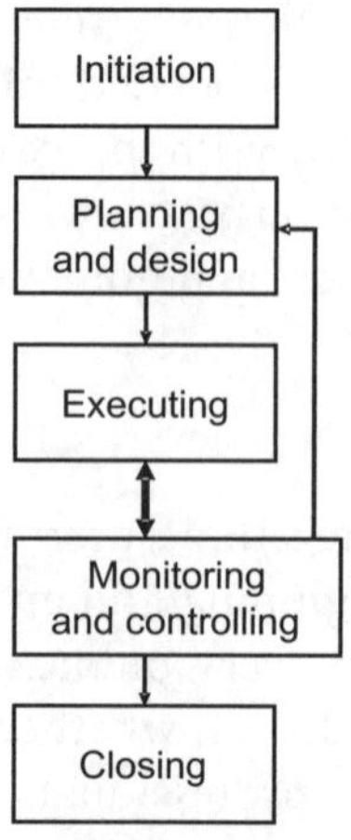

Figure 4.3 Five project management process groups and their interrelationships.

2. *Scope management*—defines the processes that ensure the project includes all the work required, and only the work required, to successfully complete the project.
3. *Time management*—defines the processes that ensure timely execution of project phases or tasks as well as completion of the project.
4. *Cost management*—defines the processes needed to ensure that project is completed within the approved budget.
5. *Quality management*—defines the processes needed to ensure that the project demonstrably satisfies all its objectives for which it was established and funded.
6. *Human resource management*—defines the processes needed for the most effective use of people supporting the project.
7. *Communications management*—defines the processes need for timely and appropriate generation, collection, dissemination, storage and use of project information.
8. *Risk management*—defines processes required to ensure that significant risks are quickly identified and rapidly mitigated.
9. *Procurement management*—describes the processes required to acquire goods and services in support of the project.

Some of the tools and techniques frequently used in project management processes include [8]:

- *Expert judgment*—recommendations from subject matter experts in the required operational or technical domains.
- *Decomposition*—subdividing project deliverables into component parts so that specific work activities or tasks and subordinate tasks can be defined for each part; all performed to a level of detail such that work and associated costs can be scheduled and controlled by the project manager.
- *Variance analysis*—study of the difference between expected project performance and actual performance, for example, the difference between the planned (or expected) quality and timeliness of one or more work tasks and the actual achieved results.
- *Rolling-wave planning*—the development of detailed project plans (e.g., detailed work tasks, schedules, and costs) for near-term requirements and environment characterized by greater certainty, while providing only high-level planning for future requirements and environment for which there is significantly greater uncertainty.
- *Dependency determination*—the analysis of "cause and effect" relationships among project work tasks to determine which tasks provide essential output before other tasks can begin, achieve a certain result, or are completed; often depicted in a dependency diagram.

One of the most important project management processes is development of the work breakdown structure (WBS). The WBS is defined as [9] "a deliverable-oriented hierarchical decomposition of the work to be executed by the project team to accomplish the project objectives and create the required deliverables." It is a common method of decomposing a project into its component tasks and subtask as well as analyzing the scope of the project according to its outcomes or deliverables. The WBS is usually provided with its associated:

- *Scope (WBS) dictionary*—detailed description of all project objectives, deliverables, work tasks, task dependencies.
- *Scope (WBS) baseline*—the detailed description of (benchmark) project deliverables, established at project inception, which can be used to determine if—or to what extent—suggested or required deliverable changes differ from the benchmark deliverables.
- *Project document updates*—description of any changes to the scope evolving as the project proceeds, capturing any changes over time and their impact.

Table 4.1 shows the WBS used to manage the writing of this chapter. The WBS is one of the four outputs from one of five discrete project management processes under project scope management.

Other process outputs in the *PMBOK Guide* include the:

- *Project management plan*—document that describes project objectives, baseline and current task schedules and budget, quality and risk management processes, and other information describing how the project will be executed, monitored, and controlled.
- *Project scope*—summary description of project deliverables and how they meet work sponsor or project stakeholder requirements.
- *Requirements documentation*—detailed description of expected capability or service, including required functions, interfaces, standards, operating environments, performance, training, maintenance, and so forth.
- *Communications management plan*—a document describing the means used to share project information with the team to maximize progress and minimize misunderstandings (e.g., to avoiding the 25 Peal Harbor–9/11–Katrina deficiencies within the project team).
- *Project schedule*—documented times or dates at which project activities, tasks and provisioning of deliverables was planned to occur, including updates showing tasks or deliverables that are on or off schedule.
- *Cost performance baseline*—usually expressed as a graphical function showing expected cumulative costs over time used to control project resources and measure variances
- *Human resource plan*—document describing how team members will be acquired, managed, and sustained to meet project work objectives.
- *Risk management plan*—document that describes how current and future risks will be identified and partially or completely mitigated as well as control of residual risks associated with partial mitigation.
- *Change requests*—formal request from project sponsor, manager, or team to change one or more aspects of the project, including deliverables, requirements, costs, schedules, and so forth.

Table 4.1 Example of a Work Breakdown Structure for Preparing Chapter 4

Task numbers	*Phase and Task Names*	*Phase and Task Narrative*
Phase 1.0	Acquire contributors	Creator/editor determines what best practices are needed for the chapter, then identifies the right experts and gains their commitment.
1.1	Define needed best practices	
1.2	Identify best-practice experts	
1.3	Solicit contributor inputs	
Phase 2.0	Manage section contributions	(1) The creator/editor provides contributors with individual writing content and schedule requirements; (2) contributors write their assigned material; (3) creator/editor collects drafts (deliverables); provides individual contributor's feedback on their inputs, then receives final inputs; finally (4) creator/editor integrates chapter with text linking beginning, middle, and end with summary to complete the chapter.
2.1	Assign sections and schedules	
2.2	Conduct weekly status meetings	
2.3	Ensure section integration	
2.4	Review drafts and provide feedback	
2.5	Integrate contributor sections	
Phase 3.0	Prepare Chapter 4 manuscript	Creator/editor (1) draws all artwork, (2) solicits and employs results form all external and contributor reviews, and (3) prepares and delivers final manuscript deliverable to the publisher.
3.1	Redraw all artwork for consistency	
3.2	Deliver manuscript for review	
3.3	Receive reviewer comments	
3.4	Review comments and edit chapter	
3.5	Prepare and deliver final manuscript	
Phase 4.0	Receives and verifies copy edits	Creator/editor makes final corrections and submits final manuscript deliverable to publisher.
Phase 5.0	Receives and verifies pages	Creator/editor makes final corrections and submits final book pages deliverable to publisher for final production.

- *Deliverable descriptions*—detailed narrative and tabular or figure-based characterization of project results meeting project sponsor or stakeholder requirements.

It is important to note that the *PMBOK Guide* processes are adaptable and that not all indicated processes and their outputs are required for all projects. For integrated interoperability projects spanning governance levels, a standard adoption of project-dependent process and input/output artifacts is recommended. For example, a project to develop standard operating procedures at a specified governance level should follow standard processes established at the next higher regional level, and so on. The following section provides examples of tailoring of process tailoring to meet project specific objectives.

4.2.3 Tailoring Guidance

The standard PMI project management best practice can be tailored to meet specific project needs. Tailoring guidelines allow the project to apply the most appropriate project management processes to meet the diversity interoperability planning needs without bypassing important tools and techniques. Tailoring allows the project to apply these appropriate processes and accompanying tools within the real-world context of project cost, schedule, performance, and quality objectives. It also accounts for the need project results, work scope, team experience, stakeholder relationships, and identified risks. Tailoring determines which process activities are needed for the project and the level of detail that these activities will be performed. To tailor the application of project management processes to achieve the desired work results, the project managers and their team should perform the steps described below.

- Step 1 *Understand*
  - Understand the scope and purpose of the project: The decision of which project management processes to employ is based on understanding of project requirements. The scope and intended purpose of the project will drive the information required, the composition of the WBS, scheduling, and resource allocation.
- Step 2 *Assess*
  - Determine the process inputs and outputs: The tailoring effort establishes process information flows, or process input and output requirements, for effective project management given the intended project scope purpose, and expected results. This information will provide understanding of what project management processes need to be performed.
  - Determine the project management processes to be performed: Based on the requirements of the project, inputs and outputs needed, the structure of the project organization, and the project risk, identify the project management processes that need to be performed to ensure project success accordingly.
- Step 3 *Tailor*
  - Tailor the project management processes: Specify information inputs and outputs to be provided form each selected project management process based on the assessments performed in the earlier steps. Add, modify, or delete process inputs and outputs—and the tasks that generate them—to reflect the scope and extent to which the selected process will be performed, and then document the reasons behind the changes to that process. For example, the WBS example in Table 4.1 showed only two levels of taking, phase (e.g., '1.0') and task (e.g., '3.3'). It was believed that this level of task detail for developing this chapter was adequate, rather than to further subdivide these tasks into level 3 (e.g., 3.2.4), level 4 (e.g., 2.3.1.2), or beyond.
- Step 4 *Review and approve*
  - Review, approve, and document the approved tailored processes: Summarize the decisions and assumptions made during the tailoring process and review the proposed tailored processes within the project management best

practice. Reach agreement on the tailored process and make adjustments as necessary across the project team and project sponsors, and ensure all tailoring results are approved by the project manager before implementation. Finally, document the approved tailored processes and use them as the foundation for the creation of the project WBS.

- Step 5 *Communicate*
  - Communicate the approved tailored processes: Communicate the result of all tailoring efforts to the project team and stakeholders, ensuring that all team members understand the processes and schedules to be followed.

4.2.4 Application of Project Management Best Practices to the Interoperability Failure Model

In previous chapters, we identified people and processes (i.e., human factors) as the principal source of interoperability failures, because the tools (technology) perform as intended given these human factors. Recent research into project success and failure has also attributed interoperability failure to these same human factors and therefore emphasizes project communications to improve success rates [10]. Disciplined and consistently applied project management best practice identifies the following criteria for effective management: clarity of communications, building trust, supporting facilitation, employing effective management processes, managing risk, and managing outcomes. These criteria are easily applied to mitigate the 25 Pearl Harbor deficiencies defined in the Interoperability Assessment Checklist in Table 3.3.

These deficiencies are mitigated in part by proper project management planning itself, because without this management best practice, lack of information sharing will likely occur on the very projects designed to improve interoperability, see Table 4.2. Again, "interoperability planning requires interoperable planning," and these processes ensure information sharing across the project team. A failure of project management planning due to these deficiencies will mean failed interoperability when there is operational need—with consequences that have been demonstrated repeatedly.

4.3 Organizational Development

People, process, and technology factors led to the 25 Pearl Harbor deficiencies cited in the World Interoperability Failure model described in preceding chapters. The organizational development discipline (Figure 4.4) can help mitigate many of the risks inherent in the people and process factors driving these deficiencies.

4.3.1 Overview

Organizational development is "a complex strategy intended to change the beliefs, attitudes, values, and structure of organizations so that they can better adapt to new technologies, markets, and challenges [12]." It can also be referred to as organizational design where "organizational design is the whole sequence of work that results in an alignment of vision/mission, values/operating principles, strategies,

Table 4.2 Mitigation of Interoperability Checklist Deficiencies Using Effective Project Management (PM) Criteria

No.	*PM criteria*	*Description [11]*	*Relevant deficiencies*	*Tools and techniques*
1	Create clarity in communications	Clarity in defining project goals and likely project outcomes is critical, specifically during the early stages of a project; otherwise many project requirements may not be identified. Incorporating theses requirements at the latter stage of the project would lead to time and cost-overruns.	PHD #2 Assumption PHD #6 Alertness PHD #11 Communications PHD #21 Relationships PHD #22 Priority	Create a project-level change management and strategic communications plan; create a sense of urgency; and establish and communicate the change vision for the project (see Section 4.3: Organizational Development).
2	Define roles and project management processes	Clear assignments of roles and responsibilities without ambiguity of overlapping responsibilities are important for conflict resolutions and productivity. This definition will lead to effective use of project team members and their expertise. It also helps functional managers to understand project requirements and provide support.	PHD #1 Organization PHD #11 Communications PHD #21 Relationship PHD #22 Priority PHD #25 Delegation	Establish project structure ensuring alignment of responsibility, authority, and accountability (ref: organizational development Section 4.3).
3	Communicate expectations	Failure to establish expectations from all the stakeholders at the project process level is a common problem that will eventually result in both perceived and actual incidences of not delivering expected results. This is especially true with stakeholders within and outside the project who are not actively involved with projects.	PHD #3 Omission PHD #9 Attitude PHD #10 Imagination PHD #11 Communications PHD #12 Paraphrase PHD #14 Disclosure PHD #15 Insight PHD #16 Dissemination PHD #21 Relationship	Communicate the change vision and establish a coalition (see Organizational Development section). Create a project charter (see interoperability strategic plan Section 4.4).
4	Employ consistent processes	Many organizations manage projects with no formal processes. Mandating consistent and formal processes would lead to improved risk management, reduced ambiguity, and increased project management maturity.	PHD #1 Organization PHD #5 Supervision PHD #19 Consistency	Establish standard operating procedures and templates for projects.
5	Establish trust	Trust is critical to knowledge sharing and teamwork. An environment of trust is established by clear definition of roles and responsibilities and it promotes team performance. An environment of trust is influenced by the organizational culture that promotes transparency, collaboration, and openness. It ultimately leads to cohesive project team and organizations.	PHD #1 Organization PHD #3 Omission PHD #9 Attitude PHD #11 Communications PHD #12 Paraphrase PHD #13 Adaptability PHD #14 Disclosure PHD #20 Protectiveness PHD #21 Relationship PHD #24 Delegation	Apply change management techniques, including empowering end-users and stakeholders and planning for and creating short-term wins (see organizational development Section 4.3).

Table 4.2 (continued)

No.	*PM criteria*	*Description [11]*	*Relevant deficiencies*	*Tools and techniques*
6	Facilitate support	Organizational willingness to help the project is an outcome of top management support, and it is critical in the event of changes in project requirements. Obtaining organizational support is a challenge in traditional organizations, where resources are usually controlled by functional managers	PHD #1 Organization PHD #5 Supervision PHD #8 Intelligence PHD #21 Relationship	Establish a coalition; plan for and create short-term wins (ref: organizational development Section 4.3). Establish project-level leadership and support (see strategic planning Section 4.4).
7	Manage outcomes	A clearly defined project mission and objectives would help develop formal evaluation of project outcomes and determine project success. It promotes performance excellence, motivation, recognition, and synergy in teams.	PHD #4 Verification PHD #7 Complacency PHD #13 Adaptability PHD #14 Disclosure PHD #16 Dissemination PHD #23 Reporting	Capture knowledge and lessons learned through feedback mechanisms to support evaluation of project outcomes (see organizational development Section 4.3).

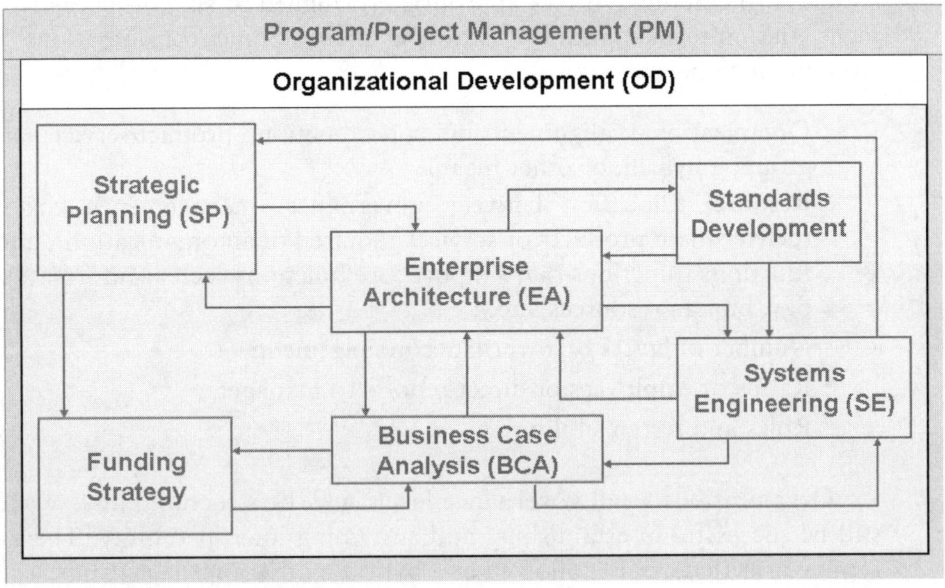

Figure 4.4 Organizational development best practice.

objectives, tactics, systems, structure, people, processes, culture, and performance measures in order to deliver the required results in the operating context" [13]. Vision/mission, values/operating principles, strategies, objectives, tactics, systems, processes, and performance measures will be addressed by other sections of this chapter. The remaining elements of organizational development/organizational design are discussed within this section with specific emphasis on people and culture—as these are fundamental tools in addressing the results of the World Interoperability Failure model on a specific public safety and emergency response organization.

Applying organizational development processes and tools, this section focuses on organizational structure, organizational change management, and strategic communications to:

- Mitigate people and process factors leading to the 25 deficiencies in the checklist;
- Enable shared, trusted understanding and collaboration within, and across, organizations to achieve a common goal;
- Increase stakeholder (in this context, those directly impacted by or who have a vested interest in interoperability success) buy-in and adoption from initial planning to implementation and overall sustainment;
- Support the six critical success (planning) factors derived in Chapter 1, namely governance, resources, approach (including organizational development), architecture, performance, and acceptance.

4.3.2 Organizational Structure

A critical first step in developing interoperability is to analyze the organizational structure to determine if it is appropriately aligned to support successful development and implementation of strategic plans. Some considerations related to structure include:

- Organizational alignment by core function, product/service, stakeholder group, a hybrid, or other means;
- Resource allocation between governance (management), core functions (deliver main products or services required of an organization), and enabling functions (functions that support core functions such as information technology, human resources, etc.);
- Number of layers of governance/management;
- Ratios of employees or direct reports to managers;
- Roles and responsibilities.

Organizations at all governance levels may be structured in several ways and still be successful in establishing and sustaining interoperability. There are many frameworks that can be followed or a hybrid/matrix approach can be selected. The *Guide to Organisation Design* by Naomi Stanford [14] outlines many frameworks that can be referenced and selected based on the type and level of governance in question. The following questions are then posed to determine which framework to employ:

- What structure supports overall interoperability?
- What structure enables fast decision and delivery of product or service?
- What structure is agile enough to adapt to changes in interoperability requirements?
- What structure maximizes information and knowledge flow?
- What structure enables two-way communications?

- How is structure impacted by the multiple levels of governance impacting interoperability?
- What structure enables the right balance between centralization and decentralization for the levels of governance impacting interoperability?
- How many layers of management/supervision are needed for oversight and control?
- What is the appropriate span of control for managers/supervisors?
- How should managers/supervisors and direct report be aligned?
- What are appropriate roles and responsibilities within the organization?
- Are responsibility, authority, and accountability aligned?

Once these questions have been answered, leaders and their planners will be better able to determine which framework should be used for organizational alignment (structuring or restructuring) to enable shared, trusted understanding and collaboration within and across organizations to achieve a common goal.

4.3.3 Organizational Change Management and Strategic Communications

All levels of governance impacting interoperability (see Chapter 3, Figure 3.1) face change related to people, process, and technology in a global environment of diverse stakeholders. Change management [15] and strategic communications are necessary to gain interoperability acceptance (see Chapter 1) and to mitigate adverse effects resulting from significant change. Well executed change management and strategic communications plans form a bridge that connects operational strategy with tactical implementation, thus reducing the risk of planning failure. For this reason, successful change occurs when leaders drive it (top-down) and end-users and stakeholders embrace and own it (bottom-up). The concept of ownership is important because it empowers individuals to take responsibility for interoperability performance.

To enable success, a phased approach to change management and strategic communications is recommended as follows: Phase 1: Prepare, Phase 2: Initiate and Implement, Phase 3: Sustain. Activities within these phases are based on industry best practices and the expertise of subject-matter experts [16]. A phased approach to change management and strategic communications:

- Accelerates implementation of change;
- Reduces resistance and addresses change at organization and individual levels;
- Increases stakeholder and end-user buy-in and adoption;
- Aligns and integrates change management and communications with the interoperability strategic plan and business initiatives;
- Eliminates adverse impact on end-users;
- Reduces turnover and loss of valued employees;
- Measures performance;
- Minimizes productivity loss.

4.3.3.1 Phase 1: Prepare

Change management and strategic communications are most effective when introduced during the initial phases of an effort. In this case, it would be introduced during program management planning and creation of the interoperability strategic plan. Initial focus is placed on providing strong leadership commitment with authority and consistent communication regarding the vision of interoperability within the organization (created during the strategic planning process—see Section 4.4). It is important to work across the entity and the levels of governance impacting interoperability to ensure these leaders and planners understand the vision, urgency, and organizational impact of reaching the desired future state of interoperability outlined in the strategic plan. Specific change management and strategic communications activities include:

- Conducting a change management and communications assessment—helps define the changes, how the changes may be accepted within the existing culture, what other major changes have occurred or are anticipated, and relevant challenges and opportunities. The assessment also includes identification of key stakeholders, a stakeholder analysis, and a summary of potential communication products and channels.
- Establishing a sense of urgency—involves communicating the need and urgency for interoperability to stakeholders and end users.
- Establishing a coalition—the coalition is responsible for leading change management activities including maintaining a sense of urgency. The coalition works hand-in-hand with the interoperability strategic plan planning team and the integrated project team or other designated group during implementation. As outlined below, there are overlaps between the planning team and coalition members to ensure consistency in messaging and success of the overall effort. Clear roles and responsibilities should be established for the coalition. Typical roles include:
    - *Champion (executive sponsor)*: Key leader defined in the strategic planning section of this chapter—empowered to make decisions related to the plan and way forward.
    - *Team lead*: Individual defined in the interoperability strategic plan section of this chapter—responsible for applying program management (Section 4.2) to support the creating, managing, and executing the project plan in addition to change management, communications, and outreach.
    - *Team members (change champions)*: Representatives from core responder-receiver and stakeholder groups defined below in the strategic planning section of this chapter.
    - *Change management lead*: Works with planning team to identify change impact and mitigation strategies, creates and maintains the change management plan and its implementation and sustainment.
    - *Communications lead*: Establishes core messages, performs stakeholder analysis, identifies communication products and channels, and creates and maintains the communications plan and its implementation and sustainment. (Note: Based on resource constraints and project size, the organiza-

tion may decide to have one lead for change management and communications).

- *Training lead*: If a change in knowledge, skills, and abilities is required based on the interoperability efforts selected, a training lead may be required. The training lead identifies training needs and creates and executes a training plan:
  - Establish a change vision (what defines success)—in most cases this is the vision from the strategic plan—the vision of a desired future state of interoperability.
  - Creating the change management and strategic communications plans—based on preferences some entities have separate change management and strategic communications plans while others include communications as a component of the change management plan. To minimize resource and time requirements to maintain the plans, we recommend creating only one plan which incorporates both change management and strategic communications activities.
  - Change management and strategic communications plan—includes the following:

    Time-phased implementation actions (refer to action plan template in interoperability strategic plan section of this chapter);
    Key messages (messages used to communicate the change);
    Training and coaching needs;
    Leadership roadmap;
    Stakeholder and resistance management plan;
    Risk management plan;
    Recognition and incentives;
    Recommended performance metrics;
    Feedback mechanisms—two-way communication is vital in overcoming any hurdles and building commitment to a new culture [17]. Table 4.3 is a communications matrix template for use in managing communications activities. Examples of communication products include workshops, team meetings, e-mail announcements, surveys, articles, brochures, posters, newsletters, and other sources.

4.3.3.2 Phase 2: Initiate and Implement

Overall change management and strategic communications support the change effort through monitoring, executing, communicating, and measuring the change

Table 4.3 Communications Matrix

Target Audience	*Message*	*Communication Product/Frequency*	*Channel*	*Responsible POC*	*Desired Outcome*	*Degree of Impact*

effort. This phase entails the hands-on execution of the change management and strategic communications plan. The use of collaborative processes should be employed whenever possible to gain the ideas and insights of the stakeholders and to establish entity-wide ownership of the interoperability strategic plan. Specific activities of Phase 2 include:

- Communicating the vision—includes designing, developing, and distributing communications and training material to communicate the desired future state interoperability vision, interoperability strategic plan goals and objectives, and initiatives/actions that will be taken to implement the plan.
- Empowering end-users and stakeholders—determining ways to involve end-users and stakeholders in strategic planning, enterprise architecture, systems engineering, and project management activities. This can include project management team meetings, working groups, coaching, mentoring, training, recognition and incentives.
- Planning for and creating short-term wins—measuring, managing, and documenting achievement of interoperability strategic plan and project-specific milestones. Data collected from end-users and stakeholders contains insights and ideas that can be implemented in the short-term.

4.3.3.3 Phase 3: Sustain

While change and communications initiatives may be successful in the short-term, the entity must take steps to ensure that the change is embedded and sustained. Activities include:

- Produce more change—capture knowledge and lessons learned including successes and failures for reuse. Employ interviews, focus groups, web polls, and surveys to obtain feedback on the progress of reaching the desired future state of interoperability. This feedback provides a means to identify best practices and areas for additional change activities which lead to updating the plan.
- Institutionalizing new approaches—focus on retaining stakeholder commitment to, and participation in, the change process. The expected end result of this effort is sustainment of a change culture—one that is responsive, agile, and resilient to adapting to needed changes to maintain interoperability.

4.3.3.4 Success Factors

The following success factors contribute to successful change management and strategic communications planning, implementation, and sustainment [18].

- Visible top management (top-down) support is a strong factor in sustaining momentum; thus the time spent by leaders giving out recognition awards, holding briefing sessions, and putting change on every meeting agenda is not a "nice to have" situation, it is a necessity.
- Improvements and change initiatives need to be viewed in the context of a total business strategy, which includes reassessing organizational structure,

business processes, team working, organizational culture and its symbols, reward and promotion structures.

- Involvement of all stakeholders (bottom-up) in any change process will ensure that they are kept fully in the picture, their ideas incorporated, and any potential reservations are taken into account.
- Check out in advance how the areas undergoing change interface with other parts of the organization and examine related processes. Seek to resolve conflicts of aims, values and interests, which may undermine progress.
- Communicate widely, internally and externally at every stage.
- Monitor and evaluate success at regular intervals, using measures agreed at the start.

The anticipated outcome of implementing a formal organizational change management and strategic communications process in interoperability *acceptance*—which includes all aspects of getting "buy in" from all levels of governmental leadership and the community, as well as responder-receiver leadership and their organizations, the "press," and the public is to establish shared (mutual) trust. Similar to project management best practice, organizational development methods are used throughout the program or project lifecycle. Organizational development processes and tools described in this section are most effective when formally planned and implemented rather than applied on an ad hoc basis.

Table 4.4 outlines organizational development processes and tools which mitigate deficiencies cited in the World Interoperability Failure model defined at the end of Chapter 3.

4.4 Strategic Planning

This section describes the critical strategic planning best practice shown in Figure 4.5.

4.4.1 Overview

A strategic plan to achieve interoperability as defined in Chapter 1 establishes an organization's interoperability mission, a desired future state of interoperability, and a roadmap to achieve the future state. A strategic plan *should* be created at each level of governance impacting operability (reference Chapter 3, Figure 3.1) but it must not be created in isolation. The strategic plan at a given governance level should always directly support (and be integrated in detail with) strategic plans of all upper governance levels, and it should be used to guide the creation of strategic plans for subordinate governance levels as well. In other words, the plans should provide well-integrated set of multiorganizational steps to achieve the common goal of preparedness. Coordinated strategic plans would be created at all levels of governance (not one alone) to convert the World Interoperability Failure model from a "worst case" calamity of deficiencies into a successful example of "best case" planning, an Interoperability Success Plan. In general, a strategic plan answers the following questions:

Table 4.4 Impact of Organizational Development Processes on Interoperability Checklist Deficiencies

Organizational Development Mitigation	*Deficiency*																								
	PHD#1 Organization	*PHD#2 Assumption*	*PHD#3 Omission*	*PHD#4 Verification*	*PHD#5 Supervision*	*PHD#6 Alertness*	*PHD#7 Complacency*	*PHD#8 Intelligence*	*PHD#9 Attitude*	*PHD#10 Imagination*	*PHD#11 Communications*	*PHD#12 Paraphrase*	*PHD#13 Adaptability*	*PHD#14 Disclosure*	*PHD#15 Insight*	*PHD#16 Dissemination*	*PHD#17 Inspection*	*PHD#18 Preparedness*	*PHD#19 Consistency*	*PHD#20 Protectiveness*	*PHD#21 Relationships*	*PHD#22 Priority*	*PHD#23 Reporting*	*PHD#24 Improvement*	*PHD#25 Delegation*
Organizational structure	●	—	—	—	—	—	—	●	—	—	—	—	—	—	—	●	●	●	—	—	—	—	●	—	●
Change management with a strategic communications plan	●	●	—	●	—	—	●	●	●	●	—	—	—	—	●	●	●	—	—	—	—	—	—	●	—
Communicating the vision	—	●		●	●	—	—	—	●	—	—	—	—	●	—	●	—	—	—	●	—	●	—	●	●
Establishing a coalition	—	—	●	—	●	—	—	●		—	●	—	—	●	—	●	—	—	—	●	●	—	—	●	—
Change management and communications assessment	—	●	●	—	—	●	—	—	●	—	—	—	—		—	●	—	—	—	—	—	—	—	—	—
Stakeholder and resistance management plan	—	—	●	—	—	—	—	●	—	—	—	—	—	●	—	●	—	—	—	—	—	—	—	●	●
Consolidating improvements and producing more change	—	—	—	●	—	—	●	—	●	—	●	●	—	●	—	●	●	●	—	—	—	—	●	—	—
Planning for and creating short-term wins	—	—	—	—	●	—	—	—	—	●	—	●	—		●	—		●	—	—	—	—	—	—	●
Institutionalizing new approaches	●	—	—	—	—	●	●	●	●	—	●	●	—	●	—	●	●	●	—	—	—	—	●	●	—
Empowering end-users and stakeholder	—	●	—	—	—	—	—	●	—	●	●	—	—	●	—	●	—	●	—	—	—	—	●	—	●
Establishing a sense of urgency	●	—	—	—	—	●	●	●	●	●	—	—	—		●	●	●	●	—	—	—	●	●	●	—
Establish a change vision	●	—	—	—	—	●	—	●	—	—	—	—	—	●	●	●	●		●	●	●	●	●	—	●

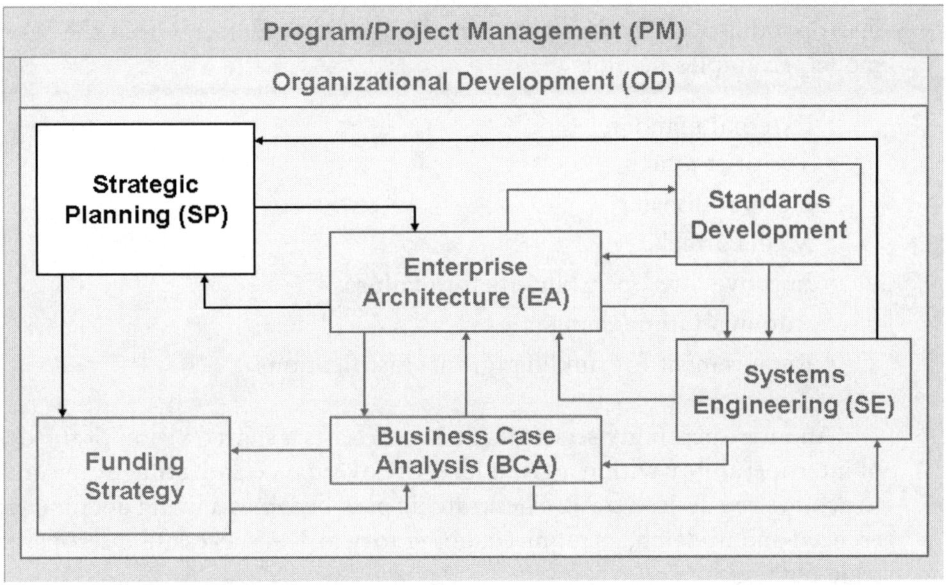

Figure 4.5 Strategic planning best practice.

- What political, social, economic, legal, regulatory, and technical issues including relevant laws, regulations, directives, and/or policies need to be met?
- What is the desired future state of interoperability we want to achieve?
- What is of most use and benefit to the stakeholders—responders and receivers?
- How do we measure our success?
- What are the milestones by which the intermediate and future results are to be achieved?

This section outlines a methodology for strategic planning based on best practices related to interoperability associated with public safety and emergency response, strategic planning, and change management (focused on human factors). The methodology harnesses the input of the varying levels of governance impacting interoperability in a collaborative method to develop a strategic plan that will not only ensure buy-in and support and be integrated with those of other governance levels, but it will also increase the probability of successful implementation and sustainment.

4.4.2 What is Strategic Planning?

Strategic planning is a systematic process of envisioning a desired future, and translating this vision into broadly defined goals or objectives and a sequence of steps to achieve them [19]. For the purposes of this book, we are referring to a desired future state of interoperability and therefore are defining an interoperability strategic plan. There are many drivers that may lead to the creation or revision of this interoperability strategic plan at any level of governance within the World Failure model. Examples include:

- External mandate,
- Terrorist attack,
- Natural disaster,
- Media attack,
- Improve interoperability before failure,
- Administration change,
- Requirement for funding (grant) justifications.

An interoperability strategic plan represents a shared vision of the desired state of interoperability within a given level of governance and a timeframe and roadmap to achieve the desired state. The strategic plan becomes a living document that is referenced and updated, as required, going forward. A successful interoperability strategic plan:

- Establishes the future vision based on an analysis of internal and external factors that may impact the future;
- Creates a framework for transitioning from the current state to a desired future state;
- Provides a broad overview of where resources will be allocated;
- Changes the way the level of governance will operate—influencing all areas of operation, becoming a part of the philosophy and culture;
- Provides a mechanism to communicate what is most important and initiate change with stakeholders;
- Establishes a baseline from which progress can be measured.

4.4.3 Methodology

The interoperability strategic plan methodology described within this section is based on a combination of United States Department of Homeland Security (DHS) SAFECOM [20] guidance and industry best practices related to interoperability associated with public safety and emergency response, strategic planning, and organizational development. DHS SAFECOM released the Statewide Communications Interoperability Plan Methodology [21] to support statewide strategic planning. Statewide Communications Interoperability Plans are supposed to be locally driven, multijurisdictional, and multidisciplinary statewide plans to enhance emergency communications interoperability. The Statewide Communications Interoperability Plan provides criteria and standards under which states could qualify for grant funding through the Homeland Security Grant Program and the Public Safety Interoperable Communications Grant Program [22]. The Statewide Communications Interoperability Plan is one framework that can be used for interoperability strategic planning, with some enhancements described in Chapter 7 to address the World Interoperability Failure model from Chapter 3.

The interoperability strategic plan methodology outlined in this section combines top-down leadership with bottom-up input from across stakeholder groups to ensure that the desired future state of interoperability conveys a realistic image of success that meets stakeholder needs. The methodology harnesses the input of the varying levels of governance impacting interoperability in a collaborative method that will not only ensure buy-in and support but also increase the probability of successful implementation and sustainment. The following strategic planning steps outlined in Figure 4.6 can be tailored, as appropriate, based on the level of governance initiating development of an the interoperability strategic plan.

4.4.3.1 Prepare

Preparing to create an interoperability strategic plan includes obtaining leadership commitment, securing resources, and creating a project management plan (see Section 4.2).

Leadership Commitment and Authority Ensuring leadership commitment and authority is the first and foremost requirement to initiate the interoperability strategic planning effort. At the each level of governance, leadership (top-down) commitment includes dedicating the appropriate priority/attention and resources (time and money) to the effort including a champion. The champion is the leader responsible for implementing and sustaining the plan and, therefore, has a vested interest in its success. For example, within a state (e.g., the first administrative division), the champion may be the state's secretary for Homeland Security. The champion imparts a collaborative vision which entails top-down and bottom-up participation in the planning process in addition to active participation across stakeholder groups to encourage stakeholder input, buy-in, and adoption of the final plan. Leadership authority implies that the leader is empowered to charter and make decisions enforcing the initiatives promoted in the strategic plan.

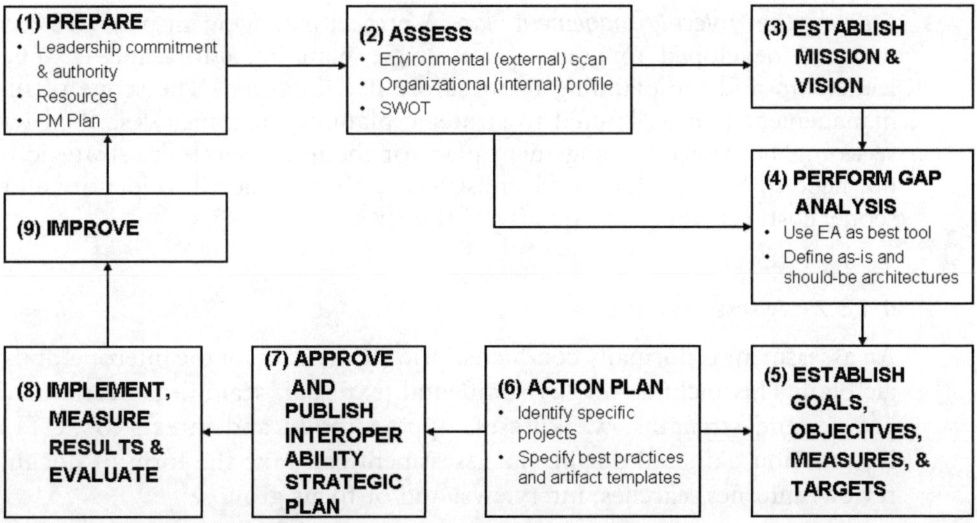

Figure 4.6 Strategic planning methodology for interoperability.

Securing Resources Project dollars and planning team resources need to be identified and gathered (see Section 4.7). The planning team is responsible for leading and coordinating the strategic planning effort. The team should be comprised of a cross-section of key stakeholders, which from practical experience, should not exceed 15 participants to remain effective. Typical planning team roles include:

- *Champion:* key leader empowered to make decisions related to the plan and way forward.
- *Team lead:* individual responsible for applying program management (Section 4.2) to support the creating, managing, and executing the project plan in addition to communications and outreach.
- *Facilitator:* leads planning team discussions, interviews, and focus groups. Ideally, this person is a neutral outsider without a vested interest in the details of the plan.
- *Meeting logistics/administrator:* responsible for coordinating meeting logistics, informing team members, and taking and distributing minutes from the meetings.
- *Team members:* representatives from core stakeholder groups, including:
  - Law Enforcement,
  - Fire Service,
  - Emergency Management,
  - Emergency Medical Services,
  - Public Works,
  - Hospitals,
  - Local Boards of Health,
  - Transportation,
  - Industry.

To support planning and implementation, organizations may want to consider including their grant administrator as a team member.

Creating the Project Management Plan A project management plan (see Section 4.2) must be developed to guide the strategic planning effort and is approved by leadership and the planning team before it is finalized. The scope of the project management plan is limited to strategic planning activities described within this section. The project management plan for the interoperability strategic plan does not need to be complex, but it must have sufficient detail to identify and manage scope, cost, schedule, and quality of the effort.

4.4.3.2 Assess

An assessment is normally conducted prior to creation of the interoperability strategic plan. This includes an environmental (external) scan; organizational (internal) profile; and strengths, weaknesses, opportunities, and threats (SWOT) analysis. Information gathered during the assessment can take the form of documentation review, Internet searches, interviews, and/or focus groups.

Environmental (External) Scan The environmental scan includes a scan or review of the environment external to the specific level of governance and equivalent organizations within the same level. For instance, an interoperability strategic plan focused on a second administrative division would conduct an environmental scan of the other five layers of governance and of other entities within the second administrative division. The scan entails a review of the political, social, economic, legal, regulatory, and technical environment including relevant laws, regulations, directives, and/or policies.

Organizational (Internal) Profile The organizational profile is an analysis of the culture and values, governance structure, demographics, stakeholders/suppliers, and operating environment within the specific level of governance. The operating environment includes products and services the entity is required to deliver, infrastructure including communication technology systems, and regional topography and geography that impact interoperability. An outcome of the profile is an assessment of the strengths that should be continued and potential weaknesses (constraints) that need to be mitigated when developing the interoperability strategic plan.

SWOT A SWOT analysis combines the results of the environmental scan and organizational profile, strengths, weaknesses, opportunities, and threats analysis. It summarizes the strengths and weaknesses of the current state in addition to opportunities for future improvement and potential threats (barriers) to implementing the identified opportunities. Some organizations conduct the assessment after the mission and vision have been established, but it is recommended that the assessment is completed ahead of time so that the planning team has a good perspective of what is impacting the organization today and what may impact it in the future.

4.4.3.3 Mission and Vision

The *mission* and *vision* are established with the planning team. The mission statement communicates the essence of interoperability (what, how, why, and for whom) within the specific level of governance. The mission is a succinct statement that articulates the organization's purpose, business, values, and beliefs as they relate to interoperability—it serves as a guide for day-to-day operations.

The vision is a compelling description of the future of interoperability within the specific level of governance—an image of what success looks like (what it will be or will achieve in the future). A vision statement should be a realistic, credible, and attractive future. It should be dynamic and should change as the environment and technologies change.

The mission and vision are vetted through leadership to ensure clarity, ownership, and a realistic image of future success. None of the subsequent steps will matter if the organization cannot articulate its purpose and where it is headed.

Ideally, facilitated workshops and collaborative decision-making processes/tools are employed, because otherwise reaching consensus on a mission and vision is difficult when various stakeholder perspectives are involved. Collaborative decision-making processes/tools enable participants to generate ideas, compare alternatives, and reach consensus in a faster, more efficient manner. Many tools allow anonymous and simultaneous electronic capture and display of individual contributions, stimulating idea generation, increasing constructive candor, and producing greater consensus.

4.4.3.4 Gap Analysis

Once a mission and vision are determined, the differences between the current state of interoperability and the desired future state must be determined using a gap analysis. It is important to gain broader stakeholder input through interviews and focus groups in order to define the "gap." The planning team works together to determine interview and focus group participants drawing from responder-receiver and other public safety and emergency response stakeholder groups. For instance, interviews may be for high-level leadership within the six levels of governance or responder-receiver groups, while focus groups would be structured based on discipline (e.g., law enforcement), geographic location (e.g., regions), or specific topics (e.g., emerging interoperability technologies). An effective focus group should have a minimum of two to five participants and no more than 10 to 15 participants based on professional experience.

Questions are created and tailored once interview and focus group structures are established. Questions are focused on how participant responses will support the newly defined mission and vision. Potential questions include:

- What aspects of existing systems "work well" in addressing interoperability issues?
- What interoperability "gaps" still need to be filled? (Use the World Interoperability Failure model at the end of Chapter 3 to determine the answer).
- What approach should be used to fill these gaps?
- Ideally, what would the future of interoperability look like within _____ (name of entity, e.g., town, province, county)?
- What may support or impede us from reaching this ideal state (based on the response to the previous question)?
- How can your organization support implementation of the interoperability strategic plan?
- Interview and focus group results are analyzed and divided into themes supporting the six critical planning factors outlined below and in Chapter 1, where possible.
    - Interoperability governance,
    - Architecture,
    - Resources,
    - Performance,
    - Approach,
    - Acceptance.

Interview/focus group results, the interoperability strategic plan mission and vision, and the SWOT created during the Assess step provide the planning team with the "gap" between the desired current and future states of interoperability. As a result of this gap analysis, the planning team will need to develop specific strategies and secure resources to close the gap.

4.4.3.5 Goals, Objectives, Measures, and Targets

Strategies to mitigate identified gaps are defined through *goals* and *objectives*. Goals are broad outcome-oriented statements that define what an entity is trying to accomplish—what must be achieved to be successful. Input from the gap analysis is used to define the goals and objectives. In contrast to goals, objectives are specific, measurable, agreed upon, realistic, time-bound (SMART) results that support achievement of the vision. Good objectives are measurable in terms of cost, performance, schedule, or other objective means. Objectives are the foundation for specific targets/measures and initiatives/actions, and they have a shorter time frame than goals.

Once objectives are agreed upon, specific *measures* and *targets* are determined for each objective. The measure is an indicator of success (how the organization will know it achieved its objective) while the target is the desired level of performance. Measures consist of two types: lead measures which are performance drivers, and lag (outcome) measures which measure success in achieving the objective. Refer to Table 4.8 in the Business Analysis section for example measures. There should be at least one target per measure focused on stretching the organization to higher levels of performance. Once targets are reached, the organization has executed its strategy.

Based on the critical planning success factors outlined in Chapter 1, Section 1.3, it is recommended that the goals and objectives are created within the following categories:

- Interoperability governance,
- Architecture,
- Resources,
- Performance,
- Approach,
- Acceptance.

The goals, objectives, measures, and targets are also vetted through leadership to ensure clarity, feasibility, and ownership. Ideally (similar to mission and vision), facilitated strategic planning workshops and collaborative decision-making processes/tools are employed, since reaching consensus on interoperability strategic plan goals, objectives, measures, and targets can be challenging when various stakeholder perspectives are involved. During this session, the action plan (described in the next section) is also created.

4.4.3.6 Action Plan

An action plan includes the steps (initiatives/actions) required to achieve the defined objectives, goals, and vision outlined in the strategic plan. An action plan represents a roadmap to implementing the initiatives in the strategic plan. Therefore, achieving an objective typically involves accomplishing a set of initiatives/actions. In some cases, organizations have an action plan for each objective rather than an overall action plan for the strategic plan. An action plan includes specifying milestones, timeline, resource requirements, and responsibility for each initiative/action. The initiatives/actions are focused on achieving the measures and targets specified for each objective. In an environment of resource (people and dollar) constraints, it is difficult to implement many initiatives at once. For this reason, it is recommended that these initiatives be tiered (tier I, II, and III, where tier I includes the first initiatives that should be started immediately) based on set criteria during the strategic planning workshop while the planning team is together. Potential criteria include:

- Ease of implementation,
- Potential cost,
- Overall impact to interoperability,
- Complexity,
- Breadth of impacted stakeholders (most likely, items that impact many stakeholder will take longer to coordinate and implement).

One potential failure of the interoperability action plan is the inability to specify the required best practices and artifacts to be employed by each initiative/project. This failure is a major risk in interoperability planning because of the critical need for close integration of action plans at all governance levels. The specification of best practices as described in this chapter and the associated planning artifact templates required to be used by the different initiatives/projects, would be undefined. This approach—specifying effectively how to achieve these initiatives/projects, much less their integration at higher and lower levels of governance—was not undertaken by the SCIP process. Thus, the detailed integration of these plans above and below the state level would be left to unspecified or random planning doctrine and conventions—therefore continuing to risk the 25 Pearl Harbor deficiencies in the World Interoperability Failure model. See Chapter 7 for an extended discussion of this point. Failure to do so is a major risk, leading to failed interoperability planning—as interoperable planning is required to achieve interoperable systems. Table 4.5 is a template of a typical action plan including specification of best practices [23].

Two of the most important categories of objectives to implement and sustain an interoperability strategic plan are the critical success factors associated with governance and acceptance. When developing initiatives related to Governance the planning team should consider:

- How should implementation and sustainment of the plan be governed? Single organization, consortium, working group, or others?
  - Who empowers this organization?
  - Should a charter be put in place outlining roles, responsibilities, timelines, and so forth?
- Who is responsible for top-down coordination?
  - How will information be shared in the future?
  - Should performance-based staff positions be put in place?

Table 4.5 Action Plan Template

Objective	*Initiative/ Project*	*Tier*	*Milestone*	*Timeline*	*Resource needed*	*Person or team responsible*	*Guidence (best practices and/or artifacts required)*
Objective 1	Initiative 1						
	Initiative 2						
	Initiative *n*						
Objective 2							
.	.	.	.	.	.	.	.
.	.	.	.	.	.	.	.
.	.	.	.	.	.	.	.
Objective *n*							

- Are there new ways interoperability will be governed within this entity?
- How will we measure and evaluate the implementation and sustainment of the interoperability strategic plan?

When establishing initiatives for Acceptance, the planning team should consider:

- How will we conduct outreach and communications? Do we need a communications (internal and external) plan?
- How will we gain buy-in from responder and receiver stakeholder groups? What activities will support this?
- How can we establish a feedback loop to ensure the strategy remains current and meets stakeholder needs?

4.4.3.7 Approve and Publish the Interoperability Strategic Plan

This step involves drafting, approving, and publishing the interoperability strategic plan using the following template:

- *Introduction and background*—Describe why the interoperability strategic plan has been created, those involved in the process, and so forth.
- *Methodology*—Provide a brief summary of methodology used to create the plan.
- *Assessment*—Summarize the Assess step (environmental scan, organizational profile, and SWOT).
- *Strategy*—Describe the agreed-upon mission, vision, goals, objectives, measures, and targets, providing context to why they were selected.
- *Action plan*—Include the action plan, including required best practices for each initiative/project and the specific integration requirements of these initiatives/projects with those at other governance levels.

- *Implementation*—Provide a description of how the interoperability strategic plan will be implemented including measurement of results, evaluation, and improvement.

The draft is reviewed by all key decision makers, feedback is integrated, and a final plan routed for approval. The approved plan should be distributed throughout the organization and integrated organizations, communicated to stakeholders, and maintained in an accessible location for ease of reference and updating as a "living" document.

4.4.3.8 Implement, Measure Results, and Evaluate

Once the plan is published, the organization needs to determine who will oversee the implementation and sustainment of the plan. It will most likely not be the planning committee, but a smaller subset considered a working group or integrated project team. For the purposes of this book, we will refer to this group as an integrated project team. As mentioned in the project management Section 4.2, integrated project teams are particularly valuable in interoperability planning because, properly done, individuals would be chosen to span the jurisdictions, governance levels, and disciplines as well as expertise needed to properly perform each initiative/project. Ideally, a hierarchical integrated project team structure would be defined in the interoperability strategic plan and each integrated strategic plan.

Communications and outreach are key to successful implementation and sustainment of the interoperability strategic plan. In addition to publishing and distributing the interoperability strategic plan, a communications and outreach campaign is essential for internal and external audiences. Wherever possible, elements of the plan should be integrated into daily, weekly, monthly, and quarterly routines. In parallel to communications and outreach, individual project plans need to be created for each of the initiatives where tier I initiatives are the first started. Refer to Section 4.2 of this chapter for best-practice guidance related to program and project management.

In addition to a project plan, funding must be secured. A business case analysis can support the investment justification by outlining the costs and benefits of the investment and potential return on investment. Refer to Section 4.6 of this chapter for a discussion on business case analysis.

A good interoperability strategic plan includes a method to monitor and evaluate the progress of the plan. Periodic evaluation of measures and targets are essential to assess progress made (or not made) toward achievement of the goals and objectives. Evaluation should also include assessment of validity (is it still current?) and feedback related to the plan. It is preferable to measure performance on a frequent basis (at least annually) to determine if the actions are supporting achievement of the objectives, goals, and vision. If not, root causes should be identified, evaluated, and mitigated through adjustments to the plan. A balanced scorecard or performance dashboard can be used to monitor progress against each of the critical success factors.

The integrated project team oversees activities described in this step and step 9 (Improve). The integrated project team meets on a regular basis (potentially biweekly) and has a charter to guide their activities. A good charter includes the integrated project team's:

- *Purpose and objective(s)*—Purpose and objectives of why the integrated project team exists and what it is supposed to accomplish.
- *Authority*—The authority provided to the integrated project team (what they can/cannot do), which should be designated from leadership.
- *Resources*—Available integrated project team resources, including any special requirements.
- *Responsibilities*—Clear delineation of member responsibilities, including whether they have to attend every meeting and what happens if they cannot.
- *Membership*—Includes names of members and their respective roles (i.e., chair/lead, etc.).
- *Meetings*—Frequency and other particulars of integrated project team meetings.
- *Deliverables*—The expected deliverable from an integrated project team, which could include business case analysis, final project selection, project management plans, project execution oversight, systems engineering plans, enterprise architecture views, IT/communication system specifications, and so forth.
- *Measures of success*—The method used by the integrated project team to measure its success.
- *Expiration/duration*—Duration of the integrated project teams lifetime on the project.
- *Effective date*—The date when the integrated project team is officially chartered to begin operation.

4.4.3.9 Improve

Any deficiencies in meeting the interoperability strategic plan measures and targets need to be improved. Based on results of the root cause analysis conducted in the previous step, the action plan is updated. If it is determined that the actual measures and/or targets were incorrect, then the overall interoperability strategic plan must be revised. Revisions should be done in a collaborative manner and must receive leadership approval before they can be implemented. In other words, the World Interoperability Failure model applies to interoperability planning as much as it does to operating information-sharing architectures. Ultimately, the interoperability strategic plan is a living document that guides achievement of a desired state of interoperability within a given level of governance. A good interoperability strategic plan created using the methodology outlined herein helps mitigate the planning deficiencies as shown in Table 4.6.

4.5 Enterprise Architecture

4.5.1 Overview

Enterprise architecture (see Figure 4.7) is an analysis and decision-support best practice which places organizational, asset, technical, policy, process, financial, and other business and tactical information into a consistent rational framework. In this

Table 4.6 Impact of Strategic Planning Processes on Interoperability Checklist Deficiencies

	Deficiency																								
Organizational Development Mitigation	*PHD#1 Organization*	*PHD#2 Assumption*	*PHD#3 Omission*	*PHD#4 Verification*	*PHD#5 Supervision*	*PHD#6 Alertness*	*PHD#7 Complacency*	*PHD#8 Intelligence*	*PHD#9 Attitude*	*PHD#10 Imagination*	*PHD#11 Communications*	*PHD#12 Paraphrase*	*PHD#13 Adaptability*	*PHD#14 Disclosure*	*PHD#15 Insight*	*PHD#16 Dissemination*	*PHD#17 Inspection*	*PHD#18 Preparedness*	*PHD#19 Consistency*	*PHD#20 Protectiveness*	*PHD#21 Relationships*	*PHD#22 Priority*	*PHD#23 Reporting*	*PHD#24 Improvement*	*PHD#25 Delegation*
(1) Prepare—leadership commitment and authority	●	—	—	—	●	—	—	—	●	—	—	—	—	—	●	—	●	●	●	●	—	●	—	●	●
(2) Assess—environmental scan	—	●	●	—	—	●	—	—	—	●	—	●	—	—	—	—	●	—	—	—	—	—	—	●	—
(3) Mission and vision—collaborative decision-making methods	●	—	●	●	—	—	—	●	—	—	●	—	—	—	—	—	—	●	—	●	—	—	—	●	—
(4) Gap analysis	●	—	●	—	—	—	—	●	—	—	—	—	●	●	—	—	●	●	—	—	—	●	—	●	—
(5) Goals, objectives, measures and targets	—	—	●	—	—	—	—	—	—	—	—	—	—	—	●	—	●	●	—	—	—	—	—	●	—
(6) Action plan—specified initiatives and projects using best practices	●	—	●	●	—	●	—	●	—	●	—	—	—	●	—	—	●	●	—	—	—	●	—	●	—
(7) Approve and publish interoperability strategic plan	●	—	●	—	—	—	—	●	—	—	—	—	—	—	—	—	—	—	—	●	—	●	—	●	—
(8) Implement, measure results and evaluate	—	—	—	●	—	●	—	—	—	—	●	●	●	●	—	—	●	—	—	●	—	—	—	●	—
(9) Improve	—	—	●	—	●	●	—	●	—	—	—	—	●	—	—	—	●	●	—	—	—	●	—	●	—

For a real-life application of this methodology reference the Massachusetts Statewide Communications Interoperability Plan (SCIP) Case Study in the included CD which represents an alignment of local, second-level, and third-level governance structures under a national (fourth-level) strategic planning precondition to get grant funds. The case study is an example of top-down structure applied to gather (and implement) bottoms-ups requirements—an important best practice in addressing the failed World Interoperability Failure model described in Chapter 3.

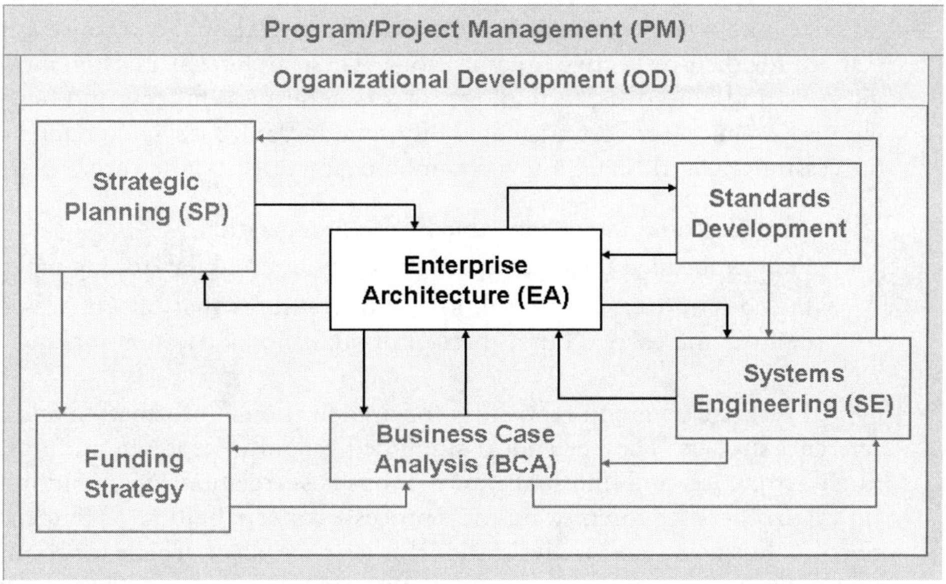

Figure 4.7 Enterprise architecture best practice.

context, we define *architecture* as "the fundamental organization of a system embodied in its components, their relationships to each other, and to the environment, and the principles guiding its design and evolution" [24]. We also accept the definition that an architecture description is "a representation of a defined domain, as of a current or future point in time, in terms of its component parts, what those parts do, how the parts relate to each other, and the rules and constraints under which the parts function" [25].

A common architecture description, in terms of a standard framework, must be employed. This framework is a fundamental tool for developing interoperability because it "provides the guidance, rules, and product descriptions for developing and presenting architecture descriptions that ensure a common denominator for understanding, comparing, and integrating Families of Systems (FOSs), Systems of Systems (SoSs), and interoperating and interacting architectures" [26].

This framework applies a common structure and language to the information and can provide a predefined set of views or representations of the collected information. In other words, it provides a simple standard approach to visualize large-scale systems of people, processes, and tools. Its use emerged from the need to create IT interoperability in businesses and government, where many individual stovepipe systems (e.g., see Figure 3.2) reduce efficiency and effectiveness. Thus, if multiple information stovepipes to be integrated depict their operational and system, architectures then interoperable interfaces will be highlighted and can then can be potentially implemented for interoperability.

4.5.2 Considerations in the Use of Enterprise Architecture

Enterprise architectures can be created at many different levels within and across governance organizations and jurisdictional or stakeholder boundaries, but these uses must be strongly coordinated. Several organizations involved in national public safety and emergency response maintain organizational enterprise architectures (OEAs) for their respective organizations. These architectures often include strategic goals and objectives, technology forecasts, business process and function models, performance, services and capability catalogs, and associated reference models. Successful OEAs all share a few common characteristics, they are:

- Accessible and understandable by the organization;
- Fully embedded into the organizations policy and day-to-day practice;
- Include a governance workflow which ensures that an OEA assessment is defined and carried out as part of organizational decision-making processes.

OEAs are a common repository for organizational information and associated reference models. They provide a simplified graphical representation of organizational structure and internal/external business relationships (and information flows) to otherwise complex business processes or capabilities. As referenced in Section 4.6: Business Case Analysis, the OEA is an excellent vehicle for routine evaluation of the business case relative to organizational goals and objectives. It supports procurement reviews, helping to ensure that planned purchases align with the organization's strategic goals and technical (interoperability) standards. The use of OEA and project-specific EAs within the system development life cycle (SDLC) ensures that system designs continues to conform to the changing organization while their operations are under development. This process of embedding enterprise architecture reviews into governance workflows is critical to sustainment of solutions identified in an interoperability strategic plan. From this point forward, we will focus on project enterprise architecture and only refer to OEA when necessary for context. The project level enterprise architecture always reaches back to the overarching OEA when it is available.

Like any other best practice, the use and deployment of enterprise architecture must be planned. Many organizations do not have an existing organizational enterprise architecture and, as such, will only be using enterprise architecture principals and methodologies on a project or program basis. Organizations with existing EAs can draw much of the background information from the organizational enterprise architecture, but will still need to follow the tenets below to develop interoperability architecture.

A key tenet for enterprise architecture is to determine the problem and purpose for the architecture and build only what is needed to address that need. Organizations should consider the use of enterprise architecture and determine if there are multiple projects or programs for which it is to be applied, or a single event. This decision affects the selection of tools and support software procured for the resulting enterprise architecture projects. Useful EAs can be built using a word processor, spreadsheet, and presentation/graphical programs common to most office computers. For organizations with severely limited resources, or only a single planned project, these minimal capabilities are adequate. For larger organizations with multiple planned projects there are several enterprise architecture tools available with varying cost and capabilities. The advantage of using a tool is the ability to easily reuse the information for future or multiple concurrent projects. By using an automated toolset the enterprise architecture information is built and refined through successive projects into a single repository of interoperability information.

The enterprise architecture repository is an excellent decision-support resource. In some national organizations, the role of enterprise architecture is evolving to include the use of databases, business intelligence reporting, and Web portals. This is an important transition for enterprise architecture from "shelf-ware" to a living knowledgebase.

4.5.3 Approach to Enterprise Architecture

There are several well-documented enterprise architecture frameworks, such as Zachman [27], Federal Enterprise Architecture Framework (FEAF) [28], and the Department of Defense Architecture Framework (DoDAF) [29]. Each of these frameworks was developed for specific purposes and provides guidance for the documentation, analysis, and display of the information contained in each enterprise architecture. The best enterprise architecture framework for tactical public safety and emergency response *operations* is a tailored version of DoDAF, version 1.5 [30]. Several enterprise architecture frameworks have been used for business enterprises and administrative systems, but DoDAF is directly applicable to the fast-paced command-and-control environment of tactical public safety and emergency response operations. In particular, DoDAF contains "descriptions that ensure a common denominator for understanding, comparing, and integrating Families of Systems (FOSs), Systems of Systems (SoSs), and interoperating and interacting architectures" [31]—directly relevant in planning the integration of independent information—sharing IT/communications stovepipes.

The use of DoDAF is of fundamental importance because it will establish a common ontology across all responder-receiver organizations, define standard views of each disparate stovepipe architecture; and provide a common means for the integration of disparate stovepipes to achieve interoperability. More specifically, this approach provides a common means of depicting and describing the dissimilar *operations*, *systems*, and *technical standards* employed ("as is")—or planned ("to be")—by each organization. Integrating these independent information stovepipes, or information enterprises, their missions and systems, to achieve interoperability is then done by integrating the corresponding architecture views and adding whatever additional operations, system, or technical standard elements are required to meet the to-be objectives.

Support for moment-to-moment and day-to-day operations, when and where possible, assures that responders-receivers are familiar with the necessary system operations, trust system capabilities and responses, and thus establishes a "culture of interoperability" across disparate organizations spanning disciplines, jurisdictions, and governance levels.

The use of enterprise architecture for public safety and emergency response interoperability planning is not a new concept. A tailored version of DoDAF v1 [32], dubbed the Public Safety and Emergency Response Enterprise Architecture Framework, was presented in 2004 to interoperability planners in Alaska [33] as well as information-sharing planners at the Massachusetts Emergency Management

Agency [34]. Subsequently, DoDAF was tailored and adopted by SAFECOM as the Public Safety Architecture Framework (PSAF) [35, 36].

> The SAFECOM Public Safety Architecture Framework (PSAF) provides an industry-validated enterprise architecture methodology to plan and develop the migration from current public safety architectures to interoperable systems. The PSAF methodology is designed to capture information about radio systems and elements to expedite interoperability analysis, along with other related functions, across various communications paths.

SAFECOM has published the PSAF in three living documents—*PSAF Volume I: Definitions and Guidelines* (2006) [37], *PSAF Volume II: Product Descriptions* (2006) [38] and *PSAF Volume III: Communications Data Model* (2007) [39].

PSAF, like DoDAF, is a formal information structure and methodology. Full implementation and documentation of all predefined architecture views in the prescribed framework formats are seldom needed for most interoperability planning projects. Instead, purposefully selected views serve as a standardized modeling toolset to depict some or all of large-scale as-is and to-be architecture. In the gap analysis performed in the Strategic Planning section above, enterprise architecture tools can be used to represent the as-is and to-be (or should-be) configurations of an architecture. The difference between the to-be and the as-is represents a gap in people, processes, or tools.

Enterprise architecture development, like interoperability, requires planning and disciplined execution for the best results. This section will discuss the planning and development of interoperability enterprise architecture. The interoperability enterprise architecture will be discussed with the following assumptions:

- A project or program level interoperability enterprise architecture is required;
- There is no existing enterprise architecture to provide information or views;
- Tools utilized are Microsoft Corporation's Microsoft Office Professional and Microsoft Visio Professional;
- Architecture products built will be based on PSAF and DoDAF.

Many elements of the enterprise architecture tie to the World Interoperability Failure model. Interoperability enterprise architecture development is broadly defined as a few basic activities:

1. Determining the purpose and scope of the enterprise architecture;
2. Selecting the architecture framework (e.g., PSAF) and architecture views, called work products, to be used;
3. Developing an enterprise architecture for the current as-is state of an architecture;
4. Capturing targeted needs and developing to-be architecture views;
5. Determining the gaps and projects necessary to transition from the as-is to the desired to-be state.

4.5.4 Determining Purpose and Scope of the Enterprise Architecture

Before construction of interoperability enterprise architecture, it is necessary to define its purpose, scope, and use to best integrate the enterprise architecture process into the suite of best practices needed to achieve interoperability. This enterprise architecture integration with project management, business case analysis, strategic planning, and systems engineering best practice is depicted in Figure 4.8. First, the purpose of the architecture must be determined—what must it help accomplish? Second, determine the scope—what are the boundaries of the enterprise architecture; that is, what is considered internal versus external to the enterprise architecture? Third, determine how the interoperability enterprise architecture will be represented—what are the architecture work-product views and outputs (e.g., gap analysis) desired from the architecture? Finally, how will a standard architecture framework DoDAF or PSAF be used?

It is important for an interoperability enterprise architecture project to be planned and scoped as any other interoperability project. Planning questions like those in Figure 4.8 determine the information collected, the manner of its representation, and the requisite level of detail required. Using a standardized framework like DoDAF or PSAF can speed the process, but like all processes, it must be tailored to the interoperability project *target* (the project purpose) architecture given the *current* enterprise architecture configuration.

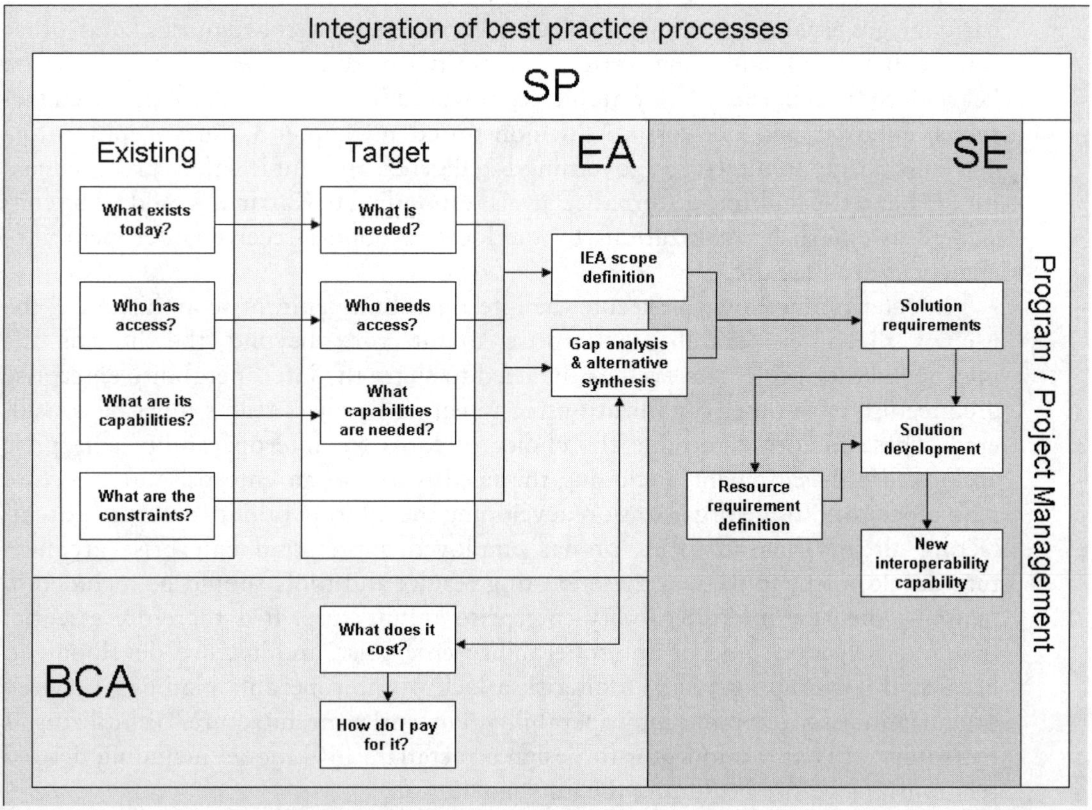

Figure 4.8 Interoperability planning from an enterprise architecture perspective.

Selection of the PSAF framework allows interoperability enterprise architecture planners to immediately employ a prescribed set of standardized architecture views to model disparate stovepipes in a common representation format. In addition, the planner will need to tailor the data collection requirements, the architecture views to be developed, and other aspects of the interoperability enterprise architecture under PSAF requirements. It is not the case that each interoperability enterprise architecture must follow the "letter of the law" defined in the chosen framework, but rather which PSAF views are required to fulfill the intent of the interoperability enterprise architecture. This tailoring process is critical to development of a focused and efficient interoperability enterprise architecture in achieving interoperability.

As is often the case, the scope of the information-sharing architecture expands as more stovepipes become known, and so can increase the level of effort necessary to complete the interoperability enterprise architecture. As a general rule, the scope of the interoperability enterprise architecture is determined by the

- Portion of an architecture controlled by each organization for interoperability planning;
- Organizations and entities that can be *influenced* to improve information sharing, and include them as *internal* organizations;
- Organizations beyond influence, which are regarded as *external* organizations, delineating the stovepipe architecture boundary.

In the case of the NPR-based Emergency Communications Architecture (ECA), national governance organizations were considered internal entities, and other state, tribal, local, and nongovernmental entities were considered external. In the Pearl Harbor scenario, a Navy architecture would have to consider Army architecture as external, and vice versa. As demonstrated in Chapter 3, the stovepipe information-sharing architectures governing intelligence and pubic safety preparedness for 9/11 and the multiple governance levels responding to Katrina, would all be considered as external organizations by the local responder-receiver interoperability enterprise architecture.

In determining how to execute the interoperability enterprise architecture, the use of the interoperability enterprise architecture beyond the life of the interoperability project, as well as the need to share the interoperability enterprise architecture with other organizations or integrated project teams must be considered. These factors determine the choice of tools for interoperability enterprise architecture development, including the need to retain an enterprise architecture data repository. If the organization developing the interoperability enterprise architecture already has an OEA, or has employed a preferred enterprise architecture-development tool, then these existing results and tools should be included in planning the new interoperability enterprise architecture. It is therefore essential that tool selection precede interoperability enterprise architecture development, because differing proprietary tools risk a lack of interoperable planning between organizations developing interoperability enterprise architecture. If utilizing a repository, it is also important to design a metadata [40] model including desired relationships before beginning the implementation.

For interoperability architecture, current information can be drawn from numerous existing sources, such as:

- Organizational charts and documents;
- Equipment and facility inventories;
- Interoperability strategic plan;
- Letters or memoranda of understanding;
- Communication plans;
- Frequency/spectrum plans;
- Existing interoperability tools;
- Policies and procedures;
- Legislative and legal constraints;
- Grant guidance;
- Other available resources, including deployable architectures;
- Existing emergency operations plans and SOPs;
- Service level agreements or mutual aid agreements;
- IT networks and systems topologies—including backbone transport and security capabilities and requirements;
- Higher-level plans such as regional, national, or international plans and agreements for example, National Response Framework [41] (NRF) or international, such as NATO's Standing Operating Procedures for the Euro-Atlantic Disaster Response Unit (EADRU) [42].

Based on the scope, purpose, and planned architecture views and products, data from the available sources listed above can be used to create the as-is architecture.

Most interoperability enterprise architecture projects will begin with collecting basic information about how organizations at all governance levels share information with each other in moment-to-moment public safety as well as emergency situations. These sources of existing information-sharing architecture include:

- Actual responders-receivers roles such as firefighter, law enforcement officer, HAZMAT specialist, doctor, and others.
- Leadership roles, such as incident commander, watch stander, fire chief, and others.
- Special units, such as search and rescue, special weapons and tactics, among others.
- Operation centers, such as PSAPs, EOCs, Joint Field Office (JFO), Incident Command Post (ICP).

Many of these organizations and facilities, either internal or external to the interoperability enterprise architecture, have operating procedures, legislation, policies, MOUs, and other existing artifacts. Once the organizations are identified and their architecture information collected, they are added to the interoperability enterprise architecture. At this point, communication needlines are added between these organizations based on *imagining* the needs under a variety of threat scenarios and documenting them in the to-be architecture. These needlines could have been defined—or implied—in the documents collected above, but frequently the specific communication requirements by scenario are not defined and, therefore, need to be vetted with the actual responders-receivers and other stakeholders to define and validate the Information Exchange Requirements (IERs). Without this definition and validation, the World Interoperability Failure model will identify deficiencies PHD #1 Organization, PHD #3 Omission, PHD #4 Verification, PHD #10 imagination, PHD #11 Communication, PHD #13 Adaptability, PHD #14 Disclosure, PHD #15 Insight, PHD #16 Dissemination, PHD #18 Preparedness, PHD #20 Protectiveness, PHD #23 Reporting, and PHD #24 Improvement, and will guarantee interoperability failures as shown in Chapter 3. The communication requirements specifying the information flows between jurisdictions, disciplines, and governance levels include:

- Information source(s);
- Information destination(s);
- Media (voice, data, video);
- Message length/duration;
- Message format/language;
- Performance (e.g., reliability and availability);
- Environment (indoor/outdoor, temperature, humidity, dust, etc.);
- Need for message acknowledgment;
- References (justification for each IER);
- Maximum allowed transmission delay;
- Priority;
- Interfaces (physical, electrical, human-system);
- Maintenance (preventative and responsive) and logistics;
- Training (operation and maintenance);
- Legacy equipment (function, interfaces);
- Security (e.g., encryption level required. authentication level required, etc.);
- Transmission environment (fixed/mobile, range, terrain, indoor/outdoor, interference, noise, etc.).

4.5.5 Architectural Views

A key shortcut for verification and validation of the current architecture, organizations and needlines is the use of graphical architecture views which depict "slices" of the as-is or to-be architecture in standard formats. These graphical views make it easier to present data to first responders and often expose gaps in data or capabilities. Without common architecture views, integration of disparate organizations and their information enterprises (either as-is or to-be) and their shared tasks and activities, information, and missions becomes problematic, ad hoc, and fraught with excessive risk. These views simplify the integration of disparate information architectures, including security (encryption) policies and procedures. Thus, the integration of disparate IT/communication architectures for moment-to-moment and day-to-day tactical public safety operations and emergency response interoperability is best performed using common architecture views. The PSAF or DoDAF explicitly define the following architecture views and their interrelationships as shown in Figure 4.9:

- Operational views (OVs), including information flows that support tasks and activities;
- System views (SVs), including hardware and software configurations;
- Technical views (TVs), including products, services and technical protocols/standards.

4.5.5.1 All Views

Not shown in Figure 4.9, the all views (AVs) provide information which span the entire architecture. The all view 1 (AV1) provides an overview of the architecture, its strategic and mission importance, who will build it, what are the artifacts, views and scope as well as noting the stakeholders for the interoperability enterprise architecture. All view 2 (AV2) is a data dictionary and contains all of the applicable terms, names, and acronyms utilized in the interoperability enterprise architecture providing a common language base.

4.5.5.2 Operational Views

The operational views (OVs) depict the tasks and activities and corresponding information-sharing requirements for responder-receivers and other stakeholder organizations, the "boots on the ground," so to speak. This ability to view the architecture from the perspective of the user is the reason that the DoDAF was chosen as the basis for the PSERAF and the PSAF. A DoDAF interoperability enterprise architecture would make extensive use of the following OVs:

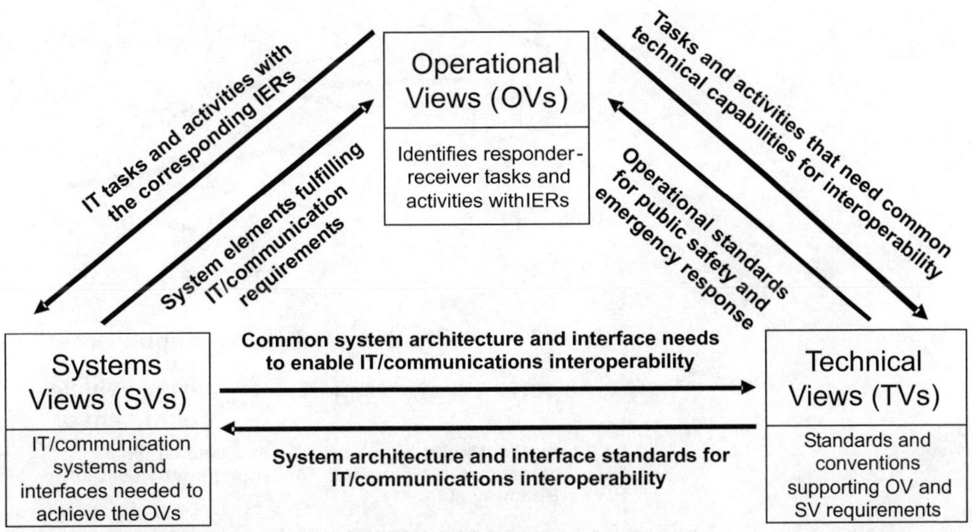

Figure 4.9 Interrelationships of DoDAF architecture views.

- Operational View 1 (OV-1)—High-Level Operational Concept Graphic: The OV-1 is a graphical depiction or cartoon of the operational organizations, their facilities, mobile units, and associated communication links and networks showing operations needed to address one or more natural or man-made threat scenarios. OV-1s would be used in executive-level presentations to depict the scope and intent of a given interoperability project. See Figure 4.10 for a sample OV-1 depicting a mass evacuation from a catastrophe—which we will title a catastrophic evacuation for brevity's sake.
- Operational View 2 (OV-2)—Operational Node Connectivity Description: The OV-2 is in an important and informative interoperability enterprise architecture view. The OV-2 depicts the participating organizations graphically as well as their various relationships (including communication needlines). OV-2s can depict a single organization's external relationships (i.e., shared tasks and activities with the corresponding IERs) in the nodal diagram as well as a larger scale view representing multiple organizations comprising a significant section of an interoperability enterprise architecture, as shown in Figure 4.11 for the catastrophe evacuation example.
- Operational View 3 (OV-3)—Operational Information Exchange Matrix: The OV-3 is a tabular or matrix view often represented in a spreadsheet. The OV-3 identifies the communication needlines and their specific characteristics—serving as a critically important interoperability enterprise architecture work product for documenting the communications for information-sharing among responders-receivers and other stakeholders. Figure 4.12 depicts an OV-3 for the catastrophic evacuation example.

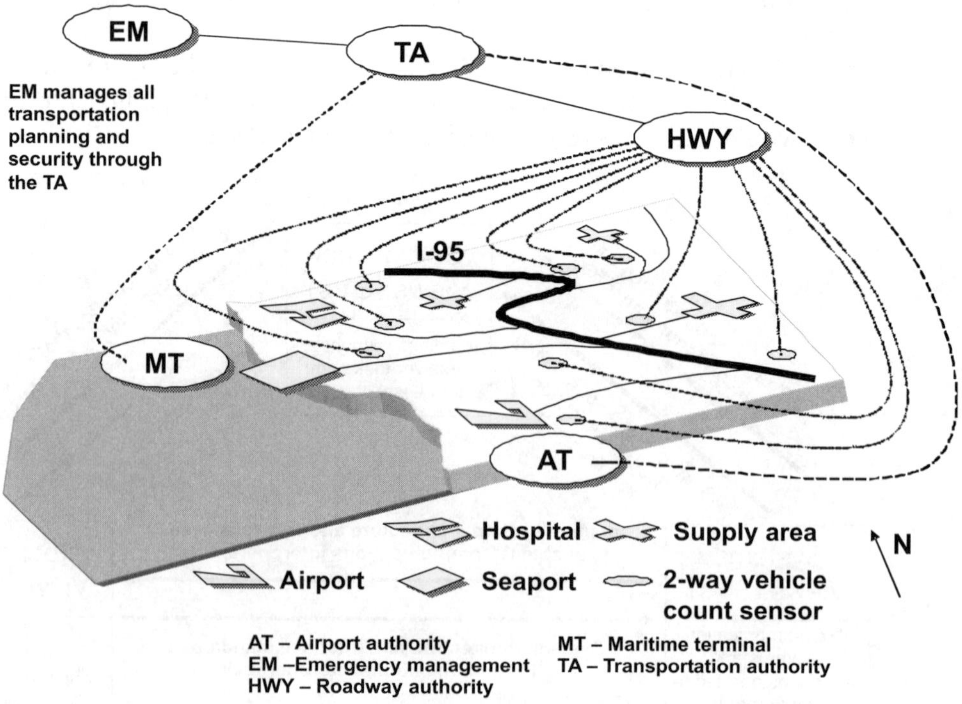

Figure 4.10 OV-1 sample for a possible catastrophe evacuation scenario.

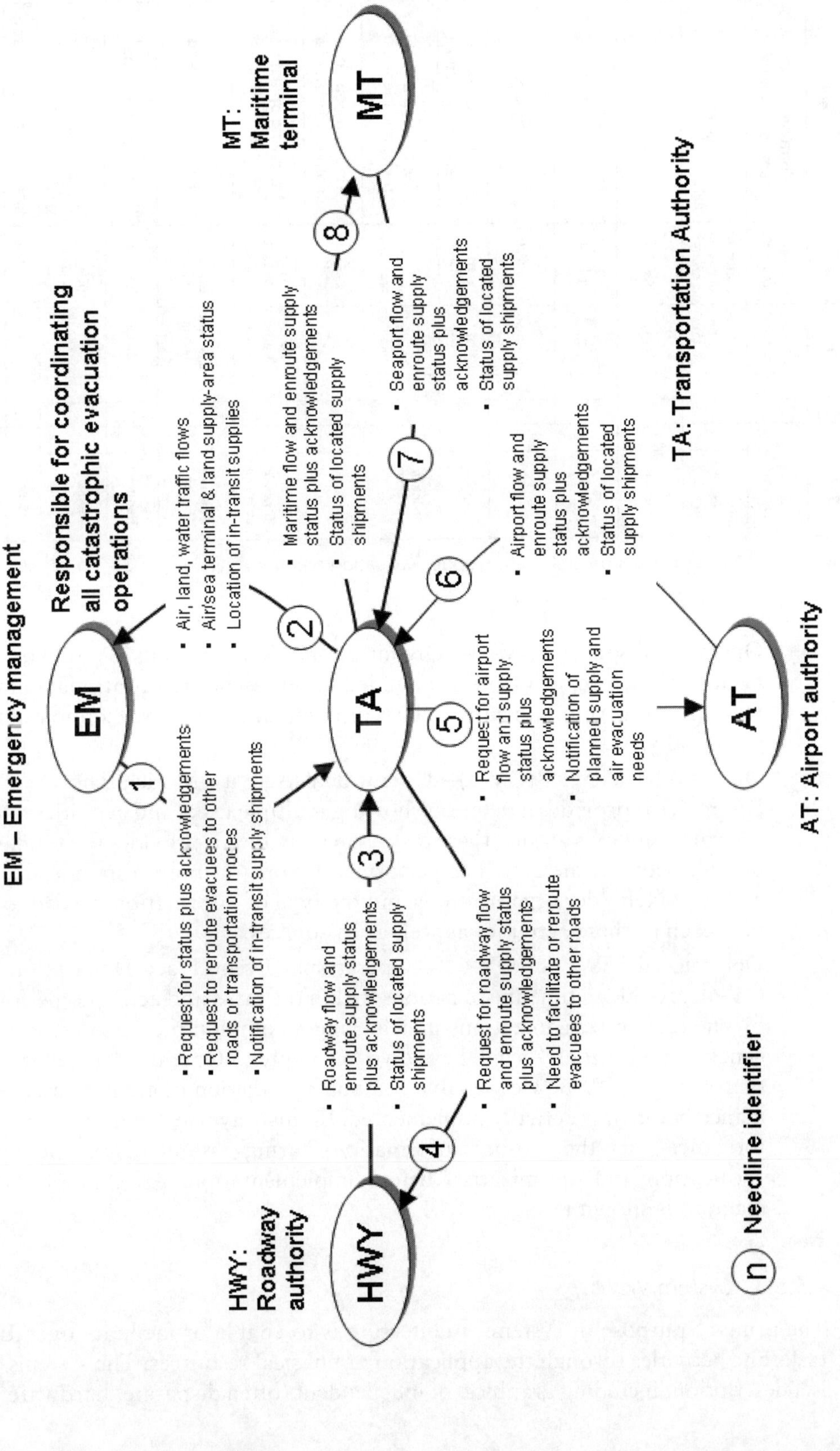

Figure 4.11 OV-2 sample for a possible catastrophe evacuation scenario.

		Information element description					Producer		Consumer		Nature of transaction				
Needline identifier	Information exchange identifier	Information element name and identifier	Content	Scope	Accuracy	Language	Sending op node name and identifier	Sending op activity name and identifier	Receiving op node name and identifier	Receiving op activity name and identifier	Mission / scenario	Transaction type	Triggering event	Interoperability required	Criticality
1	EM-TA	Status request	Status of all routes & modes	Region	1 hour	English	EM	Maintain evacuation progress	TA	Track flows & supplies	Evacuate region	Request	Periodic check	Data	Medium
1	EM-TA	Reroute request	Reroute mode traffic	Region	5 minutes	English	EM	Direct traffic reroute	TA	Force traffic reroute	Evacuate region	Direct	Flow blockage detected	Voice	High
1															
2															
...															
8	TA-MT	Status request	Status of vessel traffic	Region	1 hour	English	TA	Maintain evacuation progress	MT	Track flows & supplies	Evacuate region	Request	Periodic check	Data	Medium

Figure 4.12 OV-3 sample for a possible catastrophe evacuation scenario.

- Operational View 4 (OV-4)—Organizational Relationships Chart: The organizational picture, or OV-4, is intended to represent organizational data and relationships. An organization chart or command chart are examples of this graphical view.
- Operational View 5 (OV-5)—Operational Activity Model: The OV-5 (see Figure 4.13) provides a diagram showing multiple tasks and activities and the information flows among these tasks and activities. It provides the fundamental link between the activities performed by one or more responder-receiver and or stakeholder organizations and the type of information needed to support each of these activities as well as its sources.
- Operational View 6c (OV-6c)—Operational Event-Trace Description: The OV-6c provides a timeline of actions within the tasks and activities performed by one or more organizations in achieving a common public safety or emergency response goal. As will be shown in Chapter 5 Operational Architecture, the use of the OV-6c view enable computer prediction of architecture performance before it is actually implemented. In this way, we can measure the performance of the to-be information-sharing architecture, permitting verification and optimization before implementation. An example OV-6c diagram is shown in Figure 4.14.

4.5.5.3 System Views

The primary purpose of systems architecture is to enable or facilitate operational tasks and activities through the application of physical resources. The systems view is a description, including graphics, of independent (often disparate) hardware/soft-

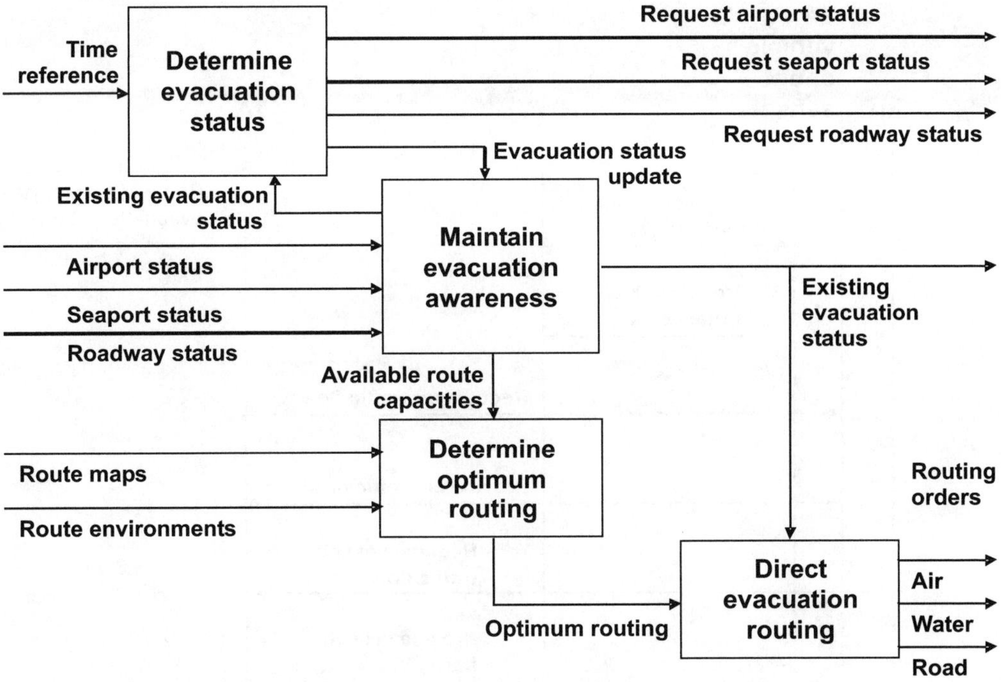

Figure 4.13 OV-5 sample for a possible catastrophe evacuation scenario.

ware capabilities enabling operational (OV) functions. For the interoperability enterprise architecture, SVs will:

- Map public safety and emergency response organizations, their mobile assets, functions, and characteristics identified in the operational architecture views;
- Identify system interfaces and define the connectivity between systems;
- Define system constraints and bounds on system performance.

The SVs show *how* multiple systems link and interwork, and also describe (if necessary) the internal construction and operations of particular systems within the architecture. For an independent network, the SV includes the physical connection, location, and identification of key nodes (including materiel item nodes), circuits, networks, mobile assets, among others, and specifies system and component performance parameters (e.g., mean time between failure, maintainability, availability). The SV associates physical resources and their performance attributes to the corresponding OVs and its requirements using the technical standards defined in the technical view (see below). In this regard, the SVs are based upon—and constrained by—the associated technical architectures.

Useful SV work products used for the as-is information stovepipes and the to-be interoperability enterprise architecture include:

- Systems View 1 (SV-1)—System Interface Description: The identification of systems and system elements, components, subsystems and their interfaces, within and among nodes.

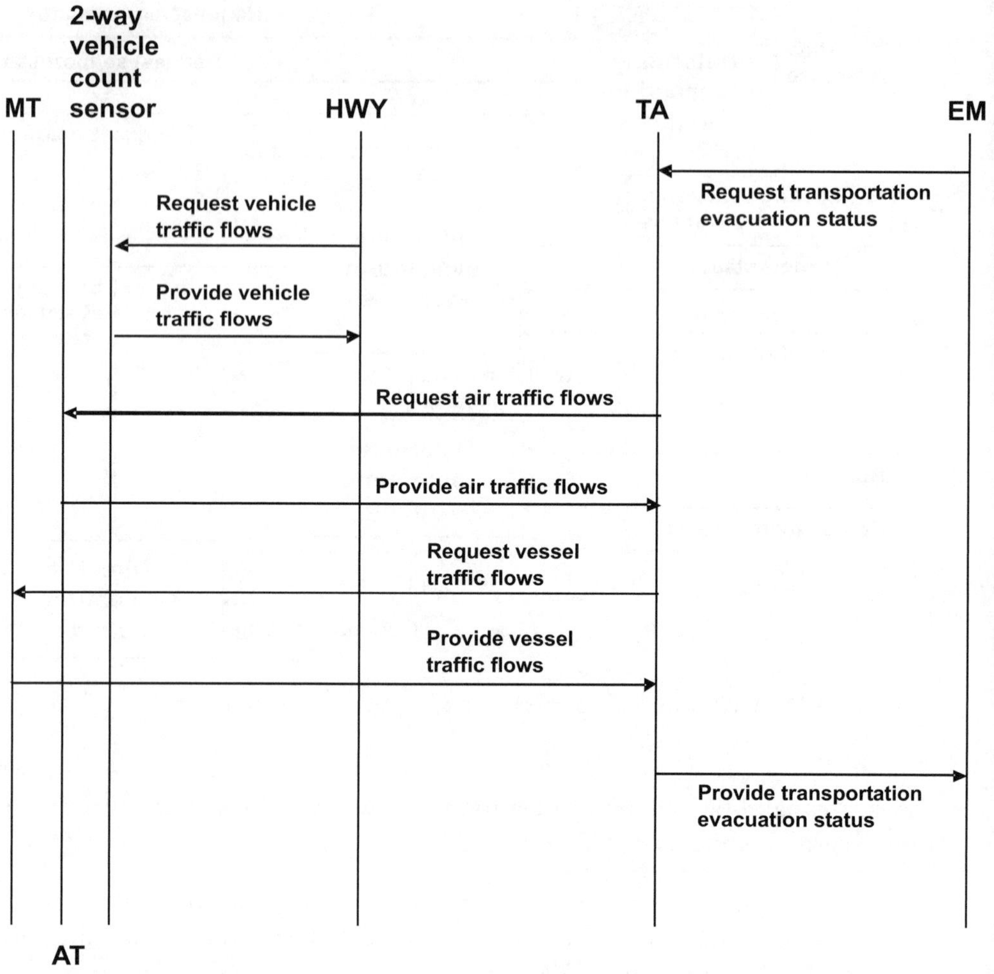

Figure 4.14 OV-6c sample for a possible catastrophe evacuation scenario.

- Systems View 2 (SV-2)—Systems Communications Description: Physical nodes and their related communications systems descriptions.
- Systems View 3 (SV-3)—Systems-Systems Matrix: Relationships among systems in a given architecture; can be designed to show relationships of interest (e.g., system-type interfaces, planned verses existing interfaces, etc.).
- Systems View 4 (SV-4)—Systems Functionality Description: Functions performed by systems and the information flow among system functions.
- Systems View 5 (SV-5)—Operational Activity to System Function Traceability Matrix: Mapping of system functions back to operational activities.
- Systems View 6 (SV-6)—System Information Exchange Matrix: Detailing of information exchanges among system elements, applications and hardware (H/W) allocated to system elements.
- Systems View 8 (SV-8)—Systems Evolution Description: Planned incremental steps toward migrating a suite of systems to a more efficient suite, or toward evolving a current system to a future implementation.

- Systems View 9 (SV-9)—Systems Technology Forecast: Emerging technologies and software/hardware products that are expected to be available in a given set of time frames and that will affect future development of the architecture.

These SV work products define the physical IT/communications interoperability enterprise architecture needed to specify the interworking and integration of the independent information-sharing stovepipes.

4.5.5.4 Technical Views

Finally, the TV is the minimal set of rules governing the arrangement, interaction, and interdependence of system parts or elements, the purpose of which is to ensure that the interoperability enterprise architecture satisfies a specified set of requirements. The TV provides the technical systems-implementation guidelines upon which engineering specifications are based, common building blocks are established, and commercial off-the-shelf (COTS) products and services may be identified. The TVs include a collection of technical standards, conventions, rules, and criteria organized into profiles that govern system services, interfaces, and relationships for particular SVs and that relate to particular OVs. These TV work products define the interoperability enterprise architecture interface standards enabling the required open architecture, ideally with a minimum of commercial intellectual property rights (IPR), minimizing single-vendor reliance. The TV work products are defined as:

- Technical View 1 (TV-1)—Technical Architecture Profile: Extraction of standards that apply to a given interoperability enterprise architecture, for example common alerting protocol (CAP).
- Technical View 2 (TV-2)—Standards Technology Forecast: Description of emerging standards that are expected to apply to the given interoperability enterprise architecture within an appropriate set of timeframes.

TVs document current use of technologies and technical standards (TV-1) and future or forecasted technologies (TV-2). Technology views are used to document and plan technologies or changes needed as technologies evolve. TVs are often the first step of governance implemented as an organization creates and begins to utilize the interoperability enterprise architecture.

4.5.5.5 Use of the Architectural Views

Not all work products may be selected for development of an interoperability enterprise architecture, because they would require a level of design detail beyond the scope of the interoperability strategic plan and available resources. The OV and SV work products can be selected and developed to emphasize many critical categories of interoperability enterprise architecture requirements, for example:

- Analyze human roles in the interoperability enterprise architecture—A human-centered supplement to the OV-4 can be used to describe the roles and responsibilities of human operators, leaders, and planners who populate the organizations. Gaps, overlaps, and unique roles for organizations and key personnel then become apparent.

- Design and evaluate security and information assurance architecture, as all chosen interoperability enterprise architecture views contain a security aspect, such as:
  - AV work products specify all security-related operational goals, strategies, and critical success factors;
  - OV work products identify the types of assets that need protecting and a rating of the assets' importance to the enterprise;
  - SV work products identify the functionality and system elements that accomplish the operational security goals;
  - TV work products define the relevant security standards, particularly across inter-system interfaces (i.e., a security standards profile (within TV products).
- Guide investment decisions, employing OV-5 to present interoperability enterprise architecture operations and relate them to information flows and the organizations and systems as mechanisms that support them (prioritizing the life-critical operations, e.g., those that reduce risk to life first, then those reducing risk to property).

The use of these interoperability enterprise architecture views will be to gather and represent interoperability capabilities, needs, and participants in an easy-to-understand format. These views will be customized to reflect the collected and required data, and to understand, document, and represent that data in common standardized views which are easily and rapidly understood.

Once the current architecture views are populated and constructed, the same techniques and views are used to define the target (to-be) interoperability enterprise architecture. The target interoperability enterprise architecture includes not only the current capabilities that are captured, but the desired information sharing and interoperability capabilities. It is very common to have multiple target architectures representing various time intervals showing a progression of capabilities toward future goals.

Once current and target interoperability enterprise architectures are developed, a gap analysis (see Figure 4.6) is conducted to determine missing capabilities and define requirements for one or more development projects to provide these capabilities. By creating the same views (OVs) for both the current and target interoperability enterprise architecture, it is simple to compare them to illustrate and analyze the variances. These variances represent new or augmented capabilities, as well as unused or inefficient capabilities that are no longer needed. The sum of these variances is the basis for the requirements definition process within the systems engineering discipline discussed later in this chapter. During the development of these requirements, there is a recursive refinement between systems engineering and enterprise architecture, where the variances are analyzed and distilled into operational and system architecture requirements. The identified requirements are then vetted and verified to ensure their alignment with the target architecture.

There are many variants of the interoperability enterprise architecture development process, including the determination of information sources and value of current architectural information. The architect/planner should decide the relevance of a given piece of information to the current or target architectures. Inputs to the current and target architectures will vary, but as a general rule, data that describes the present environment or capabilities form the as-is architecture. Useful data, such as technology forecasts, grant guidance, desired use cases, pilot results, and the interoperability strategic plan, and planned changes in responder-receiver and stakeholder capabilities will then be used to modify the current interoperability enterprise architecture baseline, forming the basis for the target interoperability enterprise architecture.

Responder-receiver organizations and other stakeholders from the interoperability strategic plan and interoperability enterprise architecture play a critical role in the development and focus of the target interoperability enterprise architecture. These participants in interoperability enterprise architecture development should have the direct experience and a vested interest in the outcomes of the interoperability strategic plan and interoperability enterprise architecture development efforts. By actively engaging these stakeholders and inviting their input using organizational development techniques, they become vested in the result and become valuable proponents for the interoperability enterprise architecture effort. Although we have focused on the methods and techniques of using enterprise architecture, the political situations in which the interoperability enterprise architecture is developed must also be considered. In this regard, a critically important value of the interoperability enterprise architecture as a strategic and analytical planning tool is that it often removes uneducated opinions (PHD #15 Insight) and personal bias (PHD #9 Attitude, PHD #20 Protectiveness, and PHD #22 Relationships) out of the picture and facilitates fact-based decisions in planning interoperability.

4.5.6 Interoperability Enterprise Architecture Views and the World Interoperability Failure Model

Enterprise architecture techniques and tools can assist in the resolution of many of the 25 deficiencies that define the World Interoperability Failure model in Table 3.3. Table 4.7 describes how each of the 25 deficiencies in the World Interoperability Failure model can be addressed using the DoDAF-PSEREA-PSAF views. The table shows how development of an interoperability enterprise architecture as described in this section can provide mechanisms to address these deficiencies. Note the necessity of including the veteran responders-receivers and other critical stakeholders from all disciplines, jurisdictions, and governance levels in the creation of these interoperability enterprise architecture views for each scenario.

4.6 Business Case Analysis

Enterprise architecture develops operational and system needs driving to-be capabilities. These needs must be used to generate a recommended solution concept from cost/benefit and risk assessment of two or more viable alternatives. The business

Table 4.7 Addressing Deficiencies with Enterprise Architecture Views

No.	*Failure of ...*	*Interoperability Enterprise Architecture Mitigation of World Interoperability Failure Model Deficiencies*
1	Organization	The OV-4 is the primary organizational view. It supports the representation of people, roles, command structures, and joint or multijurisdictional commands as represented in various incident command systems (see Chapter 5). The OV-2 shows the organizations and their communication needlines. Graphical representations of organizational elements, relationships, and priorities both clarify the chain of command and establish the roles of specific individuals and systems in scenario-based tasks and activities to mitigate this deficiency.
2	Assumption	The use of OV-2 diagrams (views) with specific purposes and focused content can identify, document, and guide planning to mitigate deficiencies and gaps in the information-sharing architecture. Creating OV-2 views showing the to-be communication paths and capabilities vetted with responder-receivers and other stakeholders can eliminate the need for assumption through proper scenario planning with imagination.
3	Omission	OV-2s and OV-4s are used to ascertain the value of specific information sharing on the roles and responsibilities of included organizations in each threat scenario.
4	Verification	Vetting OV-2 and OV-4 diagrams with experienced responders-receivers and other stakeholders can identify and eliminate the need for assumptions. Development of associated OV-3s will include verification as a matter of course, mitigating this deficiency.
5	Supervision	Augmenting an OV-4 with certification qualifications or a role-based OV-4 with associated names of qualified individuals and alternates available to fill specific roles can quickly allow supervisors to assign qualified resources to needed roles to mitigate this deficiency.
6	Alertness	OV-2, OV-3, OV-5, and OV-6c work products can be used to define and augment SOPs and CONOPS, thus defining multiple alerting paths as part of scenario-specific protocols. See ECA Case Study. Specific alert paths to multiple organizations will minimize the impact of this deficiency in the future.
7	Complacency	The immediate availability of information from alerting and intelligence sources, as depicted in the OV-4 work products, with message specificity and intercommunication in the corresponding OV-3 products, will provide the multiorganizational alertness needed to avoid this deficiency.
8	Intelligence	Creation of OV-2 diagrams specifically for intelligence information sharing in the form of actionable information that nevertheless protects sources and methods, and which are scenario-specific, is valuable in ensuring this deficiency is finally corrected. Creating the target interoperability enterprise architecture with the corresponding OV-2/OV-3 will define the specific information sharing to be vetted and implemented.
9	Attitude	By involving all relevant veteran responders-receivers and stakeholders in the development and vetting of interoperability enterprise architecture views, the development of the resulting information-sharing architecture will be based on public safety and emergency response requirements and not the limiting behavior of certain leaders and planners.
10	Imagination	Leading veteran responders-receivers and stakeholders in the rational development and vetting of target scenario-based information-sharing views will establish their experience-based imagination as vital contribution to target (to-be) interoperability enterprise architecture development.

Table 4.7 (continued)

No.	*Failure of ...*	*Interoperability Enterprise Architecture Mitigation of World Interoperability Failure Model Deficiencies*
11	Communications	Communications end points (people, teams, centers, etc) and their capabilities are defined in OV-2s, while the scenario-based information, communications flows, and associated requirements are documented in OV-3 and relevant SVs. These graphical and tabular interoperability enterprise architecture views provide an easy-to-understand representation of critical information-sharing requirements.
12	Paraphrase	Based on the views developed for (11) above, communication message sets and standards can be added to the interoperability enterprise architecture so that operational SOPs and CONOPS avoid the critical paraphrase deficiency.
13	Adaptability	Development of the target interoperability enterprise architecture for each of several relevant threat scenarios, combined with a flexible philosophy of operational tasks and activities (see Chapter 5 Operational Architecture), will mitigate this deficiency.
14	Disclosure	If a simple OV-2 had been created showing all locations or roles with access to the "Magic" intercepts (see Chapter 2), additional communications protocols could have been initiated to ensure adequate distribution of the intelligence to all organizations with the need to know. This deficiency will be eliminated by proper translation of intelligence into actionable information through structured OV development.
15	Insight	Through the process of developing OVs, governance leaders and planners are provided insight into the gaps, redundancies, and inadequacies of their current capabilities and readiness status. By involving their veteran responders-receivers and stakeholders in this development, the resulting scenario-based interoperability enterprise architectures benefit from their years of practical experience providing insight to the interoperability enterprise architecture planners.
16	Dissemination	The OVs and associated SVs will be used to explicitly—and automatically, whenever possible—disseminate the identified and detailed messages and their content type, meeting the interoperability enterprise architecture specification to mitigate this deficiency.
17	Inspection	Completion of an operational interoperability enterprise architecture makes inspection and assessment by responsible authorities and supporting staff more readily accomplished because of the complete understanding of the interoperability enterprise architecture provided by the high-level OVs.
18	Preparedness	Using interoperability enterprise architecture views in conjunction with modeling tools provides leaders and planners as well as exercise developers with the ability to experiment with "what if" scenarios by varying conditions and response elements.
19	Consistency	A "backchannel" to command elements or other public safety and emergency response personnel distorts the official announcements, alerts, plans, and so forth. Including veteran responders-receivers and other relevant stakeholders in the development and vetting of the OVs will identify the existing backchannels and determine if they need to be added to the official procedures, or if a failure of existing procedure necessitates the use of a backchannel. Either way, the lack of consistency is addressed.
20	Protectiveness	If third-party interoperability enterprise architecture planners include veteran responders-receivers and all relevant stakeholders based on each threat scenario from each discipline, jurisdiction, and governance level in the development and vetting of the OVs, a greater fostering of mutual goals and benefits of information sharing will be identified, defined, vetted, and instituted. Conceivably once out-of-the-information-loop, each discipline, jurisdiction, and governance level will then have "buy in" to the resulting target interoperability enterprise architecture.

Table 4.7 (continued)

No.	*Failure of ...*	*Interoperability Enterprise Architecture Mitigation of World Interoperability Failure Model Deficiencies*
21	Relationships	By developing vetted OVs, the true best-practice-derived information-sharing needs for optimal scenario preparedness and response are then institutionalized without the interpersonal factors circumventing or contradicting official communications paths, and finally removing this deficiency.
22	Priority	Development of the target interoperability enterprise architecture provides decision makers with a comprehensive view of all of the current and target capabilities. From this "big picture view," cross-element priorities can be impersonally assessed by multiple organizations with a proper sharing of workload to ensure the target interoperability enterprise architecture is realized.
23	Reporting	While developing an interoperability enterprise architecture cannot change willful incompleteness in personal reporting, it can develop OV-4-based reporting and command protocols, and OV-2/OV-3 definitions of reporting entities and elements can provide adequate information for impersonal assessments. By defining which organizations should be interoperable by scenario, OVs can form the basis for more detailed SOPs, CONOPS, and MOUs. In addition, SVs can define automatic reporting technologies to eliminate this deficiency.
24	Improvement	The purpose of the target interoperability enterprise architecture is to achieve interoperability in public safety and emergency response to fill all deficiency gaps identified using the World Interoperability Failure model in Table 3.3. By defining the target interoperability enterprise architecture requirements and performing a gap analysis, a roadmap for improvement can be planned with priorities established based on comprehensive information-sharing and scenario-based adaptability
25	Delegation	Specialized OV-4s can be developed showing command-and-control relationships in the target interoperability enterprise architecture. These resulting organizational interrelationships can include parameters of responsibility and authority to eliminate this deficiency. Once identified, the gaps can be addressed with SOPs, MOUs, training, and so forth.

case analysis best practice highlighted in Figure 4.15 develops and vets these alternatives into a recommended solution concept.

4.6.1 Overview

Given limited resources, it is critical that the funds available to develop the target interoperability enterprise architecture and improve interoperability at any of the governance levels described in Chapter 3 are utilized in the most efficient manner possible. The existence of new technology developments alone, or the recommendations of a well-known vendor recommending procurement of their latest products, should not be the sole driver for the prioritization of interoperability projects. The best-practice rational analysis of the cost and benefits involved with each project—as well as the projects support toward the overall strategic plan of the organization—needs to be assessed in a rational and repeatable way in order to arrive at the solution that best serves the organization. Many of the deficiencies described on the operational side in Chapter 3 also apply to the planning processes analyzed in this chapter. In particular, the business case analysis process shown in Figure 4.16 addresses planning-related deficiencies relevant to items such as PHD #2 Assump-

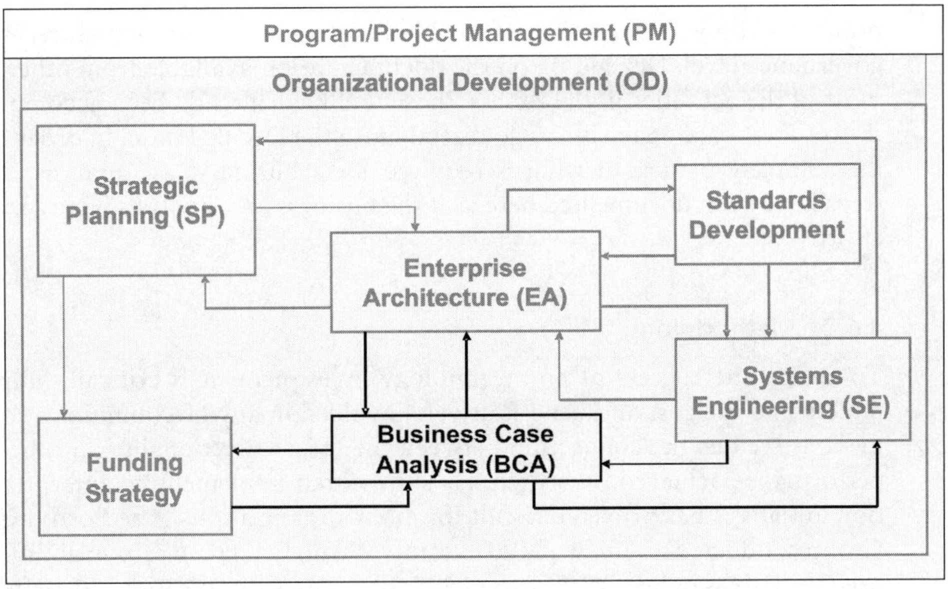

Figure 4.15 Business case analysis best practice.

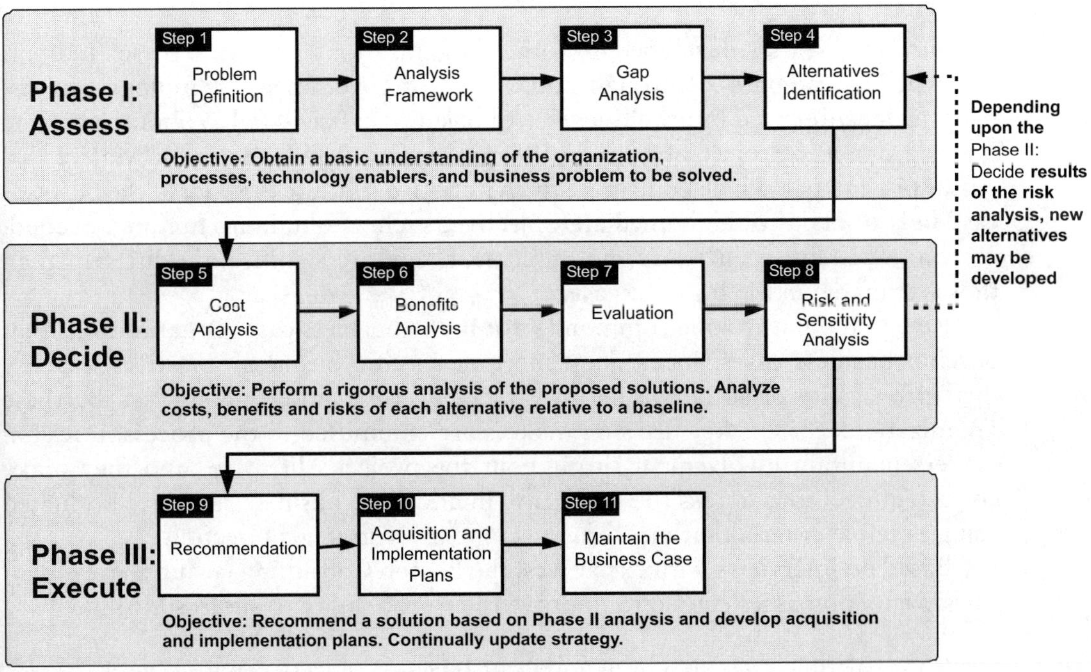

Figure 4.16 The business case analysis methodology.

tions, PHD #4 Verification, PHD #11 Communication, PHD # 14 Disclosure, PHD #19 Consistency, PHD #22 Priority, PHD #23 Reporting, and PHD #24 Improvement.

The business case analysis methodology shown in Figure 4.16 was developed mainly to analyze investments into technology at the national level; however, the principles apply to investment of any kind (e.g., training, infrastructure, etc.), at any governance level. Depending on the documentation available from other activities, most of the activities in the Assess Phase will have already been accomplished and the business case analysis can go directly into the Decide Phase. In order to provide the complete picture of what is required for a business case analysis, all process steps have been documented here as if the business case analysis were a stand-alone effort.

4.6.2 Methodology

To ensure the success of any technology investment, it is critically important to understand the cost and benefit drivers involved in any IT/communications investment as well as the connection between the target functionality and the enhanced performance achieved. Research has shown that implementing enterprise application initiatives has proven difficult for many organizations. The Boston Consulting Group conducted a survey of 100 executives who had taken responsibility for enterprise initiatives between 1996 and 1999. Based on value creation, cost-effectiveness, tangible financial impact, and goal attainment, only 33% viewed the outcome as positive [43]. Most executives felt the initiatives failed to meet their organization's business objectives, and many felt they could have achieved similar value for less money.

Similar research developed by other organizations validates these findings. Gartner, Inc. estimates, "through 2005, 70% of government technology projects that are not supported by a well-developed business case will fail to obtain adequate funding or meet project objectives (0.8 probability)" [44]. Solution Matrix discusses two factors, which contribute to an initiative's failure [45]. First, the business case fails to achieve the immediate objective, such as obtaining funding. Second, costs and benefits realized during implementation are significantly different than those estimated in the business case.

Research points to one common factor in the business case. Organizations that develop business cases, focused on meeting specific business objectives increase their probability of success. The investment decision process employed by these organizations ensures key decision makers are committed to the process. Decision makers maintain involvement throughout the project's lifecycle, and they make consistent decisions across the company. In addition, business cases are evaluated using tangible criteria that measure the company's strategic direction.

Based on interviews with executives, the Boston Consulting Group survey identifies what companies can do to improve their probability of success [46].

- Start with a strategic vision and clear business objectives (SP).
- Analyze the business capabilities required to attain the vision (EA).
- Do enough up-front work to determine the best options .
- Make decisions based on business requirements, not the capabilities of software packages.
- Divide the initiative into focused, manageable modules with quick payback (SP).
- Choose vendors carefully and manage them vigorously (PM).

- Understand the risks and manage them (SP-PM).
- Install strong, unbiased leadership to ensure the scope of the initiatives does not become too large (PM).
- Know when to stop and reevaluate the initiative.

In response to U.S. congressional legislation and federal guidance, each federal agency and DoD component developed its own business case analysis practice. In order to develop a comprehensive methodology for the public sector, business cases from many functional areas such as health care, logistics, personnel, supply, and communications, were reviewed. The analysis assessed the business case steps with respect to purpose, legislative requirements, and commercial best practices. The analysis identified 11 critical steps to conduct a business case. The business case analysis processes described in what follows contains all 11 steps designed to optimize an enterprise application investment.

The methodology shown in Figure 4.16 complies with all existing United States federal guidance and commercial best practices. It was developed after a comprehensive review of dozens of federal and commercial methodologies, including the Office of Management (OMB) standard form OMB Exhibit 300 In this context, a "business" objective translates into an "operational" objective for interoperability in each threat scenario. The key features of the business case analysis practice are focused on meeting specific business objectives, using evaluation criteria that addresses the organization's strategic direction and involves key decision makers in the process. This approach provides a constant flow of communications between the stakeholders and team ensuring that the end result meets all the clients' expectations and increases the probability of success of the investment.

As Figure 4.16 shows, the business case analysis approach is composed of 11 steps in three distinct phases: assess, decide, and execute. Phase I: Assess, establishes the framework by which the analysis is conducted. It uses the organization's mission and objectives to select appropriate measures for evaluation. In Phase II: Decide, quantitative and qualitative analyses are conducted to select an optimal alternative. This phase concludes with risk and sensitivity analysis designed to present results in a manner that accounts for potential risks. Finally, Phase III: Execute transitions the initiative from the business case analysis to implementation planning. It develops post-analysis plans for the recommended alternative and reevaluates the business case at strategic points throughout implementation to ensure success.

4.6.2.1 Phase I: Assess

Phase I refines and documents the organization's vision for the future, and forms a foundation to align key mission and IT/communication systems [47]. It requires a fundamental understanding of the organization, processes, and technology enablers. The business case demonstrates how the technology leads to enhanced value or improved service. Figure 4.17 presents the Phase I process steps.

The Phase I access process defines a problem or opportunity that aligns a proposed investment with the organization's interoperability strategic plan. It then establishes a framework to conduct the ensuing analysis on alternative architectures or project solutions. This framework defines all underlying assumptions and devel-

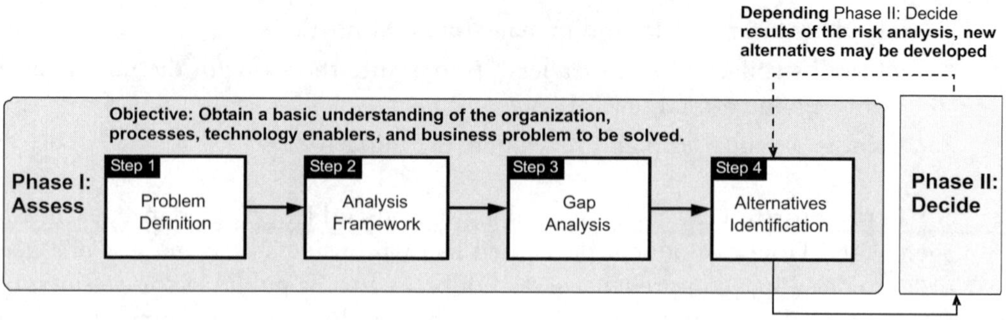

Figure 4.17 Steps in the Phase I: Assess process.

ops a baseline for the business case. It develops performance measures (see Table 4.8) to help planners evaluate the project's success at various intervals and measures differences between as-is and to-be operations to fulfill the objectives outlined in the interoperability strategic plan. Finally, it identifies alternatives that meet those objectives.

4.6.2.2 Phase II: Decide

Phase II costs and benefits are used to support decision making for the alternatives identified in Phase I. This phase utilizes a consistent cost-benefit, risk, and sensitivity analysis framework to provide planners with all the relevant information to make the best decision for interoperability. Figure 4.18 shows the Phase II steps. The key operational concept applied in all steps is that of incremental analysis. A business case is concerned with the differences between options, such as the baseline and alternative (to-be) interoperability enterprise architecture configurations or system decisions. All costs and benefits that differ between options are defined as incremental and are included in the analysis. Costs and benefits that are common to all alternatives are not relevant to the investment decision and are ignored when calculating differences.

The analysis estimates whether the incremental benefits of an interoperability enterprise architecture alternative exceed its incremental costs. The proposed interoperability enterprise architecture that maximizes incremental benefits and minimizes incremental cost is the financially preferred course of action. However, risk and strategic consideration must also be taken into account before a final course of action is selected, as will be discussed later in this section.

4.6.2.3 Phase III: Execute

The execute phase (Phase III) starts with a formal recommendation of a solution based on the framework established in Phase I (Figure 4.17) and the analysis conducted in Phase II (Figure 4.18). The financial and strategic criteria established earlier in the business case analysis process must support the recommendation, which should be fully documented to allow future reviewers to understand the rationale for selecting a particular alternative. After the committee has identified a solution, post-analysis plans should be developed to guide successful acquisition and imple-

Table 4.8 Sample Performance Measures

Type	*Focus*	*Measures*
Strategic measures	Enterprise mission goals	Percentage of mission improvement (cost, time, quality, quantity attributable to solutions and services); Percentage of planned benefits realized; Percentage of strategies fully matched to enterprise strategies.
	Portfolio analysis and management	Percentage of portfolio reviewed and disposed; Percentage of compliance to approved solution deployment; Percentage of reusability of core modules.
	Financial and investment performance	Percentage of total costs by major asset category (hardware, software, personnel, facilities, management); Budget as percentage of operational budget.
	Resource usage	Percentage of databases that can be shared; Percentage of hardware/software with interoperability capabilities.
Customer measures	Stakeholder partnership and involvement	Percentage of cooperative stakeholder and applications design; Percentage of joint development-service-level agreements; Percentage of stakeholders attending investment meetings.
	Customer satisfaction	Percentage of stakeholders reporting full use of applications; Percentage of stakeholders satisfied with application design; Percentage of stakeholders satisfied with maintenance and support; Percentage of problems resolved within target time; Percentage of products launched on time; Time taken to fulfill service requests.
Internal business measures	Architecture standards compliance	Percentage of procurement exception to architecture standards; Number of variations from standards detected by review and audit/year; Percentage of increase in systems using standard architecture.
	Project performance	Percentage of projects on time, on budget; Percentage of projects using standard methodology for systems analysis and design; Amount of backlog on enhancement and maintenance requests.
	Infrastructure availability	Percentage of computer availability; Percentage of applications availability; On-line system response time.
Training measures	Work competency and development	Number of staff trained by skill area; Percentage of staff training that uses new technologies and techniques; Percentage of staff professionally certified; Percentage of budget devoted to training and staff development;
	Responder-receiver satisfaction	Percentage of employees satisfied with the existing technical and operating environment to deliver quality products or services; Percentage of employees satisfied with training; Percentage of responder-receiver turnover by function.

mentation of the investment. Finally, the business case should be revisited at specified intervals. Figure 4.19 shows the Phase III steps.

This process allows key planners to monitor interoperability project implementation against objective criteria for meeting the target interoperability enterprise architecture. Many organizations utilize a defined software development life cycle

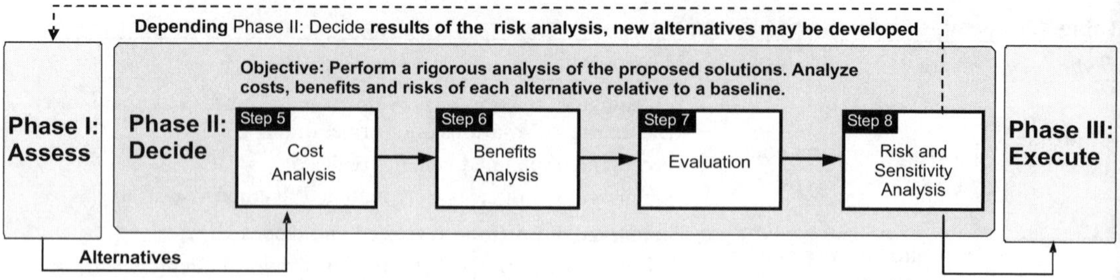

Figure 4.18 Steps in the Phase II: Decide process.

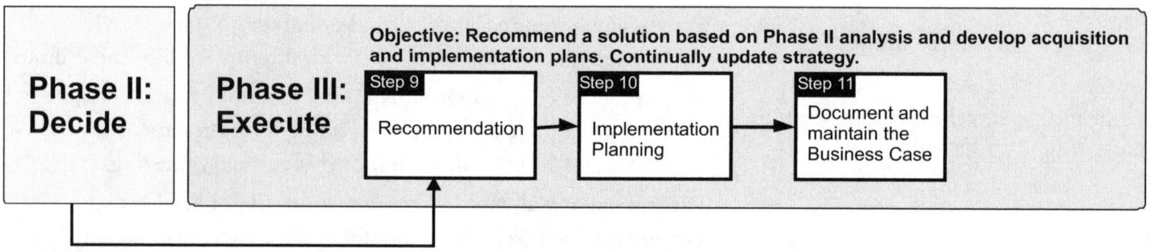

Figure 4.19 Steps in the Phase III: Execute process.

(SDLC) or systems engineering life cycle (SELC), which defines review points throughout the development and implementation of programs and projects. These review points provide opportunity for the business case, enterprise architecture alignment, and adherence to the interoperability strategic plan. If the project is not achieving desired results, the issues can be addressed early in the process and mitigation strategies can be enacted to prevent significant failures at later periods.

4.6.3 Impact of Business Case Analysis Practice on the World Interoperability Failure Model

Table 4.9 shows how the business case analysis practice addresses the potential impact of the interoperability checklist deficiencies on interoperability planning efforts. Again, interoperability requires interoperable planning.

4.7 Funding

The funding strategy process highlighted in Figure 4.20 explores various sources of potential funding for interoperability planning and development.

4.7.1 Overview

Responders-receivers and stakeholders within local and/or second-level governance are the first to respond to an incident in their jurisdiction. This public safety and emergency response community requires funding support to achieve interoperability, but funding sources must be aligned to regional strategic planning,

Table 4.9 Measures of Financial Return

No.	*Failure of ...*	*Business Case Analysis Mitigations of the Pearl Harbor Deficiencies*
2	Assumptions	Clear and well-defined assumptions are required for the business case analysis and facilitate a clear understanding of the parameters of the analysis. This is addressed in Step 2 Analysis Framework. Clearly defined and documented ground rules, assumptions, and constraints provide the analysis' basis, factors underlying the results, and provide justification for future review.
4	Verification	Step 11 Maintaining the Business Case deals with lessons learned and post-implementation review design precisely to verify that the results forecasted in the business case analysis are in fact realized in practice.
11 14 23	Communications Disclosure Reporting	These items are at the heart of the business case analysis. The purpose is to document and communicate the results of a rigorous analysis so that they can be reported to oversight agencies, disclosing the justification for the underlying investment.
19	Consistency	The stated purpose of the business case analysis is to provide a consistent, repeatable process of analyzing varying investments against a set of defined criteria in order to maximize an organization's return on investment.
22	Priority	The business case analysis performs a prioritization among competing alternatives in order to maximize the use of the limited resources available to the organization. This is done in Step 7 Evaluation, where the alternatives analyzed are compared to the baseline and ranked.
24	Improvement	As stated above, Step 11 Maintaining the Business Case deals with lessons learned and continuous improvement of the business case analysis practice.

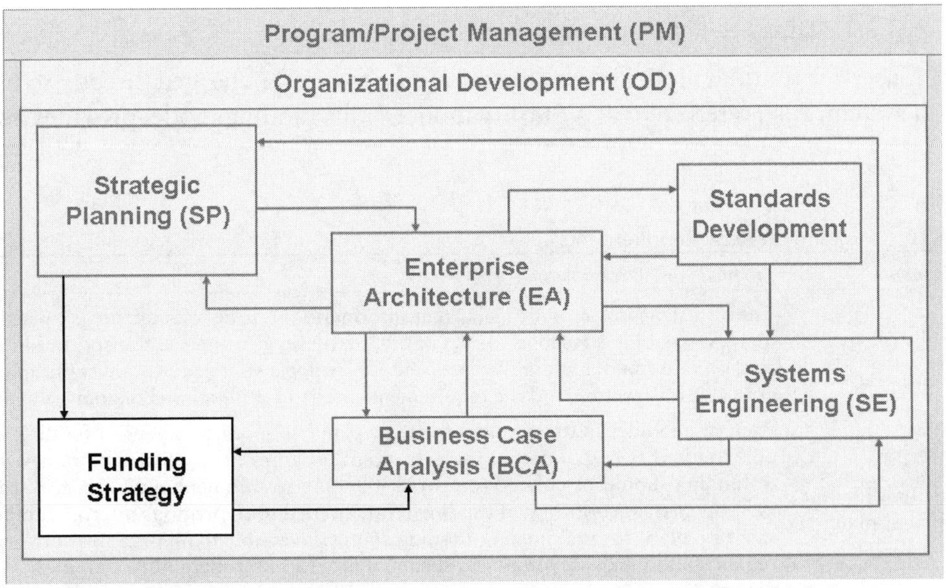

Figure 4.20 Funding strategy best practice.

funding strategies, and project management functions. Typically, the first-level of governance, such as states or provinces, would normally provide this type of resource support. However, financial support for regional coordination of responders-receivers is relatively new. As a result, planners at the associated governance levels may or may not have established programs required to support the program. Table 4.10 recommends funding strategy activities related to a region's lack of strategic plan, funding development plan, governance leaders and planners, and project management process.

4.7.2 Funding Sources

Local or second-level governance organizations typically fund interoperable communications programs through annual budgeting processes (with sources of funding from personal, property or commercial property taxes). Increasing budget constraints, as a result of the decline of tax revenues coupled with increasing homeland security mission requirements, are yielding significant funding challenges to many local/or county governance levels across the world.

With increasing demands on traditional budgets, many local or second-level governance organizations are using alternative funding models to obtain the resources needed to sustain IT/communications projects. These sources include:

- User-fee revenue,
- Municipal leases,
- Bond financing,
- Grant funding,
- Budgeting and appropriations strategies.

4.7.2.1 User-Fee Revenue

A user-fee funding model is the allocation of a tax or fee charged directly to users of a system. Also referred to as a "utility model," this funding model provides a source

Table 4.10 Funding Strategy Requirements

Requirements	*Purpose and Description*
Regional strategic plan	The infrastructure supporting current information-sharing architecture is owned by many stakeholders (such as local, county, first-level governance, transportation agencies, private companies, universities, etc.). A region's strategic plan must outline all critical infrastructure and the requirements needed for initial and ongoing investments.
Strategic funding development plan	Each stakeholder in the regional strategic plan has specific sources of funding available to him/her (e.g., internal budgets, grant programs, etc.). In addition, new models of funding should be created to provide ongoing sustainment (e.g., user fees, capital investments). A coordinated funding strategy (tied to the regions interoperability strategic plan) for maximizing funding sources available to all organizations within a region should provide long-term sustainability for interoperability activities readiness.
Governance structure	A regional governance structure is required to align the various stakeholders managing infrastructure assets and to facilitate strategic planning, funding development, and program management.
Program management for funding	Once funding is obtained for a region, its allocation must be managed to support a region's interoperability strategic plan as well as to manage the requirements for each source of funding.

of continuous investment to sustain a designated project. For example, user-fee models in the United States include:

- A tax on telephone bills that supports regional 9-1-1 systems;
- A flat fee imposed on citizen court, licensing, or DMV transactions;
- A user-fee for access to government electronic initiatives, such as obtaining government documents online, online payments, or other e-initiatives.

4.7.2.2 Legislative Strategies

Local, first- and second-level governance structures may use legislative strategies to generate new sources of funding and/or flexibility in use of funds. For example, IT appropriation legislation may request the carryover of uncommitted funds at the end of the year, the ability to reprogram savings realized for operational efficiencies gained through programs, or the ability to hire in-house staff to reduce reliance and cost for outsourcing IT services. These appropriation strategies provide for flexibility in spending funds as well as the sustainment of funds due to cost efficiencies.

The establishment of a *nonprofit* organization (an organization exempt from federal taxes because it cannot make financial profit) to support emergency services provides regions with new avenues for obtaining funding sources. For example, The All Hazards Consortium (AHC) [48] is a nonprofit organization guided by the regional states of North Carolina, District of Columbia, Maryland, Virginia, West Virginia, Delaware, Pennsylvania, New Jersey, and New York. Their mission is to help create new resources and funding opportunities for the states to support regional multistate collaboration efforts among our stakeholders from government, private sector, higher education, and nonprofit/volunteer organizations.

The All Hazards Consortium was built on the belief that state/local government is ultimately responsible for the protection of the public. Based on this assumption, the AHC sees government as the "owner of the problem." The private sector owns most of the assets, technologies, and solutions; the universities provide research and education to address the problem; and nonprofit organizations provide access to information and people who are focused on a particular segment of the problem. By bringing together all stakeholder groups into regional advisory committees, working groups, and ad hoc committees, and focusing on specific issues (with state government driving the needs), a powerful environment for collaboration is created to solve tough problems that require resources from every sector. Ultimately, however, it is the "culture of collaboration" that creates the motivation that drives the AHC and its stakeholders to work together to protect the region's citizens from all types of hazards.

4.7.2.3 Municipal Leases

Local, first and second governance-level structures may establish leasing programs where investors provide funds for up-front investment in government infrastructure systems and in return the government then makes lease payments to investors with higher interest rates. The benefits of municipal leases are that they do not count against a government's debt ceiling.

4.7.2.4 Bond Financing

Local, first and second governance levels eligible for issuing bonds may use this method for raising capital to invest in infrastructure projects. For example, in the United States, the Commonwealth of Virginia is using bond funds to finance the upgrade of a statewide emergency radio system.

4.7.2.5 Grant Funding

Many local and/or second-level governance levels rely heavily on government grant funding (i.e., provided by of the national government). Grant funding is typically provided to support a directive or mandate to incent local, first- and second-level of governance organizations to implement a particular initiative. While grants are a good source of funding, there are three key issues which pose challenges to planning and project implementation. These challenges are outlined below.

Varying Grant Sources and Development Processes There are many sources of grant funding from governments, foundations, and corporations. The ability to obtain grant funding, also known as grant development, varies greatly depending on the eligibility of an organization, eligibility of a project, and the sophistication level of the organization seeking the grant funding. Often, one locality may be better at obtaining grant funding then another; as a result there is an uneven allocation of funding for localities within a given region. If this funding is not programmed according to regional strategic plans, then the locality may purchase equipment that is not interoperable with other local and/or county government levels in the region.

Conflicting Programmatic and Administrative Requirements Grants come with specific administrative requirements, such as expiration dates, comatch amounts, and limitations on use of funds, as well as programmatic requirements, such as allocation of funding for directed federal programs. When more than one grant program is applied toward a project, there may be direct conflicting administrative requirements (e.g., one grant states equipment of a certain type is eligible for reimbursement but not another type), programmatic (e.g., one grant covers equipment purchases for interoperable communications but not for planning or organizational activities required to establish a standardized plan for equipment procurement in a region).

Sustainability Grant programs typically cover the initial cost of a system but do not provide funding support for ongoing sustainability. The individual local and/or county government must outline their own funding sustainment plan for operation of the system to cover operation and maintenance costs (O&M), including staffing, equipment maintenance, spare parts, and so forth. The ongoing O&M costs must become part of the annual budget for each organization.

In the United States, there is now a coordinated effort to address the deficiencies of grant funding in regard to regional interoperable communication programs through the new DHS PSIC grant program. This program established a broad yet coordinated set of guidelines for strategic planning, funding development, governance and funding management. While this program is a good start toward achieving a coordinated regional planning and funding approach, the long-term sustainability of regional interoperable communications programs is still left to the local, county, and third levels of governance to support via allocation of budget dollars to support O&M. With governments around the world facing fiscal challenges, the additional O&M costs for nonmission-critical government operations are often delayed or not funded. Leadership from fourth and fifth governance levels is required to establish a vision and long-term sustainment plan for this critical homeland security function.

4.8 Systems Engineering

The strategic planning, business case analysis and enterprise architecture best practices generate inputs and constraints for projects to develop real IT/communications capability that meets the documented responder-receiver information-sharing needs. Systems engineering highlighted in Figure 4.21 is the best-practice process used to develop this capability.

4.8.1 Overview

Systems engineering is engaged when it is time to focus on implementation of an interoperability project, particularly when upgrading or developing an integrated IT/communication system. At this point, preferably, organizational priorities and interoperability enterprise architecture have been established per the processes described in earlier sections of this chapter. The connection between the interoperability enterprise architecture and those "on the ground" using the system to perform their public safety and emergency response must be closely maintained to ensure that the intended information sharing is achieved. Unfortunately, there are several aspects of these types of projects that increase the risk of breaking this

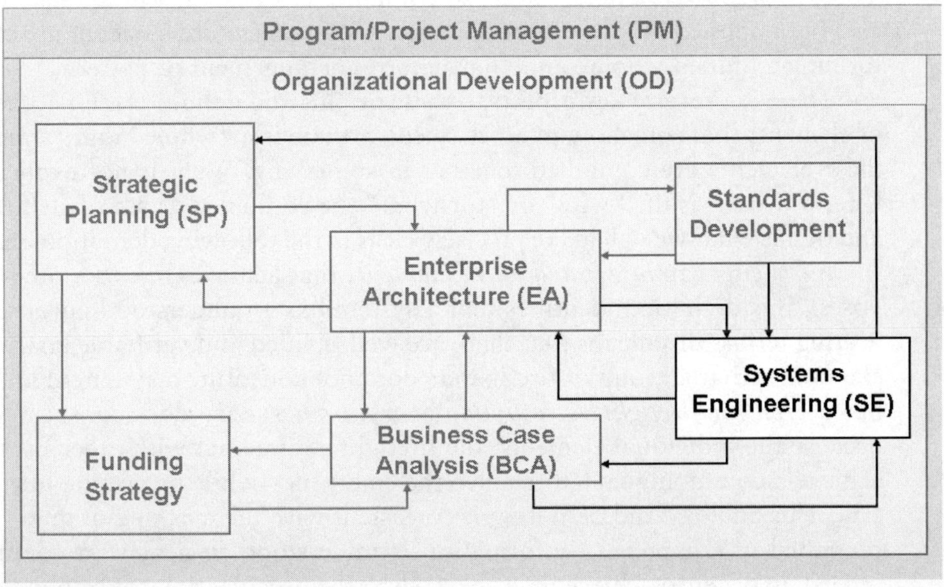

Figure 4.21 Systems engineering best practice.

connection. First, individual projects are frequently implemented by teams other than those that developed strategic plans; so there is often a gap between strategic planning and system implementation. This gap has occurred historically because people building the system are most often not the same people who developed the organizational strategic plans. Since these groups have seemingly different focuses, there is little motivation for direct collaboration, which can result in a disconnect or misinterpretation.

A second factor is that change is inevitable during project execution. Unfortunately, change can be difficult to manage well and often leads to inefficiency in keeping a project on track. This inefficiency results in unforeseen impacts downstream, to other related projects, or at a strategic planning level. When multiple projects are implemented or planned, the problems associated with lack of communication and impact of change can be amplified if cross-project analysis and coordination do not occur. The risk of encountering technical problems late in a project, when costs to correct are much higher, also increases. Finally, unanticipated technical or stakeholder obstacles may require a project to be reevaluated or modified. This often results in a situation where the projects no longer align with the goals specified by strategic plans.

All of these challenges, which are realistically part of any project, can contribute to the misalignment of operational activity with strategic vision, as well as difficulty in the realization of organizational goals. The objective of this section is to present a brief introduction to a discipline that manages these challenges at the level of *technical management* of an engineering effort.

4.8.2 Definition and Process Attributes

There are many industry definitions for *systems engineering*, but it is the meaning of the word *system* that steers an appreciation of what the discipline truly entails. The word *system* can be traced to Late Latin and Greek words meaning "an arrangement," "organized whole," and "to combine," as well as "cause to stand." It has also been defined [49] as "a regularly interacting or interdependent group of items forming a unified whole" and "harmonious arrangement or pattern."

There are several key aspects to these origins and definitions. First, there is a set of elements that somehow interact. Second, there is a "whole" that "stands" when these elements are combined together in some way. What is not explicitly stated here, however, is the "why" or "for what" the combined group of elements stand, and that is what we will make precisely clear in the following definition of system.

A system is an *integrated set of elements* that achieves one or more *stated purposes*. This definition identifies that a system has an understood purpose. In engineering terms, this means that there are well-defined and verifiable goals. A simple example is that a group of firefighters does not constitute a system in and of itself, but a *crew of firefighters responding to an emergency* does constitute a system because the individual elements, the firefighters, interact with other firefighters to achieve a common goal—to resolve the emergency while protecting life and property. This analogy and definition is consistent with the meaning of interoperability presented in Chapter 1—information is understood to achieve a common goal. Indeed, in order to achieve interoperability, the organizations of responders-receivers and other stakeholders in all governance levels, disciplines, and jurisdictions must act as a system.

The International Council on Systems Engineering (INCOSE) defines systems engineering as "*an interdisciplinary approach and means to enable the realization of successful systems.*" To make this definition truly meaningful, we must define the terms "successful" and "system." We've defined a system, but what makes the system successful is open to broad interpretation, especially from different perspectives. Systems engineers consider several key components when determining success. Perhaps the most obvious component is how well the system achieves its common purpose. However, for a systems engineer, success is not simply how well the system meets its objectives, but also about whether it was achieved in the right way, and whether the right thing was accomplished. Thus, success is defined by the right balance of three factors: (1) stakeholder expectations, (2) cost, schedule, and risk constraints, and (3) system performance.

Stakeholder expectations help define the purpose of the system—the why—as well as the how. A *stakeholder* is a group or individual affected by or in some way accountable for the outcome of an undertaking. Thus, some stakeholders have direct interest in achieving the system goals—such as the firefighter crew responding to an emergency. Other stakeholders may be affected by the system, such as those driving in the path of emergency vehicles, or those in the vicinity of the emergency. Their interests can affect the how and must also be taken into consideration in order for a system to achieve its purpose successfully. Sometimes, the goals are ill-defined because the stakeholder expectations have been misunderstood (or some stakeholders haven't even been addressed). It was pointed out in Chapter 1 that interoperability means more than just building a radio system that can communicate with radios in other jurisdictions; it is about shared understanding that drives predictable collaborative action. Therefore, simply putting a technological capability in the field will not ensure that it is used successfully. Users must know how to exchange information in a trusted way that results in coordinated behavior toward resolution of a response situation. And, they have to believe that this coordinated behavior results in more efficient resolution than what had been done previously. Otherwise, there will be no useful information exchange across perfectly interoperable radios. Indeed, building the system without taking these stakeholder perspectives and expectations into account will likely not amount to success—just technology deployed for its own sake.

The reader can relate to cost, schedule, and risk constraints. Sometimes these constraints are obvious, but often there is a grey area: how much can we deviate from the schedule and still achieve success? What is the likelihood of obtaining additional funds in order to add in a particular functionality? The systems engineer must balance development of the technical solution within cost, schedule, and risk constraints. This balance typically requires a full lifecycle view of the system and considerable attention to tradeoffs and stakeholder perspectives. Boundaries may be established as part of an organization's business case analysis initiatives, as described in Section 4.6. These factors must be addressed during technical design, and monitored throughout project implementation. Technical management of these factors is closely aligned with project management processes, as will be discussed in Section 4.8.4.

Indications of system performance are the technical measure to gauge how well a system is meeting its stated objectives. These measures are used by the systems engineer to verify that what was intended to be built has been built and that it performs in the operational environment as specified. The systems engineer ensures that the documented requirements reflect the intended operational constraints and establishes traceability early in the project. In other words, requirements must be testable, baselined, and traceable to the physical system components so they can be verified that they have been met. The intent is to identify problems, which certainly will arise, early enough in the development process to implement acceptable tradeoffs.

While system performance is often the most obvious measure of success, the systems engineer must ensure that performance remains appropriately balanced with the other factors of expectations and constraints. The systems engineering process is thus *iterative* and *interdisciplinary*. Its iterative nature enables controlled change via managed loops between specification, design, early prototyping, and verification. These loops provide the opportunity for the systems engineer to monitor and balance the success factors. During this process, the systems engineer enlists the expertise of specialists from relevant disciplines (software developers, domain specialists, etc.) to implement pieces of the project and to provide the domain knowledge necessary for technical decision-making. The systems engineer orchestrates this interdisciplinary nature of the project. Some of the technical aspects of this process are briefly examined in the remainder of this section.

4.8.3 Process Description

While there are many depictions of the systems engineering best practice, most are very similar in construct but vary slightly in terminology. Herein, we adopt the descriptive view shown in Figure 4.22. The figure illustrates the basic elements of a systems engineering approach and provides a brief description of each. While the process terminology and its visual depiction might vary elsewhere, these basic elements are always employed.

The arrows to the right in the figure depict the iterative nature of the systems engineering best practice. While they are shown between consecutive boxes, iteration can occur among any of them. Iteration through one or more process results in refinement and provides the basis for change management and traceability of requirements (from user need to operational product). Iteration occurs primarily because learning increases as the project matures. Functional analysis (B) clarifies understanding of user needs and expectations (A). Design activities (C) result in the identification of previously unspecified constraints (A), or in the identification of new opportunities for functionality (B). This type of learning then leads to the feedback loop that results in refinement of requirements and functional description. Indeed, it is during the early phase of a project, during requirements definition and design, that the greatest amount of iteration occurs as problem understanding is enhanced. The intent is that through early iteration and problem definition, one can reduce the amount of iteration and change that occurs in later steps, when it is more costly in terms of both resources and schedule.

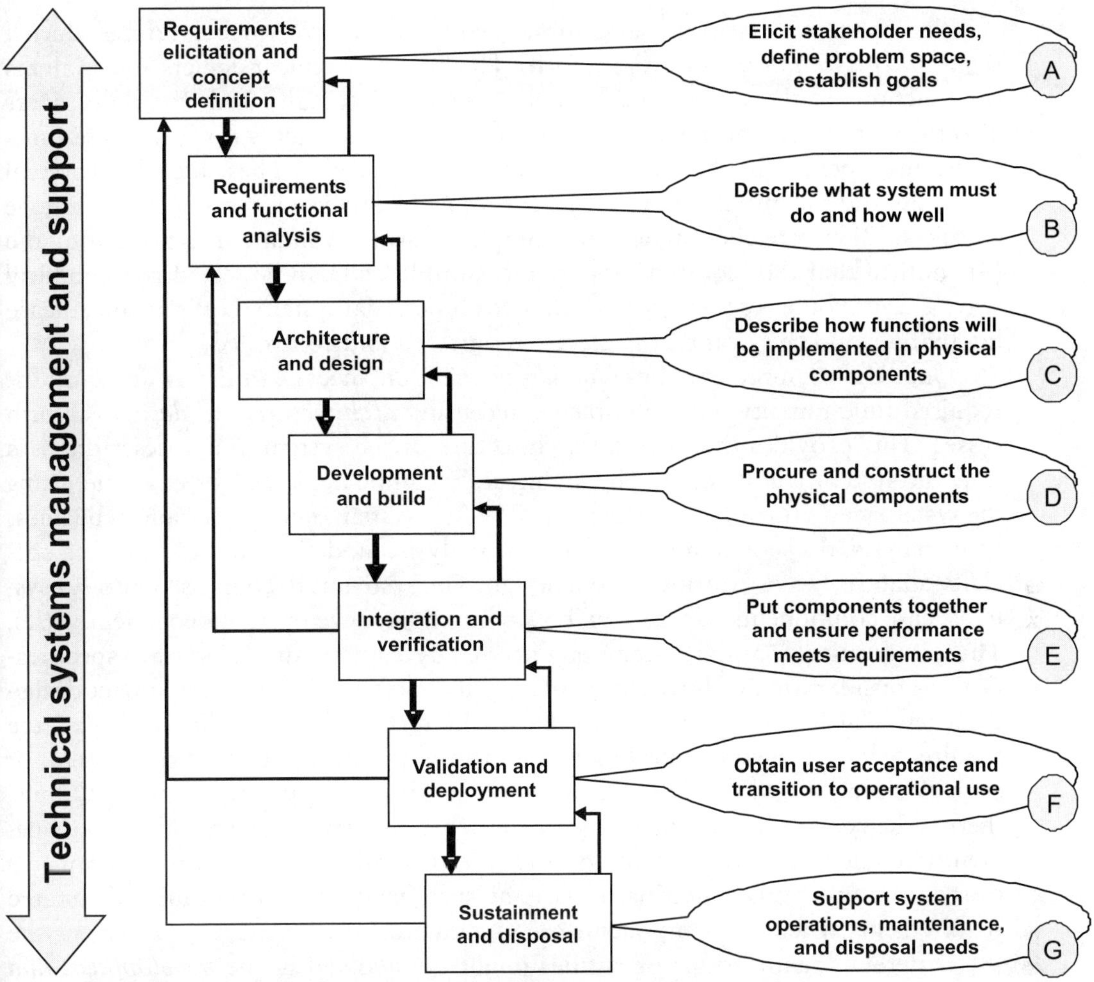

Figure 4.22 Basic systems engineering phased processes.

The primary purpose of *requirements elicitation and concept definition* (A) is to understand the stakeholders and their operational needs. This step usually involves interviews and workshops in order to elicit requirements from different perspectives. If a piece of equipment will be used by stakeholders in different operational environments for different purposes, or in different ways, then different subsets of users may have their own set of requirements or desired features. Techniques such as those described in the strategic planning (Section 4.4) and business case analysis (Section 4.6) processes described in this chapter are useful at this time. A critical and often overlooked element of this process area is to identify all the stakeholders, generally part of the organizational development process. This is also where the alignment with the strategic plan and interoperability enterprise architecture begins. Goals and missions normally provide the initial concept that must be refined at this point. Enterprise architecture activities, as described in Section 4.5, and in particular, captured by OVs, provide a critical starting point for understanding requirements.

The concept description and initial requirements are further refined during *requirements and functional analysis* (B). The systems engineer depicts and analyzes the functions of the system and how they interact internally and externally. Here also the systems engineer identifies performance requirements for how these functions must operate (how well, under what conditions, etc.). These high-level system functions and the initial user requirements must be analyzed in order to (1) remove conflicts, (2) ensure consistency and completeness, (3) validate interpretation, and (4) confirm that the requirements are measurable and can be tested in a practical way. The outcome is generally a baseline for functional system requirements, meaning that beyond this point, changes are monitored and controlled.

Once the requirements baseline is specified, engineers can determine *how* the required functionality will be implemented using *architecture and design* (C) processes. This provides the physical architecture of the system and a description of how the system will be built. The architecture can be depicted, for example, using the system views (introduced in Section 4.5). The systems engineer employs business case analysis, decision analysis, reliability analysis, modeling, and engineering specialty analysis heavily during system design. The system is decomposed into subsystems and components which can be procured or developed independently (D). Thus, it is critical that the systems engineer analyze and manage interface specifications—consideration of how components will be integrated (E) may impact design decisions. Facets of the system design might result in (1) conflict with existing requirements, (2) new constraints related to the physical nature of the system, or even (3) situations where satisfying requirements is not feasible (or desirable). Thus, there is the need for iteration, as described earlier, and structured technical management of change. The systems engineer ensures traceability of physical components to required functions, resulting in baseline specifications that will meet the stated requirements when the components are integrated.

Different organizations or entities might be involved in the *development and build* (D) of different portions of the system. Each may involve procurement, software development, manufacturing, and so forth. Given good technical management and understanding of interfaces, per processes described above, this partitioning of subsystem construction is not a significant risk. It requires, however, integrated scheduling, status monitoring, and verification planning (see Section 4.1) . The systems engineer must help to both manage change (may be necessary, but can be costly) and "gold plating" (addition of features that are not specified in requirements). This is accomplished by verifying integrity of baselines, traceability of requirements, and allocation of requirements to functions and architecture components. There are a variety of tools that are used in the analysis of architecture that help plan for how development of the physical system should best take place in order to prepare for seamless integration.

The components must come together to form the system during *integration and verification* (E). Each component is itself verified to ensure that it has been built to meet its own requirements, and then, as pieces of the system are integrated, the interfaces must be verified. The systems engineer plans for integration testing early in order to ensure that the interfaces are defined and tested in the most efficient and effective way. The systems engineer ensures that verification traces to the requirements specifications in the current baseline, as denoted by the arrows on the left side of Figure 4.22, thereby demonstrating that the system does what was stated that it must do.

Once the system has been verified to operate as planned, it is ready to transition to operational use and undergo final user acceptance and validation (F). Validation addresses not whether the requirements have been met, but whether the requirements are, in fact, ones that will result in a system that meets stakeholder expectations. This ensures that the system will, in fact, be used by stakeholders to achieve its goals. The systems engineer does not wait until deployment to perform validation, but must validate with stakeholders frequently, beginning with analysis of requirements. There are a variety of means to facilitate validation, including baseline reviews, prototypes, and incremental demonstration.

Finally, the systems engineer must address *sustainment and disposal* (G) activities. Consideration of requirements for system maintenance, upkeep, and disposal might, in fact, influence design decisions. Thus, operations, maintenance, support and disposal activities must be considered during requirements and design processes. The systems engineer must ensure that these factors have been planned for and addressed, in particular with respect to logistics, as part of life cycle planning and requirements development.

4.8.4 Systems Engineering Management

The lead systems engineer for a project is responsible for technical management of the effort, as is depicted in the vertical arrow on the left side of Figure 4.22. He or she coordinates closely with the project manager (Section 4.1) for oversight, but focuses on technical objectives and progress, and provides the analytical support that the project manager needs in order to make good project-level decisions. There are several mechanisms that a systems engineer draws on to provide technical management, and they should be directed by a systems engineering plan (SEP). The SEP defines the systems engineering effort for the project, identifies how it will be managed, and establishes how technical progress will be measured. It should be tailored for the project, addressed to the systems engineering and technical teams, and serve as a living document and central focus for direction for those teams.

Every system has a life cycle that characterizes its past, present, and future. It begins with an identified need and extends through system disposal. A life cycle is made up of stages, each of which represents a defined state of the system. Each stage has a purpose with defined inputs and outputs (e.g., artifacts, work products, decision memos), and is governed by a major approval event called a *decision gate* or *decision milestone*. At each decision gate, the objective is to determine readiness to move forward—ensuring that new activities are not pursued until current activities, on which new ones may depend, are completed satisfactorily. The systems engineer ensures that progress and direction in key technical areas meets stakeholder expectations, and that the success factors identified above remain in balance. A decision to move forward indicates that qualified experts deem the corresponding risk acceptable. Ultimately, a project's system life cycle (SLC) establishes a framework for the technical transformation of business needs into a fielded solution. There are often parallel life cycle processes for acquisition and budgeting.

The systems engineer ensures that meaningful stages and decision gates are established and defines the life cycle processes that must be employed to fulfill the purpose of each stage. In addition, the systems engineer:

- Identifies which processes occur when, how and by whom for a project within phases;
- Specifies the activities, inputs and outputs which define these processes;
- Ensures that each contributes to achieving system objectives;
- Defines approaches for resolving problems or failures.

Industry best practices and development life cycle models, such as waterfall, evolutionary, and agile are useful to help determine processes and activities that are appropriate in each stage for a particular system. The development life cycle models, often applied to software development, describe the approach to be used for development and build of system components, while the SLC provides the approach for oversight of the end-to-end system.

The SLC should be described in the SEP. Figure 4.23 illustrates a generic example of an SLC, without the details of processes and work products. The systems engineering processes described in Figure 4.23 may be implemented, as appropriate, throughout all phases of a system life cycle. The design of individual subsystems, for example, will employ their own requirements and functional analyses processes (in coherence with their development life cycle). Integration activities may involve requirements for the development of an integration environment. Transition planning, similarly, will employ systems engineering best practices to ensure that requirements for carrying out the transition are established.

4.8.5 Interoperability Deficiencies

Implementation of a systems engineering approach can result in the avoidance of many of the 25 deficiencies defined in the previous chapters. Table 4.11 provides discussion of how the systems engineering best practices can provide mechanisms to address these deficiencies (referencing process area letters in Figure 4.23). Notice that many of these types of deficiencies are addressed by systems engineering best practices that occur early in a system's life cycle. This is often where the value of systems engineering is emphasized.

Application of a systems engineering best practice to establishing the Air Warning Service (AWS) on Hawaii in 1941 would have focused on *integration* issues. Although it may appear in Figure 4.23 that integration occurs later in the SLC, it is actually a carefully orchestrated and planned initiative for which efforts begin during *requirements elicitation and concept definition* and *requirements and functional analysis*. It is early in the life cycle that system components are defined, and it is understood how they will function as a complete system in the operational environment. For the AWS, such integration planning through systems engineering and an enterprise architecture would have incorporated requirements associated with system objectives beyond the radar technology and warning capability—on how to facilitate action based on such warning. Thus, there were elements of the system that involved detection, interpretation, communication, alert, response, and inter-

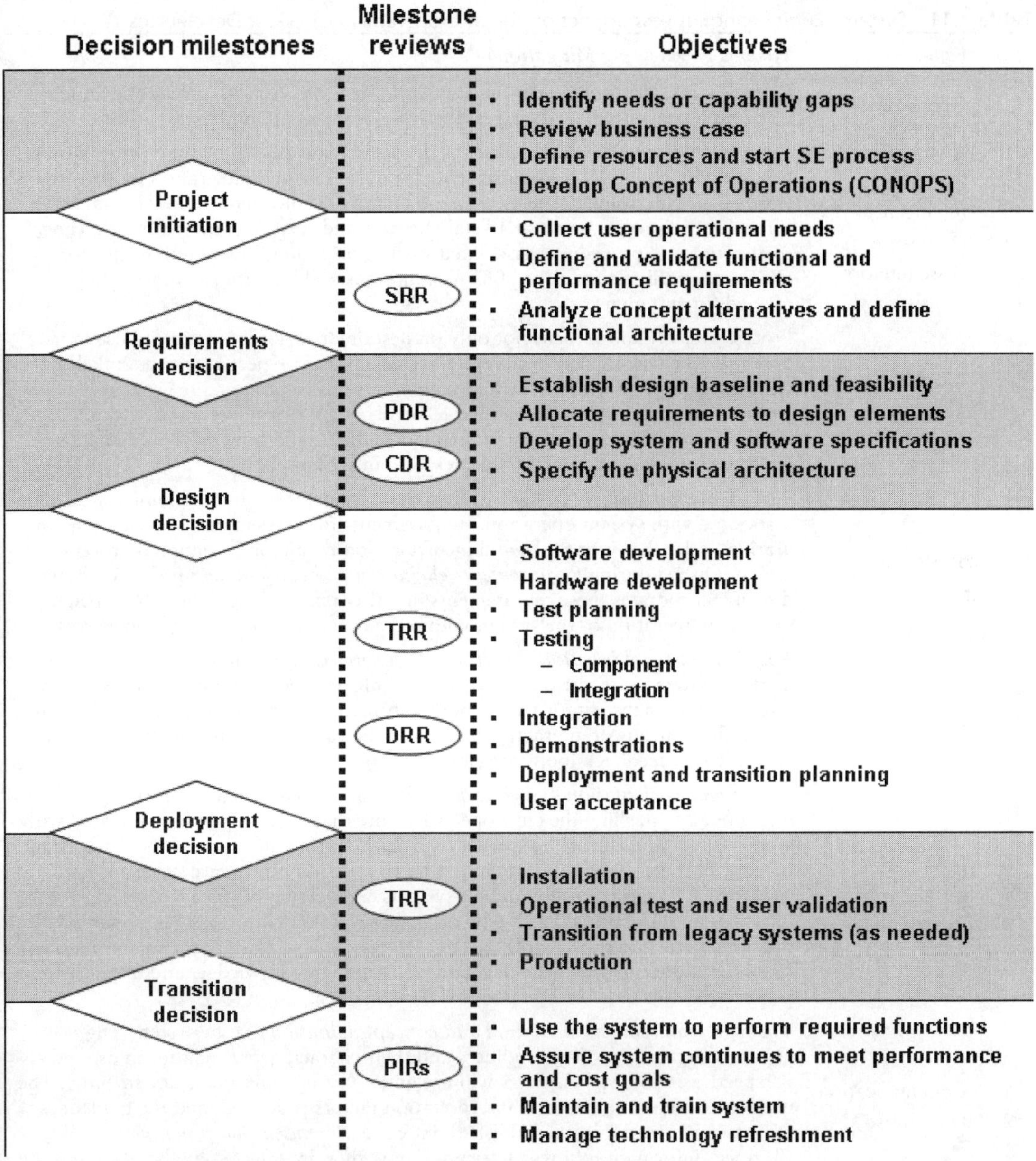

Figure 4.23 Generic system life cycle example. The abbreviations in the figure are as follows: CDR–critical design review; DDR–deployment readiness review; PDR–preliminary design review; PIR–postimplementation review; SSR–system requirement review; and TRR–test readiness review.

ception. These comprise top-level functional activities for an integrated system. Requirements and timeline analysis (see definition in Section 4.5.5.2) would then allocate performance specifications to these and lower functional elements—how quickly, what exchange mechanisms and format, what triggers, how far away, who is responsible, and so forth.

Indeed, many of the factors that resulted in the failed response to the radar detection on December 7, 1941, could have been prevented with the implementa-

Table 4.11 Systems Engineering Process Impact on the Interoperability Checklist Deficiencies

No.	*Failure of ...*	*Systems Engineering Mitigations*
1	Organization	*Requirements elicitation and concept definition* (A) establishes a baseline understanding of stakeholders, their perspectives, relationships and expectations.
2 3 8 14 16 19	Assumption Omission Intelligence Disclosure Dissemination Consistency	*Requirements and functional analysis* (B) details operational use and information flows ("who and how"). Requirements for data flow and integration mechanisms (e.g., push, pull, triggers, alerts, etc.) are identified. Information flow is first and foremost analyzed in terms of all stakeholders and their operational needs. Then, regulations and procedures associated with information flow requirements are addressed at this point in the SLC. Assumptions and limitations are well documented for system users.
4 9	Verification Attitude	Process areas A and B focus not only on describing a system's intended use in its operational environment, but on understanding the perspectives of the stakeholders who will be using the system and designing it so as to promote its use that way. Thus, understanding how command elements and subordinates approach one another and work together, including personalities, and how this could affect system operation, are taken into consideration in system planning
5	Supervision	Part of *architecture and design* (C) includes consideration and planning for training associated with system operation. Requirements for users in terms of training, integration with other user tools, and incorporation within organizational processes are established during *Requirements elicitation and concept definition (A)*. System design incorporates these requirements in a way that will facilitate, not impede, successful operation of the system within the organization's strategic objectives
6 7	Alertness Complacency	*Requirements and functional analysis* (B) ensures that thresholds and factors for alerting system users integrate and balance information from all relevant sources, interpret and analyze information both automatically and manually, are adjustable, and address means of interacting with users, elevating to command elements, and ensuring that feedback loops are closed
10 15 18	Imagination Insight Preparedness	*Requirements elicitation and concept definition* (A) requires brainstorming and intentionally "pushing the envelope" to ensure that all possibilities associated with system intent and use are addressed early in the system life cycle. This process can involve all stakeholders and often involves discussion of use case scenarios and operational scenarios. It is common systems engineering practice to consider scenarios of unlikely, as well as likely, occurrence. It is common to involve subject matter experts and those with objective points of view. The latter can serve as facilitators, but also provide historical and other domain knowledge, and interject fresh perspectives not directed by operational experience.
11 12	Communications Paraphrase	Through *requirements elicitation and concept definition* (A) the systems engineer gains understanding of stakeholders and their characteristics, relationships, and perspectives. How stakeholders will use and affect information is documented. The need to address language and interpretation factors is defined, and the guidelines for standard terminology established. Next, *requirements and functional analysis* (B) traces information flows, determines how they are affected by the users and the system, and ensures that information integrity is maintained.
13	Adaptability	*Architecture and design* (3) ensures that a system is built to be adaptable. The intended environment is analyzed, and sensitivity analysis on how the environment might change is performed. The system is designed technically for flexibility to address future change.
17 20 21 22 23 24 25	Inspection Protectiveness Relationships Priority Reporting Improvement Delegation	While these deficiencies are primarily addressed at an organizational strategic-planning level using the organizational development process, *requirements elicitation and concept definition* (A) provides brainstorming and collaboration with internal and external stakeholders. External interactions are captured well by enterprise architecture processes, and involving these stakeholders in requirements and concept definition activities provides the opportunity for understanding perspectives and building relationships. To address these deficiencies, bringing governance leaders and planners into these activities can be critical. In addition, many of the techniques associated with these processes lead to establishment of common goals and standards for defining and measuring interoperability.

tion of dedicated concept definition, requirements elicitation and analysis, definition, and architecture systems engineering activities in the consideration of a complete air defense system. As it was, the installation of radar on Hawaii resulted in simply an installation of technology. Stakeholders on the mainland driving the radar installation understood its capability, but the operational environment and stakeholder perspectives were not fully taken into account as part of a system requirements analysis. In this case, the variety of stakeholders affecting operation of the functional activities identified above contributes significantly to integration risk. For example, there were different command elements involved, even the National Park Service, all of which were protective of territory (PHD #20 Protectiveness). The concept for how the radar would be used from detect to intercept was not defined, understood, or agreed upon by all relevant stakeholders. Thus, the extent of the intended capability for an integrated air defense system was not fully understood, and those with authority did not apply an appropriate level of attention to such a system once its pieces began to arrive in the operational environment. Taking the next steps of turning the concept definition into architecture and requirements specifications provides technical understanding of subsystem interfaces—integration of the functions, from detection to intercept, in terms of both equipment and human operators. Ultimately, the air defense capability centering on the radar technology was not a *system*.

4.9 Standards Development for Interoperability

Two forms of standards development (see Figure 4.24) have served as best-practice models for the development of IT and communications interoperability standards worldwide.

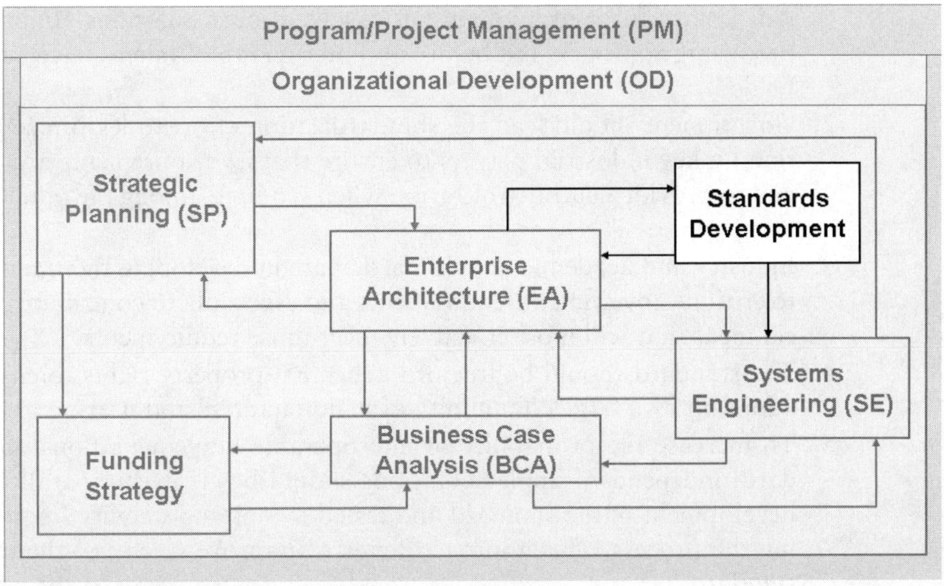

Figure 4.24 Standards development best practice.

4.9.1 Consensus Standards

Consensus standards reflect an agreed-upon approach for a given topic. These topics may concern specific technologies, or they may involve broader issues such as requirements or frameworks that influence the development, deployment, or use of a technology.

A natural extension of standards is the promotion of interoperability, given a common blueprint made available to various interested but disparate parties. We stress the term *promotion* because the presence of a consensus standard by itself does not ensure its interoperability.

Up until 1995, the Internet Engineering Task Force (IETF) required two interoperable implementations before a protocol was accepted as a standard. With the removal of this requirement, groups emerged outside of the IETF and other standards development organizations (SDO) to help promote interoperability. These groups tend to be closed to vendors, with test results kept private in order to encourage continued participation. Other SDOs, such as the International Telecommunication Union (ITU), conform to other criteria and regulations in the development and acceptance of standards.

4.9.2 Directed Standards

Directed standards are those developed under the guidance of one or more collaborating organizations to meet specific requirements. Government agencies and some industry organizations build standards for a specific need. Government standards must be free of individual or organizational intellectual property rights (IPR). During two decades of experience in developing standards for communications interoperability in the United States and NATO nations, a number of best practices have emerged:

1. An interoperability standard should be developed when experience with initial designs for a technology has resolved most questions about the most promising approach, but before noninteroperable systems have been widely fielded.
2. Government should lead the standardization effort to legitimize participation by key industrial players to ensure that its requirements are addressed and to provide neutral arbitration when required among the other contributors.
3. Industry and academia should participate in developing the standard, both to critique government requirements and especially to contribute the technical ideas that will most effectively meet those requirements.
4. The standard should be free of intellectual property rights to ensure rapid adoption by a large community of manufacturers and users.
5. To increase the probability of interoperable implementations of the standard, independent implementations should be created in parallel with the development of the standard and tested for interoperability frequently during the process. This approach tends to improve clarity of the text of the standard.

4.9.3 Types of Standards

There are three primary types of standards produced by SDOs. The most common types involve technology standards that define how a particular mechanism, be it software or hardware, must operate. In the case of the IETF, these technology standards follow two different tracks; one is experimental, and the other is realized in three different levels of maturity starting with proposed standard, continuing with draft standard (requiring two independently developed interoperable implementations), and finally ending with Internet standard.

The other body of standard's publications is classified as informational. These documents cover best common practices or present information that compliment or advance the information described in technology standards.

4.9.4 Participation

The background of individuals that participate in SDOs vary from managers, industry engineers, civil servants, system integrators (typically in the form of government contractors), and even academics. Their respective body of knowledge varies as does their roles over time, such as engineers that built systems becoming managers or system integrators and thereby broadening their understanding and exposure to the big picture. The cross pollination of disciplines can at times be contentious with engineers complaining of a lack of real-world understanding by academics, academics complaining of a lack of vision, and leading edge ingenuity by engineers. Each of these arguments has some merit, which reinforces the need to avoid totally discriminating in favor (or disfavor) of one particular type of participant. However, the measure of detail, scope, and purpose of a standard should also be a factor in which type of participant takes a leading role in the development of a standard.

The subject of participation also pertains to those individuals that take on the role of governance within the standards bodies, such as working group chairs, or area directors. These individuals directly influence the progression or acceptance of a standard on behalf of the SDO. In cases like the ITU, the role is limited to achieving consensus and providing guidance with the power of acceptance or rejection distributed equally among all participants through voting. In the case of the IETF, governance in the form of acceptance/rejection is more centralized and concentrated to a few participants. However, rough consensus versus unanimous decision determines the progression of a standard.

While the historic track record from the mid 1980s to the present of the IETF has an impressive set of accomplishments to point to in defending its current model, recent history in some areas has demonstrated a need for serious change. The Session Initiation Protocol (SIP) working group, chartered to define the means of establishing sessions (or in the case of VoIP, calls) between two end points, is a prime example of how the current system of the IETF has broken to a point of almost being dysfunctional. The time taken to advance a standard within the SIP working group far exceeds that of other working groups. The problem has become so acute that the working group chairs have promoted a complete overhaul of the group along with a related working group.

An interesting observation is that those responsible for the mess are the ones still in a position of governance or the primary leaders advocating change. In a sense, they are the inmates who constructed the asylum and the ones that continue to run it. Two solutions to address this problem are term limits and requirements for implementations that accompany the development of a technology standard. The former proposal helps promote the infusion of new ideas and perspectives. The latter helps reduce the explosion of proposed efforts that demand some measure of time and consideration by the general set of participants.

4.10 Integrated Best Practices for the Interoperability Success Plan

As shown in Figure 4.1, the employment of all best practices must be governed under the auspices of the project management best practice. These processes establish the oversight and guidance framework as well as documented artifacts used as a baseline for all projects at all governance levels. This project governance structure ensures repeatability and consistency throughout interoperability planning efforts, and also enables coordination of interdependencies among individual programs (i.e., sets of coordinated projects) and other stakeholder projects selected and planned to achieve the desired interoperability architecture. At the project level, the structured approach tailored from management processes to ensure timely and quality performance within budget are referred to as project management.

Once program/project management is established, programs and projects employ the organizational development best practice to ensure that an agency is appropriately structured with clearly defined roles and responsibilities of its individuals, departments, or divisions. In addition, they ensure that organizational changes are communicated in a structured manor to increase stakeholder buy-in and adoption from initial planning to implementation through overall sustainment. Organizational development establishes the teams of planners, stakeholders, subject matter experts, and other staff to optimally proceed with the development of strategic plans for holistic interoperability at each governance level shown in Figure 3.1. Most importantly, it establishes the governance structure needed to develop and sustain organizational commitment to interoperability programs and projects.

Once the governance structure for interoperability planning is in place, it is critical that the strategic planning best practice establish the organizational mission, vision, goals, and objectives for interoperability at each governance level. Strategic planning refines and translates these goals and objectives into initiatives that can be implemented as specific, detailed projects to achieve interoperability across multiple disciplines, jurisdictions, and governance levels. The overall process begins with the generation of an interoperability strategic plan, which gathers the policy directives and guidance from overarching governance levels to coherently define and prioritize the capabilities required for key staff and other stakeholders. It also provides a critical planning framework for other governance levels within their jurisdiction. As will be shown, coordinated interoperability strategic plans at all governance levels are needed to convert the World Interoperability Failure model from a worst-case chaos of deficiencies into holistic interoperability for public safety and emergency response.

Once programs and projects are defined as initiatives in the coordinated interoperability strategic plans, project management oversight is established over the recommended projects so solution definition can begin within a structured, performance-driven planning regimen. Within this regimen, the remaining best practices can be employed to provide the necessary guidance, methods, and tools needed to successfully define and deliver specific elements of the interoperability architecture. In this regard, one project invariably begun at the outset of any interoperability planning initiative uses the enterprise architecture best practice to illustrate both the existing and the envisioned operational environments, including detailed capabilities needed by the operational staff, organizational structures, collaboration needs, and so forth. Enterprise architecture is particularly important in representing disparate operational and system architectures in a manner such that information sharing among these architectures can be done in a common descriptive framework. Enterprise architecture also defines targeted public safety and emergency response architecture as part of an IT/communications enterprise and describes its internal and external relationships. Enterprise architecture offers the tools to assess the current as-is state of the information-sharing architecture—including all people, processes, and tools employed—as well as to postulate the should-be or, once vetted, the to-be interoperability architecture, thus enabling a *gap analysis*.

The selection of prioritized initiatives in the strategic planning practice, as well as elements of the to-be architecture targeted by this practice, is properly done using the business case analysis best practice. Business case analysis is employed to understand expected costs, prioritize targeted capabilities, and assist in providing justification for interoperability project investments. In general, the business case analysis process connects the financial realities of available funding and cost of new technology and systems to the performance improvements and resulting downstream benefits. This connection then provides the justification needed for decision makers to allocate funds in the most effective way to further an organization's objectives. In this regard, the funding strategy practice provides the necessary resources at all stages of interoperability planning; including use of best practices to develop the interoperability strategic plan as well as all other stages of interoperability architecture development. Funding strategy then matches the resource requirements estimated for the effective performance of specific programs and projects with available sources of funds and the associated timeline.

In general, each interoperability project impacting the IT/communications architecture at any governance level should be implemented according to a structured systems engineering best practice. Systems engineering ensures that the capabilities reflected in planned initiatives are accurately described by a specification of technical requirements. Systems engineering then manages design, testing, and implementation by maintaining traceability to these requirements, stakeholder expectations, and strategic planning vision. Systems engineering processes are integrated with business case analysis and enterprise architecture best practices to perform analysis of viable alternative solutions and establish operational interoperability architecture. Throughout the design of interoperability architecture, technology and technical standards must be employed to the maximum possible extent. In this regard, critical architectural standards must be defined and mandated at national and multinational governance levels to prevent the formation of information-sharing stovepipes at all other governance levels. In this regard, the standards development best practice is vital to providing common systems architecture for the subdivisions of a nation and their associated responders-receivers and other stakeholders. These national subdivisions are not—and should not—seek to establish such standards independent of their adjacent and overlapping jurisdictions. It is for this reason that national and multinational bodies have this responsibility, and on occasion, have proactively pursued such standardization (e.g., the Internet).

Those organizations that achieve success do so through the incorporation of best practices, proactively creating and updating managerial and engineering processes. However, not all of these activities are appropriate for each responsible governance level and for every type of project. These best practices can be examined for their relevance to the specific needs of a region and then tailored to fit those needs. The amount of tailoring depends on a variety of factors. For large and complex projects, significant portions of the methodology should be formally applied, while smaller projects may significantly tailor these methods to the scope of the problem being addressed. The remainder of this chapter provides an explanation of each of these best practices. Figure 4.25 illustrates how these best practices are integrated.

The individual numbered circles in the figure provide additional explanation for the information flows between base practices as follows:

1. Project management provides overarching day-to-day planning and management of schedule, resources, cost, deliverables, risks, quality, and stakeholder expectations for projects employing one or more of the best practices contained within the surrounding rectangular area in the figure.
2. Organizational development translates strategy and goals into actionable organizational change, including the development and implementation of tailored collaboration and communication techniques spanning a wide variety of stakeholders with diverse interests, needs, and motivations.
3. Strategic goals, objectives, measures, timelines, initiatives, and capability needs are provided to the funding best practice for alignment with funding sources to ensure successful implementation.
4. Organizational strategy for determination of interoperability operational and system architecture and development of high-level system requirements; create or updates the organization or project's missions, goals, values, immediate and long-range plans, and provides other overall guidance. These inputs are incorporated into the organizational enterprise architecture and represented on various reference models for use by the remainder of the organization or project team.
5. Technical capabilities and constraints from the enterprise architecture best practice provides strategic planning with strategically aligned reference models which reflect the organizational or project mission, goals, priorities, capabilities, constraints, technologies, and life cycles. This input represents both the as-is and to-be architectures providing the strategic planning best practice with the necessary information to guide decision making and strategic planning based on verifiable information and not opinion.

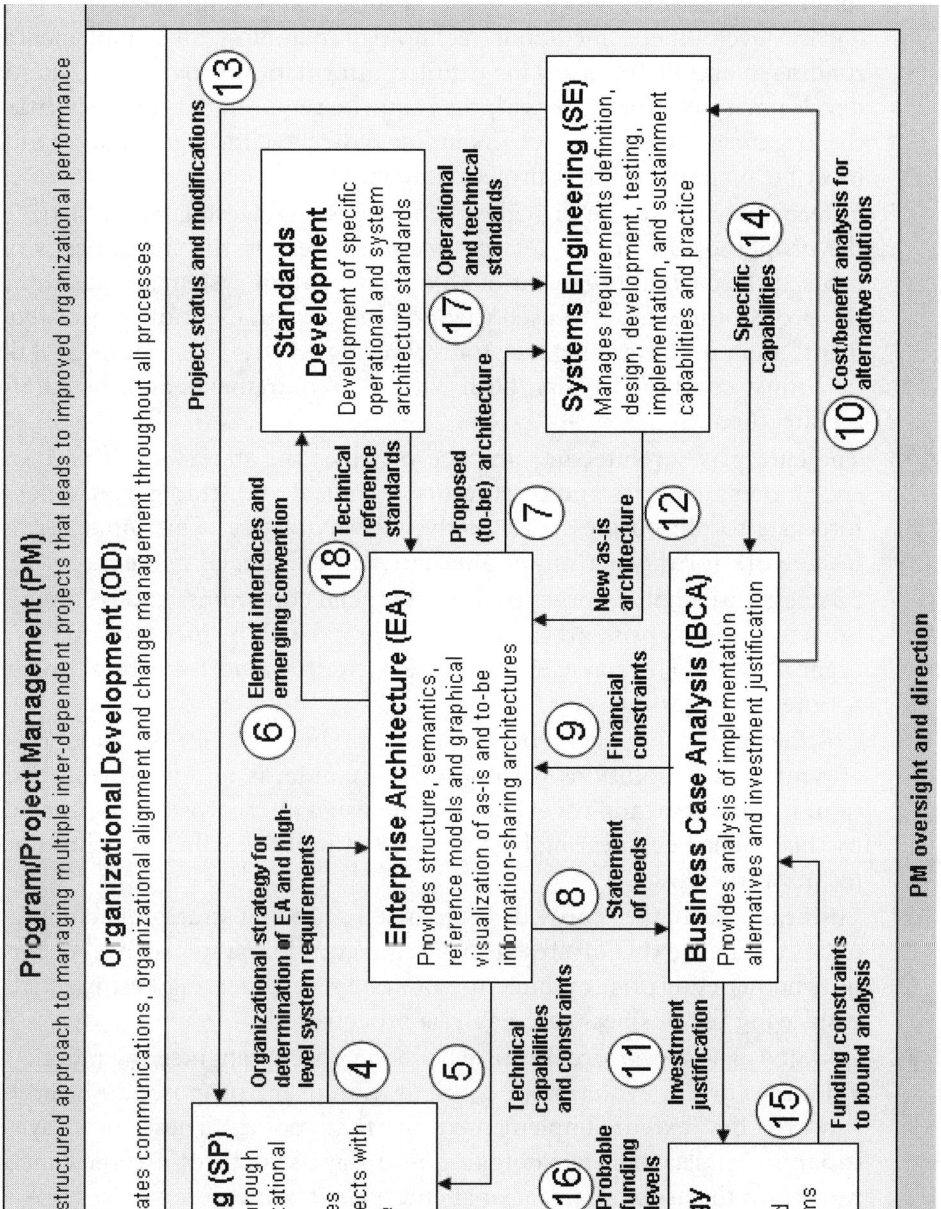

Figure 4.25 Integrated best practices for interoperability success planning.

6. Element interfaces and emergent conventions from the enterprise architecture best practice provides business, strategic, and technical drivers to standards development including technology planning and implementation roadmaps and life cycles. This detailed information provides the standards development best practice with the requisite requirements and priorities for the organization or project, including when technologies and standards must be available to meet project timelines.
7. Enterprise architecture provides the systems engineering best practice with the proposed to-be or target architecture based on the stated needs of the project; the current organizational, operational, or system architecture; and the project timeline. It is used to guide the development of project requirements and provide governance for technical reviews. The project, as it develops, must continue to align both with the requirements and the enterprise architecture.
8. The enterprise architecture practice provides the statement of need, reference models, policies and constraints, technical and strategic guidance, and forecasts to the business case analysis best practice. This input forms the framework for the analysis of alternative and financial modeling.
9. Business case analysis practice feeds financial constraints and recommendations to the enterprise architecture best practice to be incorporated into the organizational enterprise architecture as well as and transitional planning for the organization.
10. Cost/benefit analysis from the business case best practice defines and evaluates alternative solutions with particular emphasis on investment justification. It provides trade-off evaluation of design alternatives being considered by the systems engineering best practice as input to solution definition and technical review.
11. Business case analysis provides guidance to funding strategy, including estimates of likely expenditures at the architectural or systems level, depending on whether enterprise architecture or systems engineering best practices are employing the business case analysis processes.
12. Updated as-is system architecture from the systems engineering practice provides the current evolution or configuration of the proposed to-be architecture to the extent implemented at that point. These new systems, capabilities, services, technologies, and data structures are then incorporated into the as-is organizational enterprise architecture.
13. Project status and modifications from the systems engineering best practice are used to measure the achievement of goals established by strategic planning. Any modifications that occurred during systems engineering must be reflected in strategic plan initiatives and updates. This ensures that organizational goals remain aligned with implementation in the field, and also provides input for future strategic plan update cycles.
14. Specific capabilities and alternative solutions being considered by the systems engineering best practice that are identified during design activity can be analyzed using a cost/benefit analysis approach with particular emphasis on investment justification.

15. Business case analysis needs estimates of available function both to determine the range of possible alternative solutions, but also to estimate the funding risk associated with preferred alternatives when selecting recommended solutions.
16. Strategic planning uses estimated funding levels to scope resources available for priority initiatives, which in turn define the content of high-level planning for possible programs and their associated projects.
17. The standards development best practice provides both operational and technical conventions to the systems engineering practice to establish preferable technical approaches, define constraints, establish consistent methods and terminology, and guide solution design.
18. As new technical reference standards are approved, they are added to the organizational enterprise architecture in its technical reference model. Each affected technical or standard area is updated including the current (as-is) and future (to-be) configuration or forecast/roadmap. These updated reference models are used to guide technology development and procurement.

The best practices described in this chapter provide the tools needed in the World Interoperability Success Planning approach described in Chapter 7. These practices are primarily focused on building consensus across the diverse people who impact public safety and emergency response IT/communications systems. As has been shown, human factors—not technology—are the source of interoperability failure. These human factors must be managed for the greater good through considered rational thought and practice. This chapter has presented the tools for that practice.

References

[1] Select Bipartisan Committee to Investigate the Preparation for and Response to Hurricane Katrina, Failure of Initiative, *Final Report of the Select Bipartisan Committee to Investigate the Preparation for and Response to Hurricane Katrina*, Appendix 8, U.S. Government Printing Office (Washington, D.C., February 15, 2006), p. 5, available at http://www.gpoacess.gov/congress/index.html.

[2] Project Management Institute, *A Guide to the Project Management Body of Knowledge*, 4th ed., 2008 (hereafter referenced as the *PMBOK Guide)*.

[3] See www.pmi.org.

[4] For PMI background, see http://www.pmi.org/AboutUs/Pages/About-PM.aspx.

[5] *PMBOK Guide*, p. 6.

[6] *PMBOK Guide,* p. 67.

[7] An integrated project team is an establish subset of a project team with a focus on a single domain, component, element, or subsystem. Individuals on this team are also on other integrated project teams created for interworking domains, components, elements, or subsystems, so interface and other critical integration risks for the final product are greatly reduced.

[8] Refer to the *PMBOK Guide* for detailed description of project management tools and techniques.

[9] *PMBOK Guide,* Glossary, p. 444.

[10] Anantatmula, Vittal S., "The Role of Technology in the Project Performance Model," *PMI Project Management Journal*, March 2008.

[11] Derived from Anantatmula, Vittal S., Ibid.

[12] McKenzie, J., *Perform or Else*, Routledge, 1989, p. 77.

[13] Stanford, Naomi, *Guide to Organisational Design: Creating High-Performing and Adaptable Enterprises*, Chapter 1, Profile Books Ltd, 2007.

[14] Ibid.

[15] Cook, Sarah, Steve Macaulay, and Hilary Coldicott, *Change Management Excellence: Using the Four Intelligences for Successful Organizational Change*, Kogan Page, 2004.

[16] Based on the Eight-Step Model by John Kotter, professor at Harvard Business School, world-renowned change expert and author of *Leading Change and Heart of Change.*

[17] Cook, Sarah, Steve Macaulay, and Hilary Coldicott, Ibid.

[18] Ibid.

[19] Business Dictionary, www.businessdictionary.com.

[20] United States Department of Homeland Security SAFECOM Program, http://www.safecomprogram.gov.

[21] DHS Statewide Communication Interoperability Plans Fact Sheet, http://www.safecomprogram.gov/NR/rdonlyres/E7E70605-12FB-492F-ABD3-5BA708EA5FD3/0/SCIPImplementationFactSheet_FINAL_081029.pdf.

[22] Ibid.

[23] Including specific artifacts to be produced after tailoring.

[24] IEEE Standards Association, "3.0 Definitions," *IEEE Std 1471-2000 IEEE Recommended Practice for Architectural Description of Software-Intensive Systems*, IEEE, 2000, available at http://standards.ieee.org/reading/ieee/std_public/description/se/1471-2000_desc.html.

[25] DoD Architecture Framework Working Group, *DoD Architecture Framework, Version 1.0, Volume I: Definitions and Guidelines*, August 2003, p. 1–1, available at http://www.tricare.mil/jmis/download/DoDArchitectureFrameworkVersion1Feb2004.pdf.

[26] Ibid.

[27] Zachman International Enterprise Architecture, http://www.zachmaninternational.com/.

[28] Federal Enterprise Architecture Framework, http://www.cio.gov/documents/fedarch1.pdf.

[29] *DoD Architecture Framework*, http://jitc.fhu.disa.mil/jitc_dri/pdfs/dodaf_v1v2.pdf.

[30] Ibid.

[31] Ibid.

[32] DoD Architecture Framework Working Group, *DoD Architecture Framework Version 1.0, Volume I: Definitions and Guidelines*, and *Volume II: Product Descriptions*, Department of Defense, February 2004.

[33] Science Applications International Corporation, *Integrated Statewide Strategic Emergency Communications Management Plan—Technical Proposal*, SAIC Proposal No. 06-0282-71-2005-007, 14 January 2005, pp. 3–1 to 3–6.

[34] Science Applications International Corporation, *Interoperability Planning for the Northeast Homeland Security Planning Region*, Briefing, Massachusetts Emergency Management Agency, Summer, 2005.

[35] *PSAF Volume I: Definitions and Guidelines Version 1.0*, *PSAF Volume II: Product Descriptions,* February, 2006, and *PSAF Volume III: Communications Data Model*, December, 2007.

[36] Redding, C., J. Kub, J. Lee, and A. Morse, Public Safety Architecture Framework Trial Report, SAFECOM Program, January 2007, p. iii, available at http://www.safecomprogram.gov/SAFECOM/library/technology/1305_publicsafety.htm.

[37] *Public Safety Architecture Framework Volume I*, http://www.safecomprogram.gov/SAFECOM/library/technology/1251_publicsafety.htm.

[38] *Public Safety Architecture Framework Volume II*, http://www.safecomprogram.gov/SAFECOM/library/technology/1252_publicsafety.htm.

[39] *Public Safety Architecture Framework Volume III: Communications Data Model*, http://www.safecomprogram.gov/SAFECOM/library/technology/1252_publicsafety.htm.

[40] In this context, metadata is information that describes the conventions adopted in data repositories or databases.

[41] DHS National Response Framework, http://www.fema.gov/pdf/emergency/nrf/nrf-core.pdf.

[42] Standing Operating Procedures for the Euro-Atlantic Disaster Response Unit (EADRU), http://www.nato.int/eadrcc/sop/sop.htm.

[43] Archibald, N., et al., *Getting Value from Enterprise Initiatives: A Survey of Executives*, The Boston Consulting Group, (visited May 1, 2002), p. 9, http://www.bcg.com/publications/files/Enterprise_Computing_report.pdf.

[44] Gartner, Inc., *Is IT Worth It? Presenting the Public-Sector Business Case*, February 2002, p. 2.

[45] Schmidt, Marty J., "What's a Business Case? And Other Frequently Asked Questions," Solution Matrix Ltd., (visited May 3, 2002), http://www.solutionmatrix.com/WP1_A4.PDF, p. 11.

[46] Supra., note 1, p. 3.

[47] IBM, *The Business Transformation Toolkit*, September 2001, p. 1–1.

[48] See www.ahcusa.org.

[49] *The Merriam-Webster Online Dictionary* (2008), retrieved October 3, 2008, from http://www.merriam-webster.com/dictionary/system.

CHAPTER 5

Operational Architecture

Whether civilian or military, public safety and emergency response operations can be characterized as the performance of *specific* tasks and activities by responders-receivers and other stakeholders to achieve a common goal. Examples of such tasks and activities are:

- Radar operators at Opana Point detect a large flight of incoming aircraft and call the Information Center on Sunday morning, December 7th (see Chapter 2).
- A FDNY chief, learning about the collapsed South Tower, warns his subordinates over a portable radio to exit the North Tower—and a firefighter, hearing this alert, continues to help civilians escape until the collapse.
- A Coast Guard helicopter rescue swimmer is lowered to a rooftop in New Orleans, where she helps parents and children into a harness one by one, before returning to safety herself.

In each case, the actions of these people, following practiced procedures and using available tools, represent only a small part of the operations of their respective organizations. These operations are dependent on information largely collected, analyzed, and disseminated within their own organization, thus forming an information stovepipe.

What was coming to Pearl Harbor and what had arrived in New York and New Orleans requiring these tasks and activities—had been necessitated much earlier by failed interoperability among stovepipes. In each case, failed leadership and planning at many governance levels—and lacking best practices described in Chapter 4—had again led to the need for tireless initiative and self-sacrifice on the part of those heroes who must go on to account for those failures. To fix these failures, we must first define a common framework for understanding their organizations, their tasks and activities, and their information stovepipes. This common framework is like a blueprint for people, processes, and tools—including their information flows—that determine interoperability success or failure. This common framework, drawn from the best practices presented in Section 4.5 of Chapter 4, is Enterprise Architecture.

5.1 Interoperability Architecture

This section describes the process to develop an "interoperability architecture" by combining the elements of integrated, composite, aligned, and federated architectures. A brief discussion of each of these architectures follows.

Architecture is "the structure of components, their relationships, and the principles and guidelines governing their design and evolution over time" [1]. The authors of this definition considered the corresponding IEEE definitions [2]:

> IEEE STD 610.22 in 1990 defined "architecture" as "the organizational structure of a system or component." In 2000, IEEE STD 1472 provided the following definition: "An architecture is the fundamental organization of a system embodied in its components, their relationships to each other, and to the environment and the principles guiding its design and evolution."

In this case, we are interested in representations or views of *operational architecture*, which we will define in our interoperability context as "a description of the tasks and activities, operational elements, and information exchanges required to accomplish" [3] a public safety and emergency response goal. Using available tools from enterprise architecture (see Section 4), we can create architectural views consisting of [4]:

> graphical and textual products that comprise an identification of the operational nodes and elements, assigned tasks and activities, and information flows required between nodes. It defines the types of information exchanged, the frequency of exchange, which tasks and activities are supported by the information exchanges, and the nature of information exchanges.

The operational architecture therefore:

- Contains descriptions (often graphical) of the operational elements, assigned tasks and activities, and information flows required to support public safety and emergency response.
- Defines the types of information exchanged, the frequency of exchange, which tasks and activities are supported by the information exchanges, and the nature of information exchanges, in detail sufficient to ascertain specific interoperability requirements.

These activities and information-exchange requirements will span organizational (discipline and jurisdictional) boundaries as a necessary consequence of the identified tasks and activities of the collaborating responders-receivers. Operational architecture is not generally systems-dependent, that is, it is not dependent on the IT/communications system employed. Operational architecture should clearly identify the time phase(s) covered and identify whether they represents the as-is or to-be configuration.

5.1.1 Integrated Architecture

If the people, processes, and tools used in an organization were engineered "from scratch" so all the components—including the all information flows—of the resulting architecture fit together seamlessly, then an *integrated* architecture has been created.

> An integrated architecture is one in which architecture data elements are uniquely identified and consistently used across all products and views within the architecture. ... Integration of architectures enables aggregation of content to support analyses of a broader scope than what is needed or possible through a single architectural view. In this manner, integrated architectures clarify roles, boundaries, and interfaces between components of large systems of systems and act as a key tool for enterprise-level systems integration. An architecture is integrated through the mapping or standardization of terms, definitions, and relationships across the architecture to support ... decision making and systems development. [5]

In an integrated architecture, all pieces and parts are designed from the outset to work together in unison, including signal or information flows (data) between the component parts. Examples of integrated architectures drawn using simple cartoons following the operational view (OV) 1 (or OV-1, see Chapter 4) shown in Figure 5.1 are as follows:

- The Opana Point radar, the Signal Corps Radio model 270 (SCR-270) built by Westinghouse, with its antenna array, transmit and receive electrical equipment, and display scope built as one system to detect aircraft at distance—a simple field telephone is used to link the radar site with the Information Center.
- The fire department (FD) radio network, designed to provide local radio coverage at the fireground of a typical urban structure, including building-mounted or deployable radio repeaters and both mobile (vehicle or vessel mounted) or hand-portable radios, and the radio discipline needed to efficiently use the network.
- The Coast Guard (CG) helicopter itself hovering over the rooftop, with its flight controls and communications equipment, acting as a single entity in unison with the pilot, rescue swimmer, and the rescue process.

5.1.2 Composite Architecture

When multiple organizations with common integrated architectures seek to share their information, then the resulting *composite* architecture is defined.

> An architecture stemming from the integration or aggregation of content from disparate integrated architectures is known as a composite architecture. A composite architecture broadens the scope of individual integrated architectures to support joint capabilities and operational efforts. It consistently captures data elements across integrated architectures and conforms to an agreed upon "common denominator" across all individual architectures that are part of the integration effort. The

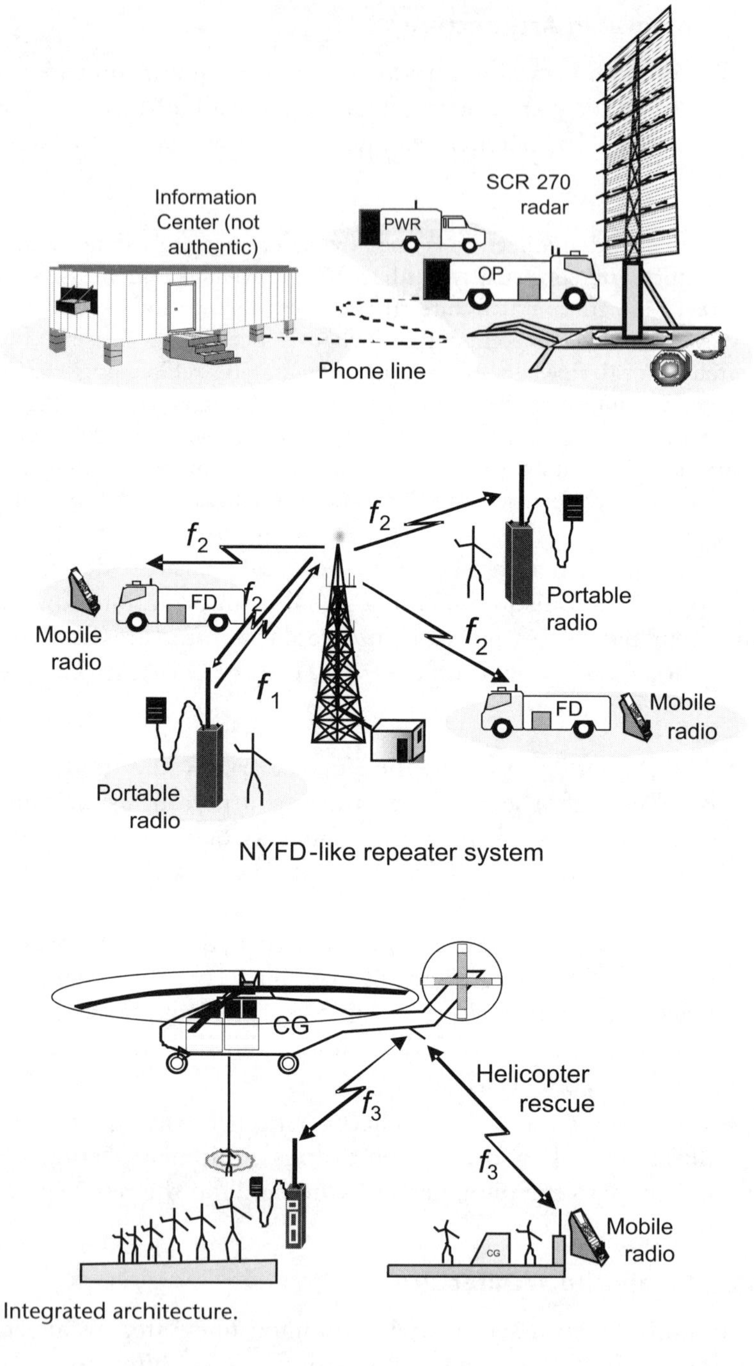

Figure 5.1 Integrated architecture.

development of a composite architecture follows the same guidelines as that of an integrated architecture, only on a larger scale. [6]

For example, if the FD radio system and CG helicopters collaborated at an incident scene, they would need a common voice channel for tactical coordination. Figure 5.2 shows that a deployable communication van with a standard audio gateway device has been deployed within radio line-of-sight of the FD radio network. The

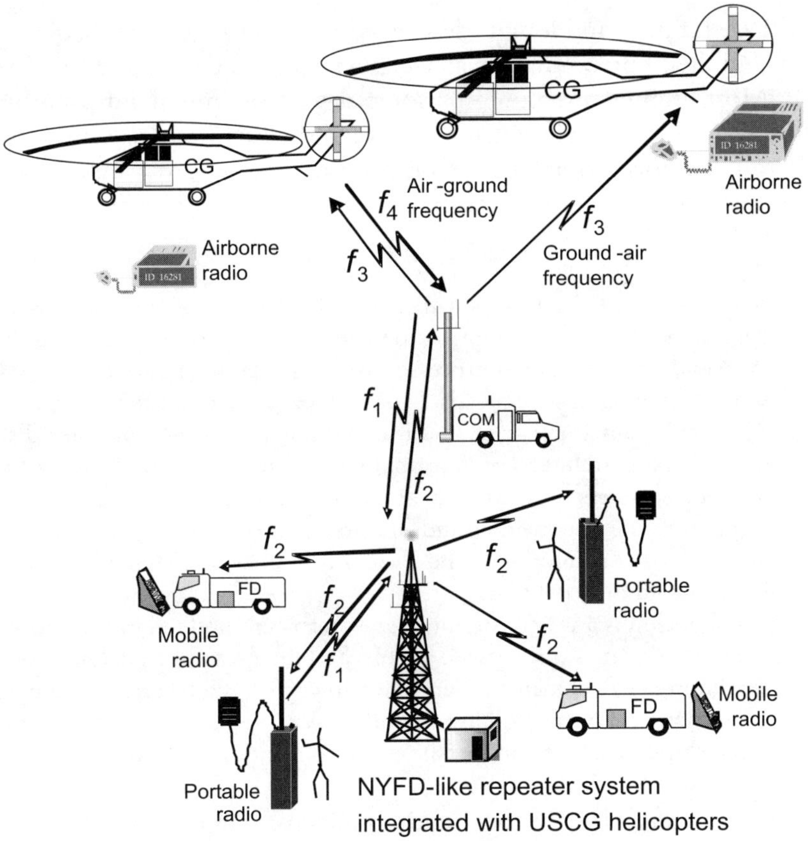

Figure 5.2 Composite architecture.

gateway is essentially a voice repeater with frequency pair (f_2, f_3), that is, it receives on frequency f_2 and transmits on frequency f_3. Thus, when FDNY units transmit on f_1, their repeater translates retransmits signal on frequency f_2. The deployable van's gateway translates the signal on f_2 to f_3 and relays the signal to the helicopters. It also translates downlink frequency f_4 to f_1 to reach the FD repeater. Communications in the opposite direction would be initiated by the helicopters on frequency f_4, translated to f_1 at the communications van for injection into the FD repeater. The integration of the air and ground systems employing a common radio frequency forms a composite architecture.

5.1.3 Aligned and Federated Architecture

One approach to sharing information between disparate architectures is to have them aligned. Architectures are said to be *aligned* when common elements between them are identified and a relationship can be represented. Consider two disparate system architectures, each containing organizational information about remote mobile units. These two systems track their own mobile units using unique unit IDs, with each unit ID mapped to a track history of latitude and longitude values. Assuming the unit IDs are transferable as simple text, the architectures are aligned by these latitude and longitude values, so each system can display the mobile units IDs and their corresponding as well as historical tracks of all mobile units.

Most architecture contains some degree of similar information which can be aligned to develop relationships between the architectures. If the two architectures share key information such as location names, acronyms, and location identifiers (e.g., geo-code, longitude-latitude), this closed relationship may be referred to as *federated* [7].

> A federated architecture is a distributed strategic information asset base, which defines the mission, the information and technologies necessary to perform the mission, and the transitional processes for implementing new technologies in response to the changing mission needs. It provides a framework for enterprise architecture development, maintenance, and use that aligns, locates, and links disparate architectures and architecture information via information exchange standards to deliver a seamless outward appearance to users. A federated architecture approach recognizes the uniqueness and specific purpose of disparate architectures and allows for their autonomy and local governance while enabling the enterprise to benefit from their content.
>
> Federation is a way to organize an enterprise's body of knowledge (architecture) about its activities (processes), people, and things within a defined context and current/future environment. Federation allows the architecture user a means to examine the enterprise from all aspects of the doctrine, organization, training, materiel, leadership and education, personnel, and facilities ... concept.

As an example of a federated architecture, imagine that the FD ground units and CG helicopters in Figure 5.2 are equipped with a common situational awareness display showing the location of ground and air units and broadband radio. An independent broadband network from their disparate radio systems is used to communicate position location and other status in a mesh network among all units, and then the result is the federated architecture shown in Figure 5.3. Their normal backbone voice network is unaffected. Although all FD and CG users of the situational awareness display have the capability of achieving a common operational picture, these assets can continue to perform their diverse organizational roles as well.

Today there are many disparate public safety and emergency response architectures in place among adjacent jurisdictions within each level of governance. This lateral disparity prevents interoperability among adjacent towns, among adjacent provinces and among adjacent nations. Lateral disparity is further complicated by the vertical disparity that exists up and down the many vertical jurisdictional levels from local to multinational. Local IT/communications systems cannot interoperate with architecture at regional governance levels, which in turn do not fit into national architecture. This lateral as well as vertical disparity of architectures characterizes the current state—the lack of true interoperability described in Chapter 3 as the World Interoperability Failure model.

In order to improve upon this current reality, it is essential to apply the concepts of integrated, composite, aligned, and federated architectures to enable holistic interoperability be implemented in a new World Interoperability Success Plan (see Chapter 7). In other words, any system being designed today must satisfy interoperability requirements so that the new system will meet World

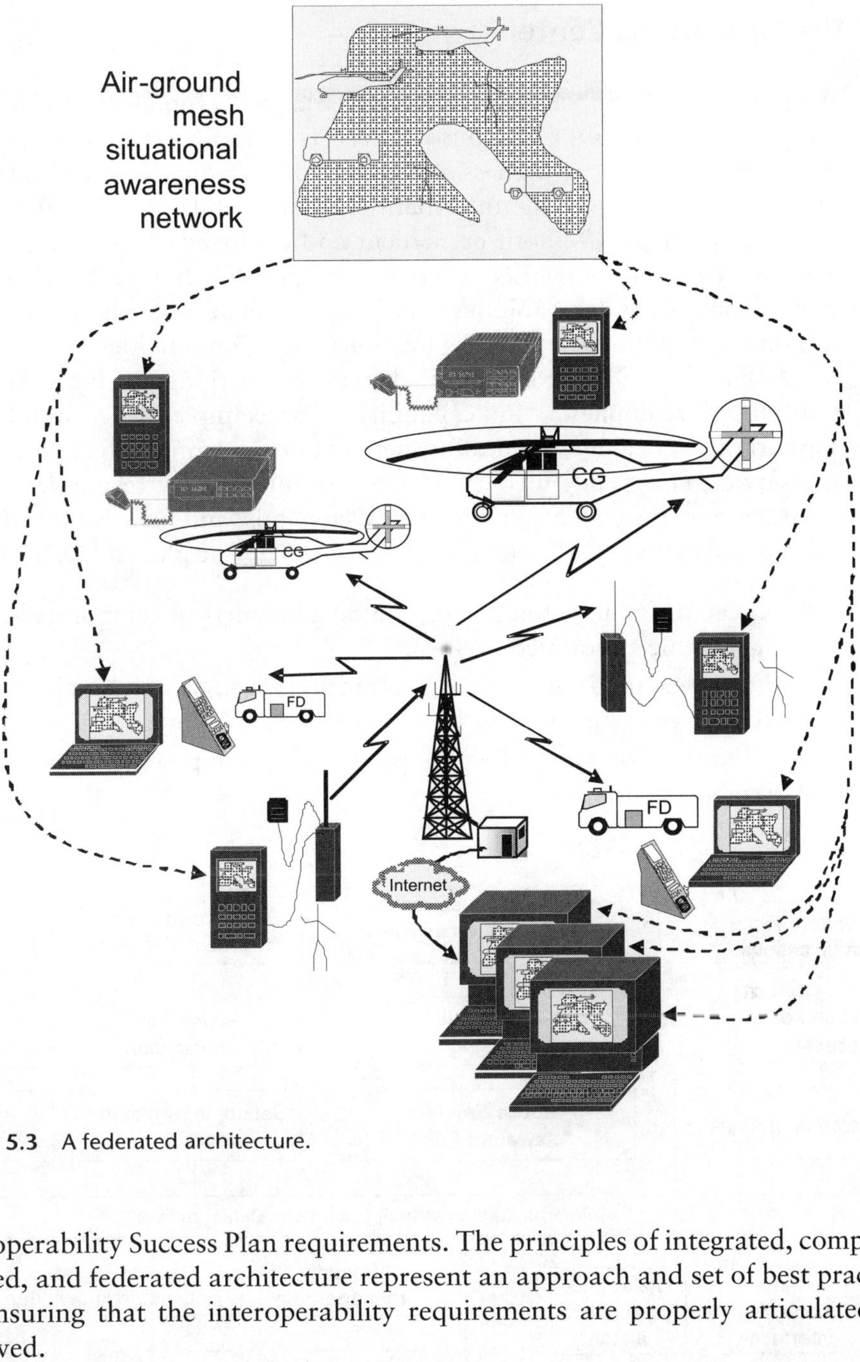

Figure 5.3 A federated architecture.

Interoperability Success Plan requirements. The principles of integrated, composite, aligned, and federated architecture represent an approach and set of best practices, for ensuring that the interoperability requirements are properly articulated and achieved.

"Interoperability architecture" refers to the proper application of these principles of integrated, composite, aligned, and federated architecture in order to assure that new systems satisfy the need to interoperate with a wide range of existing and future systems. Interoperability architecture addresses the need for future systems to interoperate with aging legacy systems, for different systems within a jurisdiction to interoperate cross-functionally with other systems in that jurisdiction, for systems in adjacent jurisdictions to interoperate laterally with one another, and for systems to interact vertically with other systems at all levels of governance.

5.2 The Operations Center

Whatever the discipline or level of governance, responders-receivers employ an Operations Center (OPCEN) of some type as a sort of consolidation or aggregation point for situational awareness and command coordination. The OPCEN at Pearl Harbor was effectively the Information Center, while for 9/11 and Katrina there were many OPCENs—both permanent and deployed. A generic public safety or emergency response OPCEN is shown in Figure 5.4. It arguably depicts all operational functions and capabilities needed by any general public safety or emergency response discipline to support its missions, tasks, or activities.

Although all OPCENs perform all of these functions at some level of sophistication, the figure implies a "full capability." For example, the general OPCEN could support the tactical interoperability needs of first responders (e.g., law enforcement, fire service, emergency medical service), supplementary responders and receivers (emergency management, public works, hospitals, and boards of health), or one or more of the many stakeholders listed in Table 3.1. The physical OPCEN could be:

- Located within a fixed (e.g., building, tunnel) or temporary structure (e.g., tent, vehicle-mounted shelter);
- Land, sea, or air deployable equipment sent to be a stand-alone capability, or replace or augment a fixed or temporary capability;
- Land, sea, or air deployable platform or equipment providing stand-alone remote capabilities;

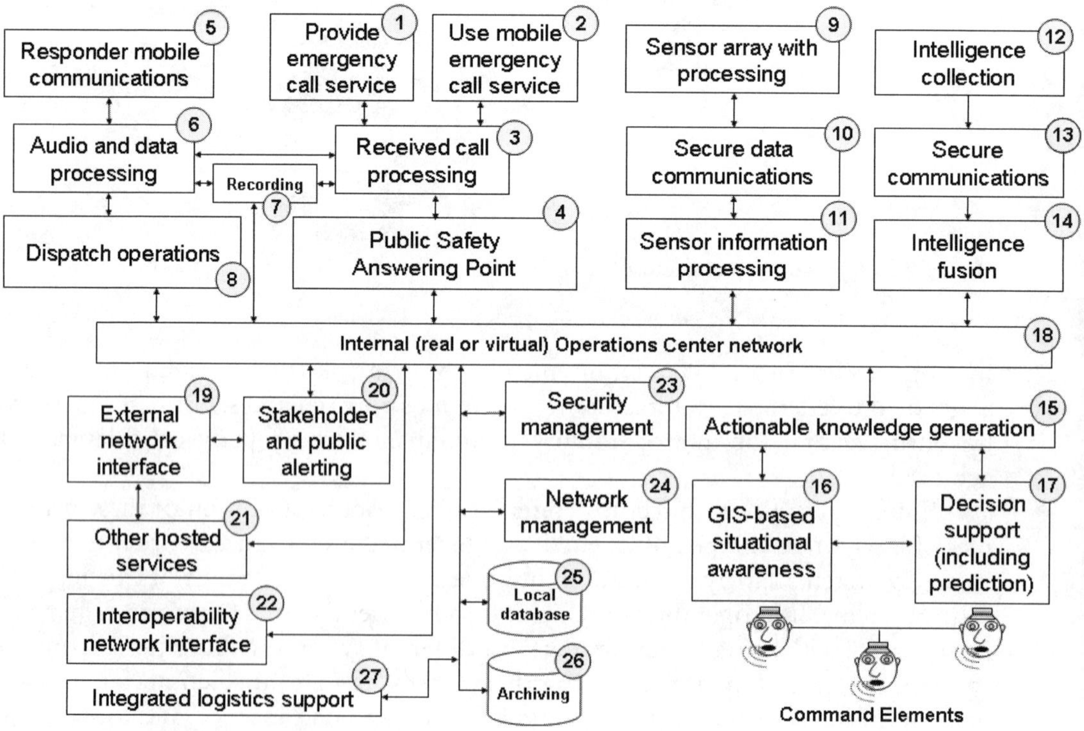

Figure 5.4 OPCEN functional architecture.

- Mobile platform (vehicle, ship, or aircraft) providing on-the-move capabilities;
- Virtual networked capability, where different functions are performed in different locations or aboard deployable/mobile platforms.

Referring to Figure 5.4, the individual functional elements numbered in the figure are defined as follows:

1. *Provide emergency call service:* The processing and routing of calls from the public for emergency service, including the routing of calls to available call-takers and retrieval.
2. *Mobile emergency call service:* The processing and routing of mobile (wireless) phone calls from the public for emergency service, including routing of calls to available call-takers.
3. *Received call processing:* The initial call taking, handling, and disposition of calls from the public for emergency service. Call processing includes assessment of the emergency, opening of an incident report, capture and tracking of incident data, and providing reassurance and guidance to the caller. In smaller operations centers, call processing and dispatch may be performed from a combined position. Included in this function is the capability for call transfer to a local public safety or emergency response dispatcher or to another PSAP.
4. *Public safety answering point:* Facilities equipped and responsible for answering emergency calls from the public for assistance. Normally associated with a local jurisdiction, such as a city or county. Typically, the calls are for law enforcement, fire, or medical emergencies.
5. *Responder-receiver mobile communications:* The various mobile communications functions for public safety and emergency response agencies to communicate with the responders-receivers under its command or coordination [8].
6. *Audio and data processing:* For audio, this processing includes removal of radio system signaling for extraction of the intelligence, formatting as required by the radio system, routing of information to the appropriate radio channels, and message switching services. Both audio and data processing may involve encryption.
7. *Recording:* Storing of audio of emergency calls from the public and radio voice calls or data exchanges with responders-receivers for immediate (e.g., several seconds of replay delay) or delayed access and archival storage.
8. *Responder-receiver dispatch:* Identification of available incident response assets, the dispatching of responders, continued incident monitoring and data capture, prosecution of emergency alarms from the field, supporting information requests from the field (where automated retrieval of the data is not available), and making/breaking of radio patches.
9. *Sensor array with processing:* The collection of raw activity or environmental information at one or more locations, such as audio, imagery, video, airborne chemicals, radiation, temperature, among others, and the possible preliminary processing of this data.

10. *Secure data communications:* The controlled and assured delivery of sensor information to the OPCEN.
11. *Sensor information (including video) processing:* Decompresses (and sometimes decrypts) and converts sensor data into usable formats. Screens received sensor data to identify occurrences of interest.
12. *Intelligence collection:* Intelligence information is collected from a variety of sources, including archived data, the Internet, electronic surveillance, reports (from other agencies), and human intelligence sources, among others. No analysis or processing is performed on this data other than by the collector in the case of a preexisting strategic alert, which would affect the priority (urgency) flagged in the data.
13. *Secure communications:* One or more media is used to communicate the captured data to provide information assurance between the collection and analysis functions. No data loss, disruption, interception or decryption is acceptable in this end-to-end transmission.
14. *Intelligence fusion:* The received information content from one or more sources is analyzed collectively to determine capability, intent, timing, and so forth, in the context of other available information (e.g., sensor data, responder-receiver reports, ongoing investigations, emergency calls).
15. *Actionable knowledge generation:* Filters, processes, and presents all available information into knowledge that can be used to determine suitable responder-receiver and OPCEN actions by the interrelated functions of situational awareness and decision-making.
16. *Geographic information system based situational awareness:* Converts actionable information into a format for effective and efficient presentation to command elements.
17. *Decision support (including prediction and forecasting):* Simplify and prioritize the decisions-making process for OPCEN command elements. In addition, predictive tools that assess theoretical consequences of hypothetical or real events.
18. *Internal network:* The internal information-sharing network within the OPCEN, able to share information among the many OPCEN functions.
19. *External network interface:* Provides access to diverse media both to deliver responder and public alerting and provide access to external hosted services.
20. *Stakeholder and public alerting:* The applications software and supporting servers that support responder and public alerting.
21. *Other hosted services:* Accesses services provided external to the OPCEN, which could include many of the functions already described in 1 through 20 above.
22. *Interoperability network interface:* Provides access to external information from other OPCENs and other sources.
23. *Security management:* Protection of the availability, confidentiality, and integrity of the associated information systems with the multilayered, end-to-end prevention, detection, and vulnerability-management solutions.

24. *Network management:* Activities, methods, procedures, and tools that pertain to the operation, administration, maintenance, and provisioning of networked systems [9].
25. *Local database:* Provides storage and retrieval of information resident at the OPCEN.
26. *Archiving:* The repository for data retained on a longer term basis useful in investigations, judicial proceedings, and training and exercises, or the recall of lessons learned in the development of operating procedures.
27. *Integrated logistics support:* Monitoring of the system architecture for status, including impending or actual equipment failures, and provides all sustainment functions including facility/equipment operations and maintenance, including operator and maintenance training, and critical parts ordering/sparing, among others.

Of course, not all OPCENs have every function depicted in the figure, or if they have the function, they may be implemented in many disparate ways. In its simplest form, the functions depicted by groups of OPCEN functions in the architecture represent the same functions performed by each responder-receiver during execution of their respective missions, tasks, or activities. For example, the public reporting enabled at the OPCEN by functions 1 through 4 in the figure, and clandestinely through functions 12 through 14 are often a natural outcome of responder-receiver interaction with the public.

The responder-receiver's equipment and monitoring devices, such as in-vehicle video, paramedic monitors, even their own senses, are reminiscent of the OPCEN's sensor capabilities in functions 9 through 11. The responder-receiver will subscribe to the mobile communications services represented in functions 5 through 8 using a portable or platform-mounted device (e.g., a vehicular, maritime, or airborne radio). They provide "on-the-ground" public alerting capability (e.g., loudspeaker, siren, speech, and even hand signals) as depicted in OPCEN functions 19 through 21. Situational awareness, decision support, recorded or archived information retrieval, and integrated logistics support (functions 15 through 17 and 25 through 27) could be provided remotely. This remote access would employ a personal communications device, mobile data terminal, or mobile computer, or at the very least, simply the capabilities of the responders-receivers themselves as guided by training and experience.

5.3 The Observe-Orient-Decide-Act Loop

The requirements for public safety and emergency response system architecture are derived directly from the operational architecture, which is a description of the tasks and activities, operational elements, and information exchanges required while accomplishing public safety and emergency response missions. These tasks and activities (see Figure 5.1)—performed by people, processes, and their tools planned with inherent interoperability—are best described as a set of well-defined, executed, and practiced actions dependent upon timely information sharing to achieve a common public safety and emergency response goal. Of course, these "actions" differ depending on the discipline, location, environment, and so forth. Rapid decisions must be made with incomplete information, and one important missing "bit" can have extensive consequences.

A radio call to a mobile unit to travel to a particular location, see a particular person, or respond to an ongoing reported incident, involves radio discipline using standard terminology in sequence that triggers the unit to respond. The predictable response is to move to the directed location to rapidly *observe* the situation, *orient* their unit relative to on-scene environment, *decide* what immediate actions to take, and then *act* as decided. Moment to moment, as soon as their action begins, the environment is affected, and so another cycle of observe, orient, decide, and act is necessary—perhaps spanning no more than a few seconds. Thus, the individual acts performed represent specific well-defined physical efforts triggered by the observe-orient-decide-act cycle, creating the classic observe-orient-decide-act (OODA) loop [10]. This OODA loop applies to all responder-receiver activities or tasks in all of their moment-to-moment, day-to-day, and long-term activities in all disciplines involved in public safety and emergency response. It provides a basic operational sequence effectively used to analyze all tactical operations in public safety and emergency response.

For example, the way in which a law enforcement officer interrogates a witness or suspect at the scene, the tone of voice analyzed for hints of deception, the physical methods used to restrain an assailant, these are all well-defined actions prompted by information sharing during their repeated cycles of the OODA loop. Paramedics observes the patient's condition from witnesses and vital signs for a rapid diagnosis, orient themselves and their tools relative to the patient, decide on what immediate steps are required to stabilize the patient, and then act as training and experience direct them. Fire units perform a similar cycle from their notification of an alarm (collecting on-scene information from an emergency call or dispatcher's relay of incident information). Upon arriving at the incident (structural or wildfire), they observe the scope and stage of a fire, orient themselves and their capabilities relative to the scene, decide how to deploy their equipment and personnel—including call for multiple alarms (backup)—relative to this observation and their training/experience, and then act (deploy) as necessary. Of course, these OODA-loops are being employed by each responder-receiver and his or her respective command elements on a continuous basis.

The OODA loop describes the basic cyclical activities of people (responders-receivers) who perform processes (e.g., activities, tasks, functions) with tools (e.g., assets, facilities, devices) to achieve a common goal. The people, and the processes they are trained and experienced to perform through their initiative, are the direct and most important element in public safety and emergency response—without them there is no response. It is therefore these people and the functions they perform that are of fundamental and primal importance in public safety and emergency response. The "tools" they employ, particularly the IT/communications information-sharing tools for interoperability, cannot be properly identified or procured for them without first having an in-depth understanding of these responders-receivers and the operations they perform. This necessity and the processes for achieving this understanding of their interoperability requirements were described in Chapter 4.

An approach applicable to operational best practice is based on a vision for incident response operations that range across all aspects of the public safety and emergency response community. Central to this vision is the concept of the OODA loop and the speed of decisions within the confines of the fastest OODA-loop cycles. The OODA loop is a central concept for public safety and emergency response operations because it describes the application of response with respect to the human or machine decision cycle. The objective for incident operations is to support effective decisions in a cycle that keeps pace or moves ahead of the threat environment based on fused observation sources, decision aids, and rapid dissemination of coordinated tasking—in other words, interoperability.

5.3.1 Observe

The observe stage of the OODA loop supports concepts for initiating action by all responders-receivers as well as their respective fixed and mobile OPCENs. There are several ways in which a responder-receiver or OPCEN operators, such as watchstander, dispatcher, operator, commander, and others would be prompted to take action in the observe phase. The observe stage employs function numbers (3, 4, 11, 16, 20) in Figure 5.4 to achieve the corresponding observe tasks and activities described in the following subsections.

5.3.1.1 Newly Launched Threshold Alerts

Event handling links the alert information from OPCEN inputs to operational task checklists, which are lists of tasks or activities triggered by an alert. Alert information can be provided directly by public emergency calls/notification in functions 1 through 4, indirectly via sensor array outputs in functions 9 through 11, and intelligence fusion outputs from functions 12 through 14. These checklists prompt the operator at the corresponding OPCEN position, including command positions associated with functions 15 through 17, to perform predefined and practiced tasks in a recommended sequence. In addition, they provide interoperable links to:

- Supporting tools, including external alerting systems in functions 19 through 21;
- Validated data-element transformations, for example, translating place names into latitude and longitude, possibly in functions 4, 8, 11, 14, 15, and, certainly, 16;
- Predefined information-sharing profiles, such as identifying a set of organizations to alert given OPCEN inputs in functions 15 through 17.

Deficiencies in operations (see Chapter 2) can be overcome by this transition to automated checklists as a key enabler of operational interoperability.

5.3.1.2 Monitoring and Correlation

Monitoring tools support receipt and categorization of structured data sources and information streams of interest, such as functions 4, 11, and 14. Events such as execution of a specific action checklist can be triggered based on user-defined rules complemented with correlation technologies, such as those technologies needed in functions 14 and 15. Correlation technology leverages semantic modeling layers that provide the operator a way to draw relationships between mission classes, threat classes, and other 'trigger' classes. This capability is supported primarily by operators in the information management (functions 4, 8, and 15 through 17) and intelligence cells (functions 12 through 14), but it may also be performed automatically.

5.3.1.3 Scheduled Watchstander's Round

This activity is dependent on a disciplined approach for reviewing status of key observation tools on an hourly (or other routine time-schedule) basis, or having them monitored by software for automatically generated OPCEN alerts. There are several categories of components that should be reviewed during these rounds, including (functions 9 through 11, 15, and 17), as described below.

Information Channels The information channels can leverage portal technologies and should be configured to use predefined channels (see functions 18 through 22). These channels, such as, Internet or Intranet portlets, should be configured to access information related to Web sites for indications of new threats or provide amplifying information associated with an existing emergency. Most of the tools in this category provide the ability to adjust refresh rate from information sources such as terrorism advisories, infrastructure protection advisories, intelligence community advisories, civil support message traffic, prevailing weather conditions, sensor feeds for critical infrastructure, and other information sources.

Near Real-Time Tracking Data Tracking data refers to various moving objects such as trucks, aircraft, ships, among others that can be kept track of using a variety of sensor equipment including radar, transponders, and so forth. A track map is a mapping application, similar to popular Web-based satellite-image mapping service maps, that is associated with this class of functions. It provides the capability to:

- Monitor air, ground, and sea tracks of responder-receiver mobile assets (see functions 2, 5, 9, and 12) or threats, such as wind-driven toxic plumes;
- Select mobile platforms of all types;
- Track individuals (functions 9 through 11) involved in geographic-based response;
- Understand the organizations that would be involved in follow-on collaboration (functions 8, 15, 17, and 22).

Collaboration and Postings Functions 19 through 22 provide the capability to collaborate synchronously between organizations as well as to provide access and ability to review new information files posted to designate virtual information sources, such as *virtual rooms* (e.g., protected Web share-space access), and the assignment of virtual rooms with associated roles and responsibilities for the rooms.

5.3.1.4 Other Sources and Indicators

Other sources or indicators may include radio, phone, sensor alarms, and others. As the operator becomes aware of new relevant information, this information should be submitted into an event log that includes sharing and dissemination capability. Operators review the shared-event log on a scheduled routine basis.

5.3.1.5 Orchestrated Observation

Automated checklists using orchestration can provide the operator or watchstander with machine-generated alarms and automatic direction based on predefined and automated process flows. These flows would be threshold, time, or exception triggered to push them to the head of the situational awareness queue.

5.3.2 Orient

The orient stage of the OODA loop supports actions in response to new events as well as consequence analysis and assessment. Threat emergence and emergency events are examples of events that trigger an operator response. The four key capability components that support the orient stage are described below.

5.3.2.1 Checklists

As described above, checklists are used to help orient the operator with action sequences to follow with respect to given events. All event types can be linked with user-defined checklists. Some system architecture tools, such as integrated situational awareness visualization displays in computer-aided dispatch systems or incident management software, provide the ability to color-code actionable events and push them to the front of the situational awareness presentation queue. The process of gathering and updating checklists is aided by operator drills and after-action reviews of real events or exercises.

5.3.2.2 Tactical Visualization

Geographic information system (GIS) maps provide the means to retrieve images, add symbology and overlays, determine geo-registered positions (i.e., located on GIS map) and save resulting products with keywords to a choice of servers or local disk drive. These tools help orient the operator by providing a means to not only retrieve geospatial products, but also receive amplifying symbols. For example, mobile units may be displayed as representative vehicle images on the GIS map using automatic vehicle location. By including standardized symbols with the vehicle icon on the GIS map, the vehicle's status, current task or activity, occupant health, and so forth may also be displayed. Depending on the OPCEN role, such as local versus regional versus state or national OPCENs, the operator may then:

- Share the image via collaboration tools to gather additional event details and contribute to shared orientation with other operators participating across the network.

- Use geo-registered, synthesized products to support mobile responders-receivers with dispatch and amplifying information, such as cordon anchor points, alert boundaries, exclusion areas (e.g., due to potential explosion or drifting plume).

The GIS display contributes to both orientation of the personnel within the mobile units and the OPCEN operators, because the information displayed by the vehicle symbol additions greatly improve situational awareness. This awareness is enhanced at the OPCEN from data captured by mobile sensors, visual reports, mobile unit status (including mobile unit personnel health status), waterborne and airborne toxin levels, critical infrastructure status, and representation of eyewitness accounts (e.g., apparent location of a launched distress flare). In addition, they reduce the need to tie up voice communication channels with location and other status chatter best left to mobile data communication systems, whenever possible.

5.3.2.3 Responder-Receiver Security Planning

These tools provide the means to visualize mobile units and assets, including logistics and supply units, which will flow through the OPCEN's area of responsibility. This function supports query and visualization of anticipated high-priority assets, such as hazmat units, food/water shipments, radio infrastructure, and so forth, which may require a heightened and coordinated security posture (pushing evolving event information to collaborating OPCENs). These tools also enable the operator to predict and visualize significant increases in vehicular and foot traffic concentrations, which may require additional coordination of traffic control units or control over traffic lights supporting public safety and emergency response.

5.3.2.4 Collaboration

Collaboration tools (see functions 4, 8, 11, 14, 15, 16, 17, 18, and 20) support several features that can help the operator become oriented with a developing situation. Preplanned assignments are posted to predefined virtual response support rooms and regional operation center virtual rooms in advance. In this way, operators then routinely visit assigned rooms to learn about scheduled collaboration conferences in support of predefined events, tasks, and activities and their corresponding schedules. The OPCEN would coordinate these schedules.

Collaboration tools navigation capabilities are provided to simplify the operator's understanding and use of orient functions. They provide the means to view the availability of representatives from collaborating units for coordination, direction, guidance, and so forth on an ad hoc basis. General guidance should be posted and updated to predefined virtual rooms to keep pace with the developing situation. Operators should review these sites frequently (e.g., hourly) during a developing situation to support the dynamic nature of response operations. Of course, automatic triggers could be established for changes of room content, either through general alerts or alerts generated by screening arriving messages for certain terms.

5.3.3 Decide

The decide stage of the OODA loop supports concepts for fusing information, structuring courses of actions, providing decision aids, and enabling the rapid dissemination of decisions. Delegation of decision authority is accomplished within the framework of interorganizational structure. Each OPCEN can assist the decide stage in collaborative action with definition of relationships and access to the associated situational awareness and virtual room repositories. Tools support decisions made at lowest command echelons possible, with higher level echelons given real-time insight and ability to override decisions on a by-exception basis. Capability components supporting this stage are decision support (function 17), tactical visualization (functions 8 and 16), collaboration (functions 4, 8, 11, 15–18, and 22), and asset management (functions 16 and 27).

5.3.3.1 Responder-Receiver Security Planning

In addition to capabilities described in the observe stage, this class of functions (11, 14, 15, and 20) also provides the capability to perform what-if analysis in support of consequence assessment and management. These tools include such capabilities as the analysis of unit hierarchy, enabling the operator to visualize the feasibility of organizing and integrating support units. These support units may include medical, logistics/supply, transportation units, and public relations, that assist local/regional resources responding to a disaster or emergency.

Responder-receiver security planning tools in a situational awareness display could include a sliding time bar that causes a unit's GIS-displayed color to change highlighting the feasibility of using that unit within the time available. Once final decisions are made regarding support to augment local or regional units, the information can be auto-exported to dispatchers and their supervisors via reporting/briefing tools, disseminated by posting to applicable collaboration virtual rooms, and perhaps delivered directly to the dispatchers themselves.

5.3.3.2 Collaboration and Command Coordination

Collaboration tools such as those represented by functions 4, 8, 11, 15–18, and 22 provide the means for decision-making conferencing and rapid dissemination of decisions. This process is improved through standard operating procedures published in a Web-based virtual room, including assignment of decision-component staff to virtual rooms, conferencing schedules, response product scheduling and briefing times.

5.3.4 Act

The act stage of the OODA loop supports rapid execution of decisions. Capability components supporting this stage are defined in the following subsections.

5.3.4.1 Tactical Visualization

These tools include functions 8 and 16, and can be used to support real-time execution response. These tools include geo-registering capabilities that enable the operator to determine coordinates of underlying maps, buildings (and their interiors), images, overlaid video, and assets. When combined with the means to collaboratively apply planning overlays such as cordon areas and tactical action symbols, the operator can use tactical voice or data communications to pass such execution data as cordon area anchor points and control lines to GIS-equipped mobile or on-foot responders-receivers.

5.3.4.2 Collaboration

As described earlier, collaboration tools (functions 4, 8, 11, 15–18, and 22) provide the means for both real-time planning and execution collaboration as well as rapid dissemination of plans and decisions. Real-time planning and execution is primarily supported using synchronous collaboration tools as shared applications, such as a digital whiteboard. This capability can be used to share such data as images, audio and video clips, planning overlays, live track pictures (of high-interest tracks), blast and plume predictions, and other artifacts. Rapid dissemination of plans and decisions are supported by asynchronous capabilities offered by collaboration virtual rooms, with associated storage space for uploading and downloading such items as maps and images with planning overlays, response plans, and so forth.

5.3.4.3 Assessment

As an adjunct to the act stage of the OODA loop, assessment capabilities actively seek to understand how well the selected course of action delivered the intended results. This analysis is supported by tools that provide both physical and functional assessments of the intended outcome built into functions 8 and 17. Additionally, results of analysis and assessment provide an additional feedback mechanism supporting further observation.

5.3.5 Dynamic Decision Model

Public safety and emergency response tasks and activities are characterized by uncertainty and risk—and no two incidents are the same, so each requires a different collection or sequence of these tasks and activities. In this regard, a decision-making model derived from Figure 1 of the United Kingdom *Fire Service Manual* is depicted in Figure 5.5 [11]. The figure shows the decision-making model with the corresponding OODA-loop phases. It is apparent that these two decision-making models are similar. A description of the model in its application to incidents—which is inherent in the OODA loop—is that [12]:

> it must be stressed that the use of the Decision Making Model should be continuous throughout an event. In the early dynamic stages it may be used many times and as the event becomes more controlled and less dynamic the model may be used less frequently.

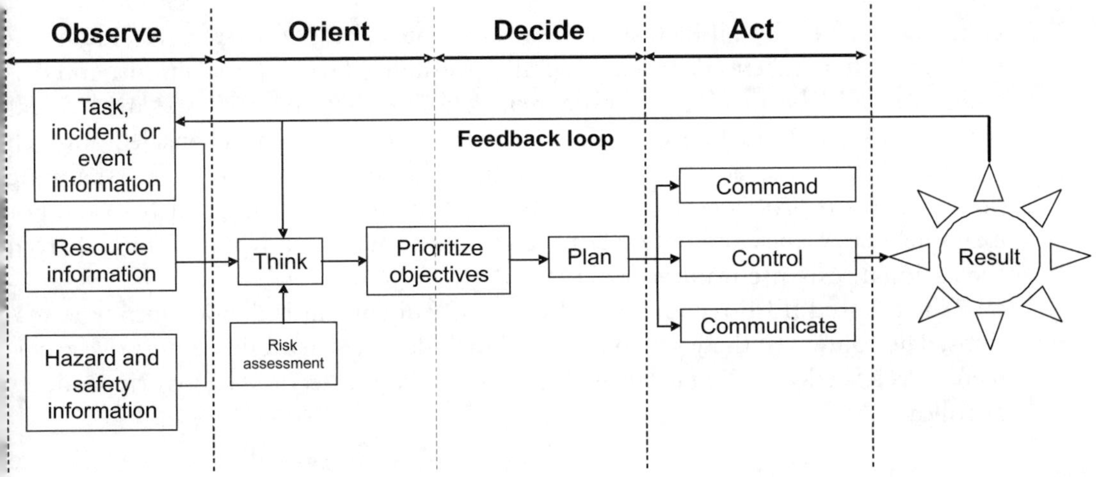

Figure 5.5 Decision-making model and the OODA loop.

This model represents the flexibility of action required by all responders-receivers and other stakeholders in dynamic incident situations. It must be effectively, efficiently, and continuously supported by the operational and system architecture developed through use of the best practices described in Chapter 4. Thus, support can no longer consist of disconnected information-sharing island stovepipes as depicted in the World Interoperability Failure model of Chapter 3. Instead, it must achieve a new dimension as a federated architecture for all responders-receivers and other public safety, and emergency response, stakeholders. To this end, plans must be developed that integrate the OODA-loop elements to best achieve this architecture interoperability environment.

5.4 Concept of Operations

Before one or more OPCENs can collaborate in a specific event, emergency, or disaster, they must establish a common understanding of *who will do what and when* for one or more hypothetical event, emergency, or disaster scenarios. This understanding must exist among public safety and emergency response organizations at every governance level that would collaborate in each hypothetical scenario, while preserving essential tactical flexibility, while ensuring interoperability. The collaborative regional approach to prevent, detect, protect, alert, respond, and recover to and from natural and manmade threats requires an overall concept of operations (CONOPS) involving all governance levels, where CONOPS is defined as "a user-oriented document that describes a system's operational characteristics from the end user's viewpoint. Synonym: operational concept description" [13]. The CONOPS then serves as a source for the development of coordinated standard operating procedures for each individual OPCEN using a standard template.

Thus, OPCENs at all governance levels should define an overall CONOPS for each scenario in which they would become involved. The best practices described in Chapter 4 must be employed in the establishment of an all-inclusive and comprehensive standard template-based CONOPS document as a specific project at each governance level. By all-inclusive, we mean that all responder-receiver organizations and other stakeholders spanning all governance layers, jurisdictions, and disciplines should be involved, if even *remotely* relevant, for each specified threat scenario (to avoid PHD #3 Omission). In this regard, the comprehensive approach requires that we consider a discrete number of representative scenarios spanning *possible*, versus *probable* (to avoid PHD# 18), natural and manmade threats. For example, review the candidate set of such threat scenarios shown in Table 5.1 from the national response framework [14].

The tailored IEEE-recommended CONOPS document is divided into nine sections. The content of these sections should include interoperability enterprise architecture (see Section 4.5) and business case analysis (see Section 4.6) work products as follows:

1. Scope (AV-1).
2. Referenced documents.
3. Current system or situation:
   - 3.1. This section addresses the current system background (objectives and scope), operational policies and constraints, description of the current system/situation (OV-1);
   - 3.2. Modes of operation for the current system (as-is OV-5/SV-4), user classes and other involved personnel (as-is OV-2/OV-4), and the support environment.
4. Nature of changes and their justification:
   - 4.1. Description and justification for changes (gap analysis), upgrades, or replacement, priorities of changes (see BCA Assess phase, Section 4.6.2.1 and Figure 4.16);

Table 5.1 Relationship of Scenario Sets to Planning Scenarios

Key scenario sets	*National planning scenarios*
1. Explosives attack—bombing using improvised explosive device (IED)	Scenario 12: Explosives attack—bombing using improvised explosive device
2. Nuclear attack	Scenario 1: Nuclear detonation—improvised nuclear device
3. Radiological attack—radiological dispersal device	Scenario 11: Radiological attack—radiological dispersal device
4. Biological attack—with annexes in *FEMA NRF* for different pathogens	Scenario 2: Biological attack—aerosol anthrax Scenario 4: Biological attack—plague Scenario 13: Biological attack—food contamination Scenario 14: Biological attack—foreign animal disease
5. Chemical attack—with annexes in *FEMA NRF* for different agents	Scenario 5: Chemical attack—blister agent Scenario 6: Chemical attack—toxic industrial chemicals Scenario 7: Chemical attack—nerve agent Scenario 8: Chemical attack—chlorine tank explosion
6. Natural disaster—with annexes in *FEMA NRF* for different disasters	Scenario 9: Natural disaster—major earthquake Scenario 10: Natural disaster—major hurricane
7. Cyber attack	Scenario 15: Cyber attack
8. Pandemic influenza	Scenario 3: Biological disease outbreak—pandemic influenza

4.2. Changes assessed but not recommended.
5. Concept definition for the proposed system (see Section 4.8.3 and Figure 4.22, and Subsection 4.5.5.2 for OV definitions):
   5.1. Background (objectives and scope), operational policies and constraints (AV-1);
   5.2. Description of the proposed system (to-be OV-1), models of operation (to-be OV-5 and SV-4/SV-5);
   5.3. User classes, other involved personnel (OV-2 and OV-4), and the support environment.
6. Operational scenarios:
   Using threat scenarios from Table 5.1, show to-be OV-1, OV-2, OV-4, OV-5 and OV-6c with SV-4 or to-be SV-5 for each scenario.
7. Summary of impacts:
   7.1. Operational impacts, organizational impacts (to-be OV-4);
   7.2. Impacts during development (the as-is/to-be transition).
8. Analysis of the proposed system (to-be OV-3):
   Summary of improvements, disadvantages and limitations, alternatives and trade-offs considered (see the business case analysis Assess phase in Section 4.6.2.1 and Figure 4.16).
9. End notes.

This CONOPS development project would be developed using the best practices described in Chapter 4, including creation of the interoperability enterprise architecture (see Section 4.5)—rolled up from all jurisdictions within each governance level. The CONOPS document would therefore include the specific detailed information exchange requirements documented as OV-3s in the interoperability enterprise architecture (see Subsection 4.5.5.2). These information exchange requirements define the critical and specific information sharing that drives the individual tasks and activities in each OPCEN, while preserving tactical OODA flexibility for the responder-receivers and other stakeholders in the field. The coordination of these scenario-dependent tasks and activities within and among responders-receivers and their respective OPCENs is formalized through the best-practices development of OPCEN standard operating procedures.

5.5 Standard Operating Procedures

5.5.1 Definition

Whether automated or manual, flexibility of action (adaptability, remember PHD #13 Adaptability from Chapters 2 and 3) in tactical situations is essential to their success as well as the safety of the responders-receiver involved, because *every incident is different*. No two house fires, traffic stops, heart resuscitations, power line outages, suspicious packages responses, or catastrophic evacuations are the same; each involves different individuals, physical or environmental constraints, weather conditions, natural or manmade threats, and so forth. For this reason, the details of each set of tasks or activities performed by a responder-receiver or other public safety and emergency response stakeholder vary with each incident or event. These tasks and activities must be dependent on the tasks and activities of collaborating OPCENs in the CONOPS (see Section 5.4). However, each task or activity is composed of OODA-loop cycles in which proven methods or techniques—largely for individual responders-receivers to perform—in observation, orientation, decision, and action can be identified, standardized, trained, executed, and the results used to provide feedback for improvement. The difference between these techniques and standard operating procedures (SOPs) has been explained as [15]:

> According to the National Fire Protection Association (NFPA), a standard operating procedure is "an organizational directive that establishes a standard course of action." In other words, SOPs are written guidelines that explain what is expected and required of fire service personnel in performing their jobs. A comprehensive set of SOPs defines in significant detail how the department intends to operate.

SOPs don't describe how to do the job (technical skills), they describe the department's rules for doing the job (procedural guidance). An example might help to clarify this point. Operating an emergency vehicle requires both technical skills and procedural guidance. Some possible components of each are listed below:

Technical Skills	*Procedural Guidance*
Location of vehicle controls	Use of seat belts
How to activate warning systems	Permitted uses of warning systems
Operation of communications equipment	Communications protocols (e.g., radio discipline)
Vehicle driving skills—(accelerating, braking, turning, etc.) speeds, passing other vehicles	Procedures at intersections, permissible
Parking skills	Vehicle placement, use of traffic cones

Other important examples of technical skills in communications would be manually selecting specific trunked channels or frequencies on a specific portable or mobile radio; selecting organizations on a situational awareness display to connect into a common ad hoc talk group [16]; locking the stanchions on a portable antenna tower and using the pneumatic pump to elevate the antenna; or the specific protocol (i.e., exchange of specific words or phrases) used to achieve radio discipline on a voice channel. All of the functions described for the OPCEN in Figure 5.4 include similar technical skills, which are increasingly automated to gain the benefits of assured action while avoiding deficiencies of relying on human initiative whenever possible.

In other words, the "technical skills" describes *how to physically perform* specific actions in each phase of the OODA loop; these actions do not form the content of standard operating procedures. A standard operating procedure governs *if and when* one or more of such technical actions are appropriate or required within the scope of a common task or activity, where these tasks and activities involve [17]:

- Expanding organizational missions—emergency medical care, hazardous materials response, technical rescue, fire prevention/public education, and terrorism incidents.

- Increasing legal and regulatory requirements—safe work practices, public and employee right-to-know, equal opportunity (race, gender, age, disability), performance standards, employee relations, and much more.
- Increasing complexity in emergency response techniques and equipment—personal protective measures, chemical safety, infection control, building and industrial codes, information management, training systems, and so forth.
- Increasing coordination and reporting requirements with other groups [interoperability]—emergency response agencies, community managers and planners, mutual aid organizations, federal and state governments, member associations, and others.

The SOP will reference skills, best practices, or standards of practice [18]. For these reasons, SOPs are increasingly necessary to provide [19]:

- Explanation of performance expectations—SOPs describe and document what is expected of personnel in the performance of their official duties. As such, they provide a benchmark for personnel, an objective mechanism for evaluating operational performance, and a tool for promoting a positive organizational culture.
- Standardization of activities—SOPs identify planned and agreed-upon roles and actions. This information helps standardize activities and promote coordination and communications among personnel. SOPs also simplify decision-making requirements under potentially stressful conditions.
- Training and reference documents—Written SOPs can provide the framework for training programs, member briefings, drills, and exercises. These activities, in turn, improve the understanding of work requirements and help identify potential problems. A comprehensive SOP manual also serves as a self-study and reference document for personnel.
- Systems analysis and feedback—The process of researching and developing SOPs provides opportunities for managers to compare current work practices with the state-of-the-art in the emergency services. Feedback from outside groups, technical experts, and staff can help to identify potential problems and innovative solutions.
- External communications—SOPs clarify the department's operational philosophy and recommended practices. As such, they may prove useful in communicating organizational intentions and requirements to outside groups, or enhancing the public's understanding of the ... [organization].

Formally invoking SOPs through memoranda of understanding, standard operating procedures are critically important as they are the embodiment of trust among collaborating organizations in the form of predictable collaborative action. Thus, great care is taken in their development, that is, tailored use of the best practices described in Chapter 4.

5.5.2 Needs Assessment

Thus, standard operating procedures govern "external communication" needed to provide "increasing coordination and reporting requirements" with other organizations. These standard operating procedures govern the tasks and activities performed in the OPCEN as well as in the field and imply an integrated combination of manual (e.g., human actuated) and automatic actions. For interoperability, they define the types of information to be shared, the protocols with which to share that information, and the order of steps/actions required to share it within the domain of one or more regional threat scenario-based CONOPS. Standard operating procedures then define what information will be shared, how it will be shared, that is, the mutually agreed upon procedures for transmission, reception, acknowledgement, and verification.

Applying the Interoperability Assessment Checklist from Chapter 3 to our World Interoperability Failure model (step 1 in Figure 5.6) would demonstrate a significant lack of scenario-based information sharing among public safety and emergency response organizations and other stakeholders at all governance levels. Increased information sharing will require achieving public safety and emergency response interoperability culminating in an interoperability enterprise architecture spanning jurisdictions, disciplines, governance levels, and possible manmade and natural threats. The interoperability enterprise architecture (Section 4.5) defines information exchange requirements (see Section 4.5.5.2) that detail the trusted coordination of the people, processes, and IT/communication system tools in the OPCENs (see Figure 5.4) of collaborating organizations. These information exchange requirements define the specific information transmitted and

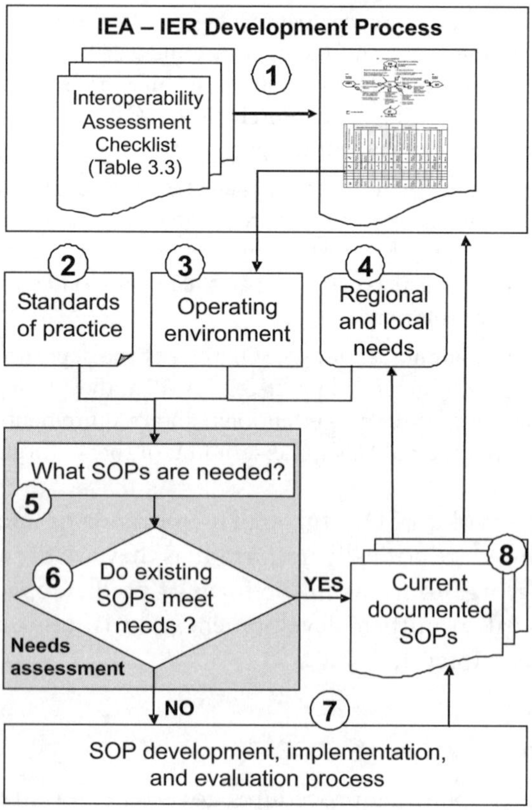

Figure 5.6 Standard operating procedure assessment process.

received by each OPCEN based on the tasks and activities it performs. Therefore, these information exchange requirements drive the needs assessment for one or more standard operating procedures with built-in interoperability for the participating OPCENs.

The questioning process used to examine the operating and legal environment probably identified some areas that are not subject to outside authorities where SOPs need to be changed. Laws, regulations, and standards are very useful, but they cannot possibly address every local need. Examples of department-specific or community-specific factors that might prompt the modification of existing SOPs include [20]:

- Target hazard occupancies (e.g., factories, nursing homes, hospitals, pipelines, tank farms, railroads, scrap tire piles, high-rise buildings, etc.);
- Unusual road conditions (e.g., load-limited bridges, steep hills, major road construction, etc.);
- Mission (e.g., coastal or water rescue responsibility, location along airport flight path, proximity to nuclear power plant or industrial site);
- Prevailing weather conditions (e.g., subject to tornadoes/hurricanes, extreme hot/cold weather, etc.);
- Significant increase or decrease in incident volume;
- Staffing increase or decrease;
- Changes to community plans or mutual-aid agreements.

Perhaps the most common reason for initiating a review of SOPs is a bad incident outcome, such as a firefighter injury or death, an EMS patient injury or death, or a catastrophic property loss.

Referring to Figure 5.6, OPCEN coordination is developed for both natural and manmade threat scenarios (step 3) to the population and critical infrastructure in the associated geographic region (step 4) using best practices (step 2). This development (step 5) will lead to the identification of which standard operating procedures should be created or updated by each public safety and emergency response organization with other stakeholders. In general, these standard operating procedures will define the time, authorization, sequential actions, and methods to be used in ensuring successful interoperability while performing the requisite tasks and activities. The standard operating procedures are developed in coordination with all collaborating organizations and stakeholders for each threat scenario through the CONOPS developed for each threat scenario.

Once the regional interoperability enterprise architecture has been developed—preferably founded upon national and multinational architecture—individual organizations and stakeholders are then capable of defining the scope and capabilities required in their individual standard operating procedures. These standard operating procedures would follow a common national, or at least regional, template using standard terminology and interoperability enterprise architecture information exchange requirements derived from regional planning efforts. These information exchange requirements (depicted in the example OV-3 shown in Figure 4.12) would be defined for the same manmade and natural threats used to develop the CONOPS (see Section 5.4). For each scenario, planners would produce integrated information sharing requirements among all other relevant public safety and emergency response organizations in the region. The processes for defining standard operating procedures to be created have been defined as follows:

- *Step One: Develop organizational support for conducting the needs assessment.* Support can take many forms, from providing appropriate resources (personnel, time, meeting space, etc.) to demonstrating the commitment ... managers to make changes based on the results of the study. The purpose of the needs assessment should be clearly stated and understood by all members of the department.
- *Step Two: Develop a plan of action.* Assign responsibilities, divide tasks, and create a realistic schedule. Conducting a thorough needs assessment can be an intense and time-consuming process. Creating a plan of action helps maintain interest, promote accountability, and help the group stay focused.
- *Step Three: Review current SOPs.* Develop familiarity with existing SOPs. Look for previously identified problem areas and inconsistencies among SOPs. Consider potential impacts of the department's operating environment on existing SOPs.
- *Step Four: Gather information on internal factors affecting SOPs.* Examples include other department documents (see Chapter 2), post-incident analyses, exercises, drills, surveys, and interviews.
- *Step Five: Consider external factors affecting SOPs.* The impact of laws, regulations, and standards should be considered here. Information sources may include literature reviews, professional and trade organizations, government agencies, other departments, accreditation manuals, and the Internet.
- *Step Six: Develop a list of required SOPs based on the needs identified in steps 4 and 5.* Consider the impact of internal and external factors, as well as the operating environment and standard of practice.
- *Step Seven: Analyze existing SOPs based on the information gathered in prior steps.* Compare existing SOPs to the list developed in step 6. Identify areas where new SOPs are needed, or existing ones should be modified or deleted. Realistically prioritize the results in light of department resources. SOPs required by laws or regulations must receive a high priority, as should those addressing critical health and safety issues.
- *Step Eight: Create a formal needs assessment document.* Summarize findings and provide recommendations. Explain priorities and provide a specific rationale for adding, modifying, or deleting SOPs. [21]

Thus, this best practice for developing standard operating procedures refers to the use of organizational development expertise (see Section 4.3) and development of detailed operational requirements in a needs assessment. This needs assessment is similar in scope and intent to the *requirements elicitation and concept definition* (step A) shown in Figure 4.22 of Section 4.8. It provides the detailed specification for the necessary standard operating procedures to be created for a public safety and emergency response organization, and it is founded on information exchange requirements developed from the regional interoperability enterprise architecture and CONOPS for each threat scenario.

5.5.3 Developing Standard Operating Procedures

The implementation of standard operating procedures by OPCEN personnel, mobile responder-receivers, and other stakeholders will vary considerably, including: a physical checklist taped to a wall next to their radio console; electronically presented to them by their computer-aided dispatch system or emergency management software; or automated in the orchestration application of an advanced situational awareness and decision-support capability. Whether or not tasks and activities performed in the field or in an OPCEN involve a "human in the loop," it is necessary to standardize the sequence of specific steps to be taken for a variety of scenarios represented in the regional interoperability enterprise architecture.

No matter the form of implementation, the standard operating procedures will be developed from specification based on regional threat-based interoperability enterprise architecture and derived information exchange requirements. Given these information exchange requirements, a standard operating procedures development process [22] can be tailored to ensure the necessary interoperability. This process, depicted in Figure 5.7, is driven by the need to improve holistic interoperability and avoid the expanded Pearl Harbor–9/11–Katrina deficiencies in the Interoperability Assessment Checklist. This approach employs the best practices of Chapter 4.

5.5.3.1 Step 1: Select Project Manager and Establish Integrated Product Teams

A responsible project leader/manager must first be selected. This project manager will be skilled in project management best practices, and establish a project management framework for the standard operating procedures development work [23].

> Within the team, a variety of project management techniques can be used to monitor and facilitate the process. These range from the use of specialized software appli-

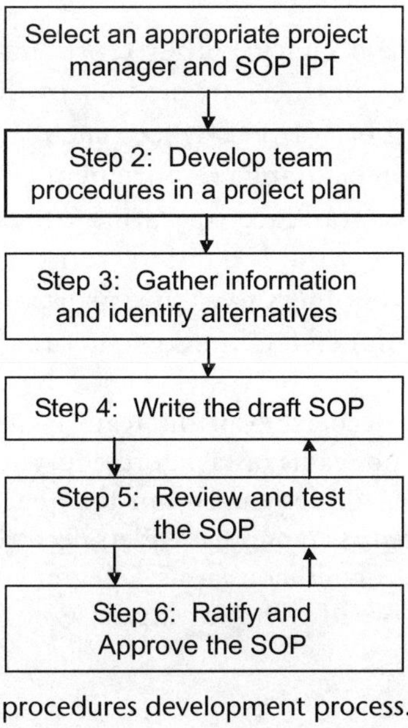

Figure 5.7 Standard operating procedures development process.

cations to more "low-tech" solutions like checklists, milestone charts, process diagrams, and flowcharts.

In addition to proper project management oversight for the standard operating procedures development project, organizational development facilitation can be used to help a diverse the project team build consensus and ensure organizational buy-in for their results.

This organizational development facilitation would establish a governance structure for the project and employ elements of change management to support the participants in this process. As shown in Figure 5.7, the project manager can employ organizational development expertise (see Section 4.3) to establish the project team. Members of this team could include [24]:

- Operational or line emergency responders, who carry out SOPs and have the greatest stake in their success;
- Safety program representatives, such as the health and safety officer, incident safety officer, operational medical director, department physician, and safety and health committee members;
- Managers, who have a broad organizational perspective that includes budgeting, public relations, political considerations, and other factors;
- Representatives of labor unions and employee associations, who understand contractual issues, articulate the viewpoints of their members, and provide insight on alternative procedures;
- Mutual-aid and regional response agency representatives, to help integrate and coordinate multijurisdictional responses;
- Representatives from other local government agencies that may interface with department operations, such as a third-service EMS agency or municipal health agency;
- Legal counsel or other personnel familiar with the content and implications of related laws, regulations, and standards.

Depending on the scenario, other representatives from the many stakeholders listed in Table 3.1 would be participants in the project team as well. Integrated product teams would be formed, where each team is assigned a specific functional area (e.g., information exchange requirements integration, command coordination, situational awareness, standard operating procedure draft writing, etc.) Each integrated product team would contain representatives of all relevant responders-receivers and stakeholders to ensure proper domain knowledge with established organizational governance facilitated structure [25]:

> The structure and leadership of the team is also important. Possibilities include assigning special sub-teams or task forces for one-time studies, and creating standing sub-committees for designated functional areas (e.g., health and safety, EMS). Personnel appointed as committee leaders need good interpersonal and group facilitation skills to help build consensus, integrate competing concerns, and facilitate decision-making. Broad technical expertise in a range of functional areas is also helpful.

Ideally, one or more of these participants would have been involved in establishing the regional interoperability enterprise architecture and associated scenario-dependent information exchange requirements for the region.

5.5.3.2 Step 2: Develop Team Procedures in a Project Plan

The organizational development facilitation support can help the project manager develop the functional or operational plans for the project team [26].

> Committee members should quickly establish the "rules of the game." In particular, individual roles and the mechanics of decision making—voting, negotiation, consensus-building, delegation, comment procedures, etc.—must be determined and communicated to all participants. Related considerations might include meeting dynamics, quorums, rights of appeal, representation of minority opinions, and other procedural aspects of the group decision-making process. Team members also need to understand the process for implementing the new SOPs. For example, does the chief or agency head have authority to alter the SOP without further input from the team?

The organizational development facilitator will make sure none of the deficiencies impact the SOP-development process, which is reminiscent of the business case analysis (Section 4.6) and systems engineering (Section 4.8) processes in the standard best-practice sequence of develop requirements, define alternatives meeting requirements, recommend and vet a preferred solution.

These procedures cannot be properly developed without a common standard operating procedure template developed at the multinational or national governance levels—in order to ensure standard operating procedure interoperability at all governance levels. Developed using best practices, a preliminary common standard operating procedure template has been recommended [27]:

> Regardless of their intended use, however, several items are usually included in any SOP:
>
> - Numbering system—Important for reference, usability, and integrating individual SOPs into the overall manual.
> - Effective date—Date the SOP is officially adopted for use in the field. (This may be different from the date of issue. The effective date may be purposefully later than the date of issue to allow for all department members to be informed and/or educated on the new SOP.)
> - Expiration/review date—Important for ensuring the currency of SOPs by establishing a date for periodic review and revision, if needed.
> - Title—For ease of reference and usability.
> - Description of purpose or rationale statement—Describes the purpose of the SOP, why it is needed, and what it intends to accomplish.
> - Authority signature(s)—Indicates that the SOP was properly created, reviewed, and approved by the fire chief or agency head, department attorney, medical director, and/or other responsible authorities. Some departments include an indication of endorsement by member organizations such at an IAFF local.

- Scope—Describes situations for which the SOP was created and the intended audience.
- General procedures—The "body" of the SOP, sets forth broad procedural guidelines for responder-receiver operations.
- Specific procedures—Specific actions necessary under the SOP to safely mitigate a situation.
- References—Source material used to create the SOP or useful in future evaluations.

The specific standard operating procedure template to be used multinationally, or at least nationally, would be developed, vetted, and finalized by a critical best-practices-driven project at this overarching governance level. Use of a mandatory template approach has been successfully demonstrated when connected to funding sources [28] and spectrum use.

5.5.3.3 Step 3: Gather Information and Identify Alternatives

The detailed standard operating procedure needs assessment is the fundamental document to be understood by the project team and its subordinate integrated product teams defining the goals of its mission. In addition to this needs assessment, the project team should also understand and benefit from its parent regional CONOPS document with its included regional interoperability enterprise architecture and information exchange requirements that must directly integrate with the standard operating procedure to be developed. Just as in the systems engineering process, alternative concepts for providing the general and specific procedures must be identified before developing draft procurement documents such as statement of work [29].

> Members of the SOP Development Team should familiarize themselves with the rationale for changes described in the Needs Assessment, along with any recommendations or comments by the Needs Assessment Team. Various approaches can be used to familiarize committee members with this information. Methods range from individual review of written materials to formal briefings by department managers or other experts. Whatever approach is taken, the goal is to avoid duplicating the work of the Needs Assessment Team.
>
> Additional research undertaken by the SOP Development Team at this stage should focus on gathering information, answering questions, resolving problems, and identifying alternative procedures for consideration as SOPs.
>
> Input from focus groups, drills, or exercises could form the basis for a new procedure. Specific language from regulations or standards can help identify procedural or technological options for the department to consider. Alternatives might be generated within the Development Team, using information gathered during the research effort, or based on input from other department members. Brainstorming [through imagination] within the group can be particularly helpful in this process; no alternative should be immediately discarded, however unorthodox it may seem at first.

The "alternatives" represent potential portions of one or more draft standard operating procedure taken from the indicated sources, particularly feedback from after-action reports from real and exercise events.

5.5.3.4 Step 4: Write the Draft Standard Operating Procedure

The potential draft standard operating procedure inputs identified in step 3 above are integrated by the appropriate integrated product team into one or more draft standard operating procedures. During the writing of this document, several important guidelines have been identified—particularly preventing standard operating procedure from becoming training manuals instead of organizational guidelines defining the step-by-step tasks and activities required for collaborative action.

The following suggestions for writing SOPs will enhance the "user-friendliness" of the final product, a key variable in determining success or failure during implementation [30].

- Level of detail—Generally speaking, SOPs should provide only broad procedural guidelines, not specific details of task performance. For example, a new SOP may include guidance to "vertically ventilate the roof of a building with a suspected backdraft condition prior to committing interior forces." However, too much detail is provided if the SOP describes related job tasks, such as, "(1) Check for overhead wires and obstructions prior to raising ladders to the roof; (2) Raise at least two ladders to the roof, ensuring that 3-5 rungs are protruding above the roof line; (3) Take tools to the roof including saws, axes, hooks, ropes, etc.; and (4) Cut at least a 4' by 4' hole and enlarge as needed." Remember: SOPs are not training manuals; they are broad organizational guidelines for performing tasks that members have already been trained to accomplish safely and effectively.
- Clarity and conciseness—SOPs should be clear, concise, and written in plain English. While simply excerpting regulatory language is easy for the SOP Development Team, such language is often difficult to understand and apply to operational situations. Clear and simple statements are the best way to describe actions in SOPs. Using an "outline" or "bulleted" style instead of a continuous narrative simplifies the presentation of information and helps clarify relationships among different components of the SOP.
- Target audience—Write for the majority of the department. Some members may require more help understanding the SOP than others. Others might already be highly experienced in the subject area. Generally, SOPs should be written to address the needs and educational level of the majority of department members, using language they can easily understand.
- Flexibility and ambiguity—To be effective, organizational guidelines must be unambiguous. At the same time, SOPs should provide enough flexibility for the on-scene commander to make decisions based on the situation at hand. Balancing the need to reduce ambiguity while maintaining flexibility can be difficult. Department SOPs should be precise but inherently flexible, permitting an acceptable level of discretion that reflects the nature of the situation and the judgement [*sic*] of the incident commander. (This concept should be explicitly stated at the beginning of the department SOP manual.)

These draft standard operating procedures must be developed with proper configuration control according to a detailed project schedule. Internet or other collaborative tools should be used to minimize the number of face-to-face meetings required following the mandatory first all-hands project "kick-off' meeting needed to establish relationships. The drafting/writing integrated product teams produces a draft standard operating procedures according to the schedule established by the project manager for review by the entire standard operating procedures project team.

5.5.3.5 Step 5: Review and Test the Standard Operating Procedures

A draft review cycle across the entire standard operating procedures project team must be facilitated by the project manager. Collaboration and voting on specific standard operating procedures content elements can be professionally facilitated—perhaps through Web-enabled collaboration tools—so consensus is reached in the final draft document.

> Conducting a "peer review" of draft SOPs can be extremely informative for team members and department managers. The process involves sending copies of the draft SOP to all or selected department personnel affected by the SOP and others with pertinent knowledge or experience in the subject area... Drafts might also be sent to mutual-aid or other agencies that are affected by the SOP but not represented on the Development Team.
>
> Comments received during this review often contain valuable insights on the feasibility of the SOP, helping to identify problems before they occur. Conversely, they may simply serve to verify the efforts of the Development Team, also beneficial information. For hard data, formal critiques, drills, and exercises can be conducted to test SOPs. Indeed, there is perhaps no better way to safely determine the effectiveness of a new SOP than by conducting a realistic, real-time exercise [31].

The exercises to be used for test and improvement of the standard operating procedures must be realistic in terms of true capabilities, with no *assumed* (PHD #2 Assumption) capabilities such as flawless communication. Again, the completed standard operating procedures must be seen as a "living document" with a configuration control process established by the project manager—with an inherent improvement cycle accounting for regional interoperability enterprise architecture and its improvement cycle.

5.5.3.6 Step 6: Ratify and Approve the Standard Operating Procedure

Finally, the completed standard operating procedures must be officially backed by command elements or leaders with premiere public safety and emergency response domain knowledge and experience. Otherwise, the ratification replicates PHD #15, Failure of Insight, provided by those with no expertise in the domain but authority over it. In our notional worst-case World Interoperability Failure model, none of the leaders and planners at any governance level meets these criteria [32].

> The individual or office approving the SOP should have expertise and clearly defined authority in the subject area. ... This person may have taken an active role throughout the development process or may limit involvement to this final phase. If authority derives from persons with no emergency response expertise (e.g., political appointees), a mechanism for a meaningful ratification process must be established. In any case, organizational leaders must fully understand the rationale for, and the implications of, adopting the SOP.
>
> Depending on the complexity of the SOP, it may be necessary to give department decision-makers a detailed briefing on its contents. Team members should stand ready to explain the process used to identify and select the procedural alternative that forms the basis of the SOP. [Project] Members should also be prepared to quickly evaluate and provide feedback on recommendations or revisions requested by ... [key domain authorities].

The completed standard operating procedures must be thoroughly trained and adapted if necessary for the greater benefit of achieving holistic interoperability.

5.6 Emergency Response Frameworks

Step 2 in Figure 5.6 showed that the standard operating procedures needs assessment must include consideration of standards of practice. Multinational and national standards of practice in public safety and emergency response require detailed information-sharing architecture to mitigate or eliminate the deficiencies identified by the Interoperability Assessment Checklist. We will assume that no such detailed information-sharing architecture exists, so the deficiencies remain prevalent. Nevertheless, high-level frameworks for such architecture have been established, but they *do not* constitute information-sharing architecture. Nevertheless, such architecture should be developed within these frameworks.

5.6.1 The Resilience Framework in the United Kingdom

The United Kingdom resilience framework describes the roles and responsibilities of responder-receivers and other stakeholders at different governance levels in the post–9/11 world. In this new framework, more agencies have public safety and emergency response roles:

- *Local and regional structures*—Nine regional Government Offices provide interface between local responder-receivers and the national government. Each Government Office has a regional resilience teams that "have been set up in each of the Government Offices to co-ordinate the response of the whole Government Office area, and to act as the first point of contact for any resilience issues *in normal working hours* [Italics added]." [33].
- *Government liaison team*—Communications with His Majesty's Government is through the strategic coordination group, the lead government department, and the government liaison team. The government liaison team is normally headed by the regional resilience team lead (the regional resilience director), unless and until the incident is determined to be terrorist related, in which case

the government liaison team is headed by "a senior Home Office official" [34].

- *Local resilience forum*—The local resilience forums are statutory bodies composed of the Category 1 responders [35] identified in the United Kingdom Civil Contingencies Act of 2004. It is the "basic mechanism through which local government, emergency services and utility services co-operate under the Act. The Act requires that Category 1 responders attend the local resilience forum, and the guidance offers model terms of reference for it, which includes ensuring joint policy positions, risk assessments, planning, exercising, etc." [36]. The local resilience forum is usually, but not always, chaired by a police representative. (In Scotland, "the local resilience forum is called the strategic cocoordinating group, which if not properly understood gives potential for misunderstanding with the English model.") [37]. The local resilience forum, however, is made up of responder-receivers engaged in emergency response after an incident occurs, in which case the forum effectively becomes the strategic cocoordinating group.
- *Regional resilience forum*—The regional resilience forum is a subsidiary committee that, in the absence of an ongoing emergency, provides planning for regional coordination such as "compiling a regional risk map; considering policy initiatives in the area of civil protection that emanate from the government; facilitating information sharing; including lessons from exercise and incidents at home and abroad, and co-coordinating multi-agency exercises and training." [38].
- *Regional civil contingencies committee*—In times of emergency, the regional resilience forum can convene instead as the regional civil contingencies committee "to co-ordinate the response to, and recovery from, and emergency at regional level in England." [39].
- *Cabinet office briefing room*—At the central government level, the cabinet office briefing room (COBR or "Cobra") provides situational awareness and advice from stakeholders for key ministers. COBR responds with increasing activity to the three emergency levels in the central governments CONOPS, namely Level 1—"significant," Level 2—"serious," and Level 3—"catastrophic," the latter being chaired by a senior cabinet minister or the Prime Minister [40].

The military can support civil authorities according to military aid to civil authorities (MACA) rules, but as "the Armed Forces maintain no standing forces for MACA tasks, and assistance will be provided on an availability basis only. Therefore it is essential that responding agencies do not base plans upon the assumption of military assistance." [41].

In the United Kingdom, His Majesty's Government has adopted the incident command system as defined by Volume 2 "Fire Service Operations: Incident Command" of the *Fire and Rescue Manual*. As defined in the manual, the United Kingdom incident command system [42, 43]:

constitutes the doctrine of the Fire and Rescue Services (FRS) in the context of operational incident management and control processes that flow from it.

The Incident Command System is the recognized "nationwide safe and effective system for managing operations." It presents the key elements of effective incident command in three functional areas; these are:

Organization on the Incident Ground—this gives the Incident Commander a recognised system from which to work when organising and using resources at an Incident.

Incident Risk Management—the principal consideration of Incident Commanders is the safety of their personnel. Therefore, prior to deciding upon the tactics an assessment of risk must be performed. The Iincident Commander must identify the hazards, assess the risks, and implement all reasonable control measures before committing crews into a risk area.

Command Competence—considers the skills and understanding required by an Incident Commander and the importance of maintaining such competencies.

The United Kingdom incident command system defines a three-tiered command structure as follows:

- Operational [Bronze]—which is the level at which command of immediate "hands-on" or task level work is undertaken at the scene of an incident. If the incident develops with several agencies needing to work effectively together, each sector would become a Bronze Command in line with other agencies. It must be emphasised that even if the incident is "multi-agency", the operational level role would not be formally labelled "Bronze" unless a tactical or silver level was in place, and usually when a multi-agency "Silver" group has been formed.
- Tactical [Silver]—which is the level of overall command on the incident ground and which ensures that the Operational levels are supported, and if there are several sectors operating to ensure their operations are coordinated to achieve maximum effectiveness. Again, if the incident develops along multi-agency lines this role would become Silver Command. An incident may be formerly structured with multi-agency Bronze and Silver Commands without a "Gold" being in place.
- Strategic [Gold]—may be invoked where an event or situation may have significant impact on resources, probably involving a large number of agencies or has impact for an extended duration. Thus will be referred to as Gold Command and will normally be implemented as a multi-agency group ...bringing together Gold Commanders from relevant organisations.

The incident command system manual contains a variety of OV-2-type graphics (see Section 4.5.5.2) showing the various command roles for multiple sectors, where one or more sectors may have the same function. These diagrams clearly show the chain of command (e.g., Bronze-Silver-Gold), but no information exchange requirements are defined between these command elements and their units. In addition, command support roles are identified for larger incidents as needed, where various officers could be assigned a variety of support roles.

At larger incidents-specific areas of resource control may be delegated to appointed officers. Such areas may include:

Operational Sectors	*Decontamination*	*Safety*
Command support	Water	Communications
Marshalling	Foam	Press Liaison
BA Main Control	Relief Management	
Logistics	Crew Rehabilitation and Welfare	

> Each of these functions may be assigned by the IC [Incident Commander] to support sectors if the scale of the operation demands it. If this occurs, the support Sector Commander will report in the line of command to the Command Support Officer, who has similar command responsibility level to an Operations Commander [44].

The incident-command system guidelines consider information sharing in the context of "span of control" [45]. The number of lines of communication, implying radio or telephone talks groups (i.e., independent work groups and their communications), will be dependent on the scope and complexity of the incident response. It will be increasingly complicated as multiple agencies are required in support. The guidelines state that five such lines of communication may be nominal, dropping to two or three during crisis periods and increasing to six or seven under more stable conditions. The specific types of messages, language used, and other communication requirements are not provided by the guidelines and are to be developed locally or regionally.

5.6.2 Framework in the United States

5.6.2.1 Incident Command System

The national response framework references the national incident management system [46] and the tactical command hierarchy defined by the incident command system (ICS) in the United States. The United States incident command system is different than the United Kingdom incident command system (see Figure 3.1) [47]. In the United States, the

> ICS was developed in the 1970s following a series of catastrophic fires in California's urban interface. Property damage ran into the millions, and many people died or were injured. The personnel assigned to determine the causes of these outcomes studied the case histories and discovered that response problems could rarely be attributed to lack of resources or failure of tactics. Surprisingly, studies found that response problems were far more likely to result from inadequate management than from any other single reason.

The incident command system defines the organizational structure at the incident command post, which determines the units that may need to be supported by information sharing for each threat-based scenario. The incident command system organizational structure for a large incident, which may be reduced in scope and extent as it is tailored for smaller incidents [48], is shown in Figure 5.8.

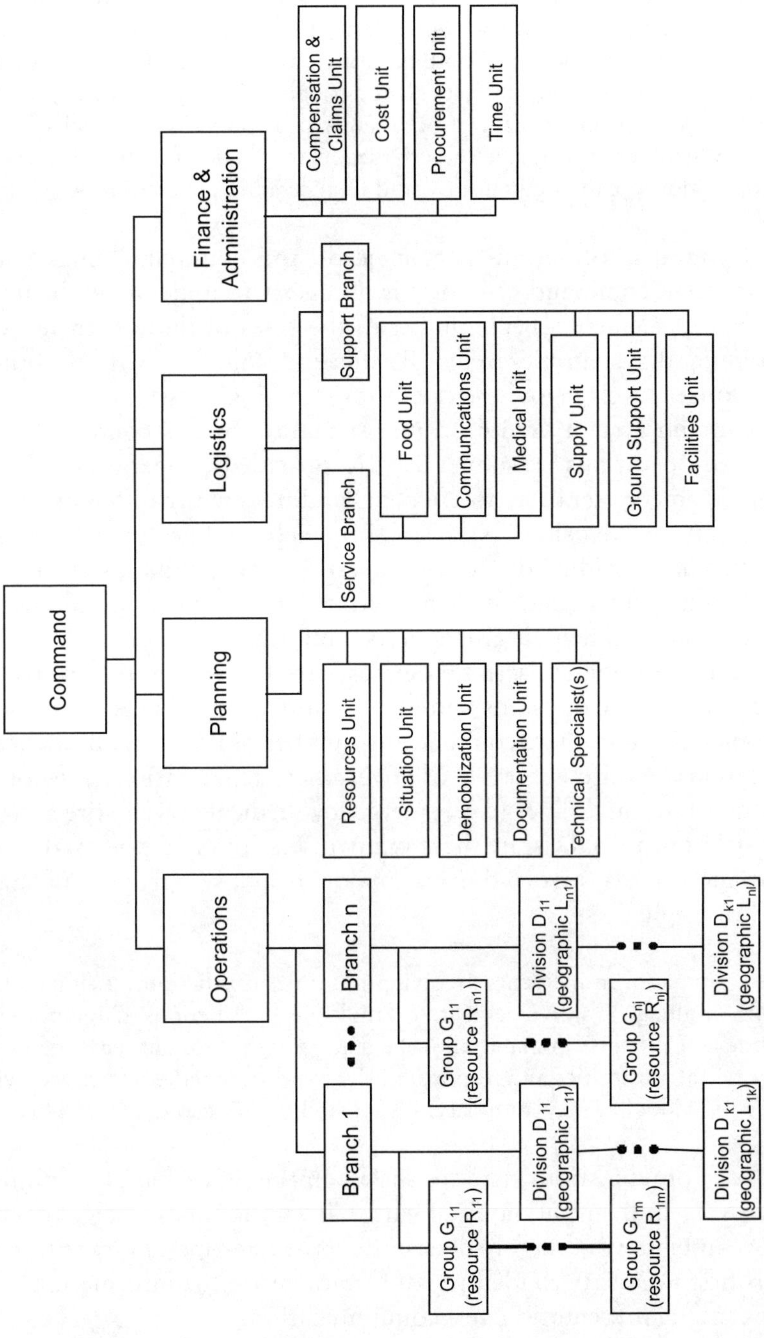

Figure 5.8 Typical incident command system organizational structure for a large-scale incident.

> The ICS is a management system designed to enable effective and efficient domestic incident management by integrating a combination of facilities, equipment, personnel, procedures, and communications operating within a common organizational structure, designed to enable effective and efficient domestic incident management. A basic premise of ICS is that it is widely applicable. It is used to organize both near-term and long-term field-level operations for a broad spectrum of emergencies, from small to complex incidents, both natural and manmade. ICS is used by all levels of government—Federal, State, local, and tribal—as well as by many private-sector and nongovernmental organizations. ICS is also applicable across disciplines. It is normally structured to facilitate activities in five major functional areas: command, operations, planning, logistics, and finance and administration.

To understand information-sharing needs and communication requirements within the incident command system, it is necessary to understand the information flows between its elements, beginning with the roles of these elements within the incident command system hierarchy. To this end, Figure 5.8 shows branch, division, and resource levels below the operations and logistics functional areas of the incident command system organizational structure. In this context, branches are subdivisions of operations personnel used to provide the operations chief with an incident management hierarchy to effectively address many different (spaced) geographic locations, a "division" is assigned to each, and groups, unique personnel resources, such as individual discipline, expertise, or capability, such as search and rescue (SAR) units, emergency medical services, special weapons and tactics units, air units, psychological help, logistics units, and others.

In the figure, a double subscript index resource R_{11}, such as the one used to show there could one or more divisions and one or more groups in each branch. The '...' between branches shows there could be any number of branches. Branches could be used to differentiate different levels of government (e.g., national, regional, local), of perhaps different disciplines (e.g., emergency medical service, fire service, police service, board of health). Despite these many disciplines and potential jurisdictions and governance levels involved in a major response, proper management is maintained through:

> Unified Command: In incidents involving multiple jurisdictions, a single jurisdiction with multiagency involvement, or multiple jurisdictions with multiagency involvement, Unified Command allows agencies with different legal, geographic, and functional authorities and responsibilities to work together effectively without affecting individual agency authority, responsibility, or accountability [49].

Each of the organizations that are shown in the figure has certain information input, processing, and output requirements that vary in time, much like the OPCEN shown above in Figure 5.4. Before determining the interoperability enterprise architecture, it is first necessary to identify and characterize the information required to best understand its presentation and communication.

> Communications/Information Management
>
> - Integrated Communications: Incident communications are facilitated through the development and use of a common communications plan and interoperable communications processes and architectures.

- Information and Intelligence Management: The incident management organization must establish a process for gathering, analyzing, sharing, and managing incident related information and intelligence[50].

A useful source for describing the range of potential incident command system group definitions involved in a major incident response is the automated exercise assessment system developed by the National Guard Bureau as a training and tabletop exercise tool. The automated exercise assessment system was developed through the solicitation of input from over 140 public safety and related organizations nationwide. Through that effort, generic functions were defined that are well aligned with incident-command system group definitions. These functions were defined such that they could be combined to represent any local configuration of services, such as affiliating emergency medical service with the fire service or, in other communities, with the police service. Table 5.2 provides a list of the 40 public safety and incident-related functions that might define independent groups in the incident-command system operations organization—spanning all incident command system organizational units and functional groups.

Operational response divisions could be defined that span a large geographic area, perhaps many tens to few hundred miles in extent. This geographic expanse contains islands; coastal inlets, estuaries, and harbors; and landscapes of heavily wooded rolling hills and plains. This situation reflects current conditions, as many communities have their own emergency manager.

Many of the information-sharing requirements within the incident command system come from the planning section. As Figure 5.8 showed, the planning section has the following units:

- *Resources*—provides arrival logging and status monitoring of all the physical assets available to incident management, including personnel, teams, facilities, supplies, and all deployable equipment for incident response; overall, information that represents a certain volume of data that "initializes" situational awareness inasmuch as the physical assets for incident management are concerned.

Table 5.2 Public Safety and Incident-Related Functions

Administration	*Fatality Management*	*Public Health*
Animal control	Food and water	Public information
Chaplainry	HAZMAT	Public works
Communications	Incident Command	Resource management
Debris management	Information and planning	Search and rescue
Donations management	Law Enforcement	Scene safety
Elected official	Logistics	Scene security
Emergency management	Military support	Trade unions
EMS	CISM/mental health	Transportation
Energy	Medical facilities	Veterinary medicine
EOD	Mass care	Voluntary organizations
Fire	Operations	

- *Situation*—maintains situational awareness by collecting, processing, and organizing resource and incident status information, providing summary status information, and incident forecasting capability to provide forewarning of possible future incident related events; overall, providing the status—and potential future status—of all relevant aspects of the incident and the physical assets available to the incident commander to manage the incident response.
- *Documentation*—provides an important logging function for all incident management decisions and actions—in the incident action plan, as well as providing all document handling (e.g., copying, storing) requirements for the incident response; essentially, the unit is responsible for all record keeping of data stored as information to support incident decision making as well as a source of future after-action "lessons learned," thus providing a short-term and archival capability for the incident response, and distribution of the incident action plan to divisions, their groups and units, that may be geographically separated.
- *Demobilization*—creates the incident demobilization plan, which is documentation that defines when and how all incident response and management personnel and assets to eventually be "decommissioned" from the incident response once they are no longer needed, so they can return to their normal preincident "ready" status or their normal day-to-day activities; the incident demobilization plan must be created, reviewed, and distributed to divisions, their groups and units, that may be geographically dispersed over a wide area.
- *Technical specialists*—personnel with special skills, assigned to any incident-command system organizational element as required, but professionally capable of the unique expertise for which they've been called to an incident, such as meteorology, environmental impact, resource use and cost analysis, flood control, water-use, explosives, structural engineering, firefighting, medical/health care, medical intelligence, pharmacology, veterinary services, agriculture, toxic substances, radiation health, intelligence, infectious disease, chemical or radiological decontamination, law enforcement, attorney, industrial hygiene, transportation, and other scientific expertise; overall, a complete array of incident and hazard-related professionals available to provide the right information in every stage of prevent, detect, protect, alert, respond, and recover.

The logistics section provisions the incident-command system response [51].

> The Logistics Section provides for all the support needs for the incident, such as ordering resources and providing facilities, transportation, supplies, equipment maintenance and fuel, food service, communications, and medical services for incident personnel.

The component units in the logistics section that were identified in Figure 5.8 perform the following functions:

- *Supply unit*—responsible for ordering, receiving, storing, and processing all personnel, equipment, facilities, tools, and supplies needed for incident command system operation; thus, it must support information exchange, processing, and display with potentially distant sources of supply.

- *Facilities unit*—charged with establishing all shelter and sustainment capabilities (e.g., food and water service, sleeping, sanitation and showers, and staging) for the incident command post, base of operations, and camps, using all available existing structures, tents, trailers, and all other forms of shelter with the proper collective protection for the nature of the incident; thus, except for supporting status of facilities and their capabilities to geographically separated divisions, primarily local coordination of structural provisioning is required.
- *Ground support unit*—provides maintenance and repair for all tactical equipment, vehicles, and mobile ground support equipment; records usage time and condition, supplies fuel, provides ground transportation, and create/executes the incident traffic plan; thus, coordination and information sharing are required across geographically dispersed divisions and intransit (ground) assets.
- *Communications unit*—responsible for development and execution of the communications plan and supporting all aspects of radio and wireline communications in support of the incident-command system as well as all its supporting functions, including "... planning the use of radio frequencies; establishing networks for command, tactical, support, and air units; setting up on-scene telephone and public address equipment; and providing any required off-incident communication links. Codes should not be used for radio communication. A clear spoken message—based on common terminology that avoids misunderstanding in complex and noisy situations—reduces the chances for error. The use of common terminology allows emergency management/response personnel to communicate clearly with one another and effectively coordinate activities, no matter the size, scope, location, or complexity of the incident. [e.g., in support of the supply unit]" [52]. (As of 2009, no common terminology has yet been defined and accepted nationwide.)
- *Food unit*—tasked to provide proper all food services for incident personnel, wherever they are and under whatever conditions they must be fed, so the food unit must closely coordinate with the planning section to locate people and animals to be fed, the facilities unit to arrange for space to serve food, the supply unit to order food, the ground support unit to transport food, preparers, and servers, and air operations for remote transport; so the food service unit must exchange information from several incident-command system organizations to function properly.
- *Medical unit*—develops and executes the incident medical plan as well as provides all medical services to all incident personnel, including procedures, medical care (e.g., vaccines, occupational health care, prophylaxis, and mental health services), medical transportation (for incident responders only), patient tracking, and paperwork assistance for injuries and deaths; thus, the medical unit must interface will all other incident command system units as they're all staffed with personnel subject to direct or indirect injury in and around the incident scene.

Finally, the finance/administration section includes the:

- *Time unit*—responsible for tracking hours worked by incident command system personnel for compensation, tracking these hours if necessary for their respective agencies, and also tracks equipment usage time collected by the ground support unit and the air operations support group (if used) for aircraft; thus, only periodic information exchanges are required, and arguably at a lower priority than other incident command system requirements.
- *Procurement unit*—responsible for all financial matters relative to vendor contracts, coordinating "with local jurisdictions to identify sources for equipment, prepares and signs equipment rental agreements, and processes all administrative requirements associated with equipment rental and supply contracts"; thus, the unit must work closely with the supply unit in the logistics section, but arguably can then support the supply unit's interaction with vendors indirectly through the supply unit's information exchanges with those vendors—requiring a level of priority just below tactical operations and transportation, but nevertheless higher than the time, compensation and claims, and cost units for these exchanges given the importance of resupply during extended incidents.
- *Compensation and claims unit*—charged with handling all injury (and other) compensation and claims by incident personnel, requiring close coordination with the medical unit; thus, periodic information exchanges with the medical unit are required, but arguably not at the same priority level of other incident command system functions.
- *Cost unit*—maintains records regarding the usage of personnel and equipment for the response, provides cost estimates to the planning section, and thus must have access to accurate cost data; however, the information exchanges necessary to support the cost unit can be periodic, tied to other unit exchanges with vendors (say, through the procurement unit) and agencies (through planning unit exchanges with jurisdictions and agencies providing the incident command system with personnel and equipment), and therefore a lower priority than tactical or incident command system support exchanges.

The information exchange requirements among these incident command system organizations define the information processing, display, and associated communication requirements of interoperability enterprise architecture for regional incident response. To specify the appropriate information processing, display, and exchange capabilities of a fixed or deployable emergency operations center (OPCEND), it is first necessary to characterize the associated information exchange requirements of these incident-command system sections. Several of the intra- incident command system information exchange requirements implied by these unit descriptions are depicted in the incident command system OV-2 shown in Figure 5.9.

The figure also shows that there are three primary external information-sharing interfaces to the information-exchange requirements applied to an incident or set of closely related incidents, including:

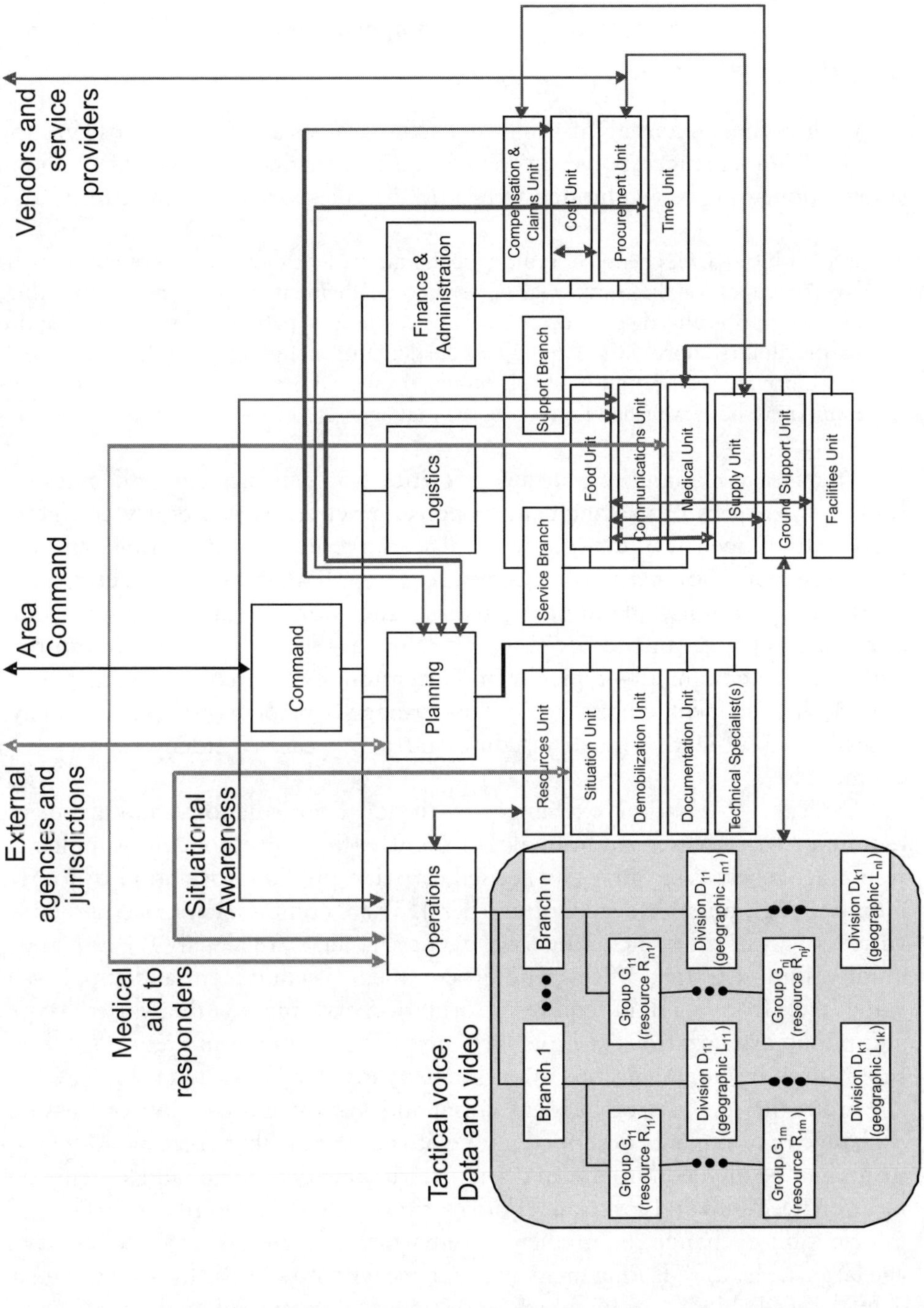

Figure 5.9 An OV-2 for the United States incident command system.

- *Area command*—to provide coordination with wide-area authorities supporting the response to multiple, perhaps uncorrelated, incidents;
- *Local and regional agencies and jurisdictions*—to support coordination with local and regional first and supplementary responders-receivers;
- *Vendors*—to provide resupply of equipment and materials used in the incident response.

Each of these external information exchanges has a unique purpose in supporting the local incident response. In some large-scale incidents, the incident command system alone may be too limited to respond to the scope of the incident [53].

> Acts of biological, chemical, radiological, and nuclear terrorism represent particular challenges for the traditional ICS structure. Incidents that are not site specific, are geographically dispersed, or evolve over longer periods of time will require extraordinary coordination among all participants including Federal, State, Tribal, local, private sector, and NGOs [non-governmental organizations]. An area command may be established to oversee the management of such incidents.

The area command establishes priorities between incidents, allocates critical resources, ensures proper incident response practice, avoids policy conflicts with participating agencies, identifies critical resource needs to OPCENs, and ensures transition from short-term to long-term recovery. In that regards, all command sections of operations, planning, logistics, and finance/administration could be affected by area command decisions. For this reason, the exchange of information with the area command—typically and intentionally located as close as possible to the specific incident command system-managed incidents—must be timely and accurate, as now the incident commander for each incident reports to area command.

Critical information exchanges with local jurisdictions and agencies are required to access additional incident (first) responder supplementary responder-receiver resources as needed, provide public information to local communities (e.g., possible evacuation orders). These communications could be with a variety of authorities, including local agency facilities, the local OPCEN, and community administrative offices. The procurement of equipment and supplies (food, water, etc.) from vendors requires information exchanges with a variety of private and public organizations storing such supplies. As shown in Figure 5.9, both the supply and procurement units are principally involved in these exchanges.

From the analysis of incident command system information processing and exchange requirements described above, it is apparent that corresponding requirements can be divided into internal and external incident command system communication needlines and associated information-exchange requirements. The external information-exchange requirements are particularly important to effectively manage large-scale, long-lasting incidents or those with no specific locations (e.g., a pandemic, see Table 5.1)—in which case an area command will oversee incident command system management practice where it is applied. These priorities (see rightmost column in Figure 4.12) affect the selection of IT/communications equipment to provide the corresponding situational awareness and command coordination as well as the corresponding OPCEN CONOPS and standard operating procedures.

Some of these organizations involved in an incident, such as those listed in Table 5.2, have the own emergency facilities and communication capabilities for information sharing, while others are limited to publicly available means (e.g., telephone and Internet). Current differences in capabilities and standards, the use of disparate databases and naming conventions, the use of different and incompatible information processing and display capabilities, and other obstacles to interoperability must be addressed by the development of unified interoperability enterprise architecture for public safety and emergency response extending beyond the national response framework and incident command system frameworks.

5.6.2.2 National Response Framework

The national response framework provides an overarching hierarchical operational structure for emergency and disaster response. It defines the high-level roles, responsibilities, and actions of local, regional, and national authorities in emergency preparedness and disaster response [54].

> The Framework is intended to strengthen, organize, and coordinate response actions across all levels. The doctrine of tiered response emphasizes that response to incidents should be handled at the lowest jurisdictional level capable of handling the work. The vast majority of incidents are, in fact, managed locally.

The Framework is focused on incidents of all types, including acts of terrorism, major disasters, and other emergencies. For the purpose of this document, the term "incident" refers to an actual or potential occurrence or event.

The framework specifies several tasks for each of the three phases of emergency management it addresses, namely prepare, respond, and recover, by defining a series of steps for each phase a follows [55–57]:

- *Prepare*—defining the "... six essential activities for responding to an incident: plan, organize, train, equip, exercise, and evaluate and improve. ..."
- *Respond*—"Four key actions typically occur in support of a response: (1) gain and maintain situational awareness; (2) activate and deploy key resources and capabilities; (3) effectively coordinate response actions; then, as the situation permits, (4) demobilize. ..."
- *Recover*—"Short-term recovery is immediate and overlaps with response. It includes actions such as providing essential public health and safety services, restoring interrupted utility and other essential services, reestablishing transportation routes, and providing food and shelter for those displaced by the incident. Although called 'short term,' some of these activities may last for weeks."

5.6.2.3 National Incident Management System

The national response framework incorporates the national incident management system, which describes the headquarters-level organizations (see the OV-2 diagram of Figure 5.10) at the federal joint field office (JFO) and the 15 emergency support functions (ESFs) described in Figure 5.11. These emergency support functions are the functional operating organizations at all governance levels represented at the joint field office. Although the individual CONOPS of the emergency support functions have been developed, the scenario-dependent interoperability enterprise architecture with documented and vetted information-exchange requirements does not exist. No detailed information exchange requirements (OV-3s) were established for the indicated needlines in this OV-2 diagram, nor were operational views (e.g., OV-5 and OV-6c) developed.

The figure does not show the emergency response teams deployed to prepare for the establishment of the JFO. The figure shows the stable JFO positions with partnerships spanning state and local OPCENs as well as private sector organizations:

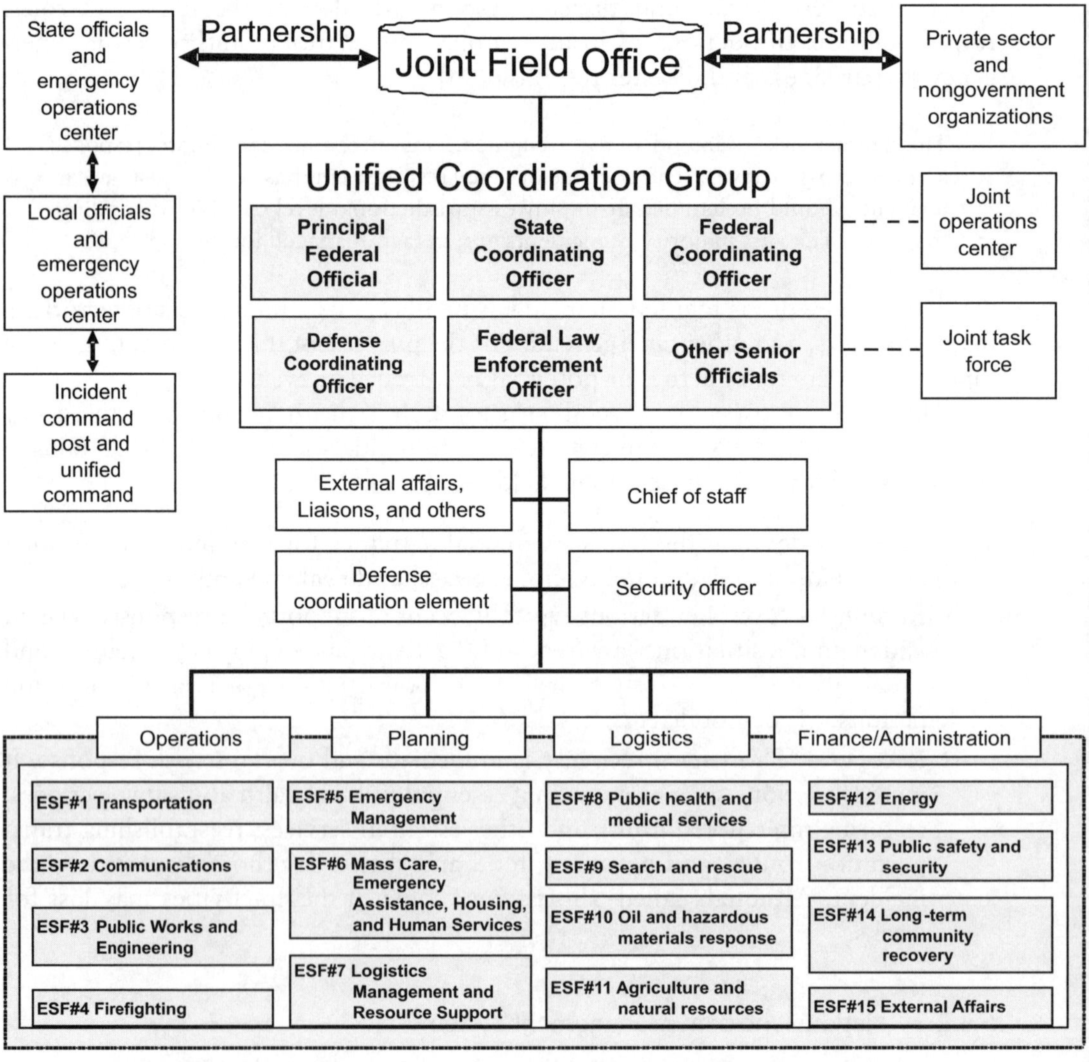

Figure 5.10 OV-2 for the joint field office in the National Response Framework [58].

ESF#1 Transportation
Coordinator: Department of Transportation

- Aviation/airspace management and control
- Transportation safety
- Restoration and recovery of transportation infrastructure
- Movement restrictions
- Damage and impact assessment

ESF#2 Communications
Coordinator: National Communication System (Department of Homeland Security)

- Coordination with telecommunications and information technology industries
- Restoration and repair of the telecommunications infrastructure
- Protection, restoration, and sustainment of national cyber and information technology resources
- Oversight of communications within the national government incident management and response structures

ESF#3 Public Works and Engineering
Coordinator: Army Corps of Engineers

- Infrastructure protection and emergency repair
- Infrastructure restoration
- Engineering services and construction management
- Emergency contracting support for life-saving and life-sustaining services

ESF#4 Firefighting
Coordinator: Forest Service (Department of Agriculture)

- Coordination of national firefighting activities
- Support to wildland, rural, and urban firefighting operations

ESF#5 Emergency Management
Coordinator: Federal Emergency Management Agency (FEMA)

- Coordination of incident management and response efforts
- Issuance of mission assignments
- Resource and human capital
- Incident action planning
- Financial management

ESF#6 Mass Care, Emergency Assistance, Housing, and Human Services
Coordinator: FEMA

- Mass care
- Emergency assistance
- Disaster housing
- Human services

ESF#7 Logistics Management and Resource Support
Coordinator: General Services Administration and FEMA

- Comprehensive, national incident logistics planning, management, and sustainment capability
- Resource support, such as facility space, office equipment and supplies, contracting services, etc.

ESF#8 Public health and medical services
Coordinator: Department of Health and Human Services

- Public health
- Medical
- Mental health services
- Mass fatality management

ESF#9 Search and rescue
Coordinator: FEMA

- Life-saving assistance
- Search and rescue operations

ESF#10 Oil and hazardous materials response
Coordinator: Environmental Protection Agency

- Oil and hazardous materials (chemical, biological, radiological, etc.) response
- Environmental short- and long-term cleanup

ESF#11 Agriculture and natural resources
Coordinator: Department of Agriculture

- Nutrition assistance
- Animal and plant disease and pest control
- Food safety and security
- Natural and cultural resources and historic properties protection
- Safety and well-being of household pets

ESF#12 Energy
Coordinator: Department of Energy (DOE)

- Energy infrastructure assessment, repair, and restoration
- Energy industries utilities coordination
- Energy forecast

ESF#13 Public safety and security
Coordinator: Department of Justice (DOJ)

- Facility and resource security
- Security planning and technical resource assistance
- Public safety and security support
- Support to access, traffic, and crowd control

ESF#14 Long-term community recovery
Coordinator: FEMA

- Social and economic community impact assessment
- Long-term community recovery assistance to states, tribes, local governments, and the private sector
- Analysis and review of mitigation program implementation

ESF#15 External Affairs
Coordinator: Department of Homeland Security (DHS)

- Emergency public information and protective action guidance
- Media and community relations
- Congressional and international affairs
- Tribal and insular affairs

Figure 5.11 Emergency support functions in the National Response Framework [59].

Principal Federal Official (PFO). By law and by Presidential directive, the Secretary of Homeland Security is the principal Federal official responsible for coordination of all domestic incidents requiring multiagency Federal response. The Secretary may elect to designate a single individual to serve as his or her primary representative to ensure consistency of Federal support as well as the overall effectiveness of the Federal incident management. ...[60]

Federal Coordinating Officer (FCO). The FCO is a senior FEMA official trained, certified, and well experienced in emergency management, and specifically appointed to coordinate Federal support in the response to and recovery from emergencies and major disasters [61].

State Coordinating Officer (SCO). The SCO plays a critical role in managing the State response and recovery operations following Stafford Act declarations. The Governor of the affected State appoints the SCO, and lines of authority flow from the Governor to the SCO, following the State's policies and laws [62].

Defense Coordinating Officer (DCO). The DCO serves as DOD's single point of contact at the JFO for requesting assistance from DOD. With few exceptions, requests for Defense Support of Civil Authorities (DSCA) originating at the JFO are coordinated with and processed through the DCO [63].

Senior Federal Law Enforcement Official (SFLEO). The SFLEO is an official appointed by the Attorney General during an incident requiring a coordinated Federal response to coordinate all law enforcement, public safety, and security operations with intelligence or investigative law enforcement operations directly related to the incident [64].

Defense Coordinating Element. Defense Coordinating Element consisting of a staff and military liaison officers to facilitate coordination and support to activated ESFs [65].

Participation of the region is sometimes specific, but often implied (nonspecific). It is assumed (PHD#2 Assumption) that the ability of the states and federal agencies to provide material support and personnel draws primarily from regional resources. In this regard, communications related to provisioning this support are essentially back-copied to national-level agency OPCENs as a formality and not as a first course of action in order to expedite processing and speed much needed resources to meet identified needs. Communications within the region should not be affected, but the link between nodes in the state/local impact zone response and regional nodes may suffer due to shortcomings in the state/local impact zone (communications links represented as thin dashed lines).

The national level provides federal oversight, funding, resource support, and situational awareness activities. Communications are assured (communications links represented by use of solid lines). We assume that coordination will take place according to the national response framework, which should provide for centralized situational awareness at regional and national OPCENs. All communications between nodes will be restricted by any problems with the weakest participant, that is, communications between national level and state/local impact zone will be restricted by state/local impact zone limitations regardless of national-level communications capabilities. It is anticipated that these imitations will be expressed primarily in terms of quality and availability.

The national response framework does not define the information-sharing architecture (i.e., scenario-dependent interoperability enterprise architecture with information exchange requirements), to be used at any governance level. Instead, this determination is relegated to subordinate jurisdictional levels as follows: local, tribal, and state actions. local, tribal, and state governments can address the inherent challenges in establishing successful information-sharing networks by:

- Creating fusion centers that bring together into one central location law enforcement, intelligence, emergency management, public health, and other agencies, as well as private-sector and nongovernmental organizations when appropriate, and that have the capabilities to evaluate and act appropriately on all available information.
- Implementing the *National Information Sharing Guidelines* to share intelligence and information and improve the ability of systems to exchange data.
- Establishing information requirements and reporting protocols to enable effective and timely decision making during response to incidents. Terrorist threats and actual incidents with a potential or actual terrorist link should be reported immediately to a local or regional Joint Terrorism Task Force.

However, neither the national response framework nor the national information-sharing guidelines [66] provide specific direction to address the "the inherent challenges in establishing successful information-sharing networks" [67] using best practices. The National Information Sharing Guidelines describe fostering culture, supporting efforts, developing training/awareness—these actions will maintain nationwide information-sharing stovepipes for the foreseeable future. In other words, best-practices methodology to define and achieve regional interoperability enterprise architecture and information-exchange requirements template guidelines needed to engineer holistic interoperability *are not employed* in these documents. As with the national response framework, these national information sharing guidelines relegate the details of information sharing to the states—continuing to support the establishment of information-sharing stovepipes at multiple governance levels nationwide [68] (see Figure 3.6).

> Each State will be encouraged to define and document how it intends to carry out intrastate efforts to gather, process, analyze, and disseminate terrorism information, homeland security information, and law enforcement information. This process is commonly known as the "fusion process." Defining this process should include the following:
>
> 1. In those States where there exist multiple fusion centers, one fusion center, with the demonstrated capacity to serve as the statewide center or hub, should be designated as the primary interface with the Federal Government. This statewide fusion center should also coordinate the gathering, processing, analysis, and dissemination of homeland security information, terrorism information, and law enforcement information on a statewide basis.
>
> 2. The Executive Agent of each Urban Area Security Initiative (UASI), as well as the applicable State's homeland security advisor, must work together to determine the most effective manner in which to incorporate the UASI into the statewide information sharing framework.

3. In those instances in which the UASI has established a regional fusion center, the activities of the major urban area fusion center should be incorporated into the statewide fusion process.

4. Each State and major urban area fusion center is encouraged to coordinate with the appropriate Federal authorities to develop synchronized protocols for sharing information with the private sector.

The national information sharing guidelines goes on to state that [69]:

> The Federal Government, working in partnership with State, local, and tribal authorities, will seek to define the current national information sharing capability that exists through the existence of existing State and major urban area fusion centers. State and local authorities will be asked to support these efforts by assessing and documenting the baseline level of capability of their existing fusion centers.

The implication from these words is that the federal government will *not* provide an overarching information-sharing architecture, but rather *will* "seek to define" this architecture based on the myriad of capabilities and conventions adopted by the independent states. Instead of a proactive architecture design for states to emulate and with which to integrate, the national government will "define the current national information sharing capability that exists through the existence of existing State and major urban area fusion centers."

The principal information-sharing approach addressed in the national response framework is to colocate staff from different organizations in a common location [70], referred to as a

> Multiagency Coordination System (MACS). The primary function of MACS is to coordinate activities above the field level and to prioritize the incident demands for critical or competing resources, thereby assisting the coordination of the operations in the field. MACS consists of a combination of elements: personnel, procedures, protocols, business practices, and communications integrated into a common system. For the purpose of coordinating resources and support between multiple jurisdictions, MACS can be implemented from a fixed facility or by other arrangements outlined within the system. Examples of multiagency coordination include a State or county emergency operations center, a State intelligence fusion center, the National Operations Center, the Department of Homeland Security/Federal Emergency Management Agency (FEMA) National Response Coordination Center, the Department of Justice/Federal Bureau of Investigation (FBI) Strategic Information and Operations Center, the FBI Joint Operations Center, and the National Counterterrorism Center.

Multiagency coordination systems were used in response to Katrina, and were defined in the predecessor to the national response framework, known as the national response plan. So their existence alone does not ensure interoperability, as more detailed planning is needed. As the national incident management system states [71]:

> Written agreements [e.g., standard operating procedures] allow agencies within the system to conduct activities using established rules and are often self-defined by the participating organizations. A fully implemented MACS is critical for seamless

> multiagency coordination activities and essential to the success and safety of the response whenever more than one jurisdictional agency responds. Moreover, the use of MACS is one of the fundamental components of Command and Management within NIMS, as it promotes scalability and flexibility necessary for a coordinated response.

However, no standardized CONOPS or standard operating procedure templates are provided. The implication is that the unifying methodology for such development is left to those responder-receiver organizations that have the time and processes to do it as they choose. Thus, no holistic interoperability will result from this ad hoc approach and the deficiencies will continue.

5.6.3 Assessment of Emergency Frameworks

The differing United Kingdom and United States incident command system frameworks vary in their terminology, organizational structure, coordination, and integration with higher command authorities, reminiscent of the national-level stovepipes shown in Figure 3.6. Both frameworks provide local-to-regional-to-national organizational response hierarchies, and both consider planning, response, and recovery phases. The United Kingdom guidelines emphasize the importance of leadership and tactical/operational decision making, a critical factor in mitigating the Pearl Harbor deficiencies. By contrast the greater emphasis in the national response framework is in organizational structure by discipline and support function. Nevertheless, the philosophy of unified command and localized action supported by regionalized coordination is apparent.

Although both architecture frameworks preserve local flexibility by leaving further details of planning and implementation to the local authorities, they provide no interoperability enterprise architecture templates to these authorities. These templates are standard information-sharing formats, such as the operational views described in Chapter 4. These OVs would define the information-sharing requirements needed to address each of the Interoperability Assessment Checklist deficiencies in the World Interoperability Failure model. By failing to refine these information-sharing requirements, the information sharing implied—but not detailed—by these frameworks, allow many of the deficiencies to remain. As has been shown, information-sharing definition and detail captured with interoperability enterprise architecture templates, provides basis for achieving interoperability. Failing such detailed planning, small amounts of information may be missed—with the same consequences that on-scene responder-receivers must make up for incomplete planning with their own initiative and self-sacrifice. Despite the many systems, plans, and frameworks, the underlying deficiencies remain.

5.7 Architecture-Centric Engineering

5.7.1 Problem

The development of public safety and emergency response architecture has been shown to be critically dependent upon achieving interoperability, that is, trusted understanding and predictable collaborative action (e.g., information-driven tasks and activities) towards a common goal. These architectures must be derived from multiple threat-based scenarios, spanning frequent and mundane as well as rare but possible catastrophic events. The classic "just-buy-something" approach to architecture development, while avoiding PHD #15, Failure of Insight, and PHD #24, Failure of Improvement, is ineffective given the:

- Complexity of diverse operational scenarios facing public safety and emergency response organizations;
- Sophistication of existing and emerging technological functions and capabilities;
- Number and complexity of existing responder-receiver and stakeholder SOPs;
- Need for comprehensive joint operations among multiple governance levels and their responder-receivers as directed by national response frameworks;
- Need to make architecture improvements to an operational capabilit, such as an existing OPCEN, which must be available and functioning on a 7×24×365 (e.g., continuous) basis;
- Existence of emergency operations frameworks (e.g., differing national response frameworks) that leave nonlinear opportunities for information-sharing failure with tragic consequences.

Often, these tasks and activities need to be performed among OPCENs as depicted in Figure 5.12.

5.7.2 Innovative Approach

An innovative architecture development approach, called Architecture-Centric Engineering (ACE) has been developed that meets these needs [72]. It provides an accelerated yet low-risk approach to operational and system architecture development, while mitigating or eliminating the Interoperability Assessment Checklist deficiencies (see Chapter 3). The ACE methodology is built from an optimal combination of modeling and simulation and experimentation as depicted in Figure 5.13. It is based on the interoperability enterprise architecture OVs described in Chapter 4, and depicts catastrophic evacuation example.

The ACE approach takes doctrine and CONOPS documents to create OV-1 use cases, and OV-5 activity models. These lead to simulating key mission scenarios assuming the modeled public safety and emergency response organizations perform those activities. The dynamic models and executable simulations—see Figure 5.13—use version 2 of the Universal Modeling Language (UML 2.0) [74]. UML provides an interoperable visual object-based modeling tool enabling exchange of model information with agreement on semantics and notation.

The numbered elements in Figure 5.13 are defined as follows:

1. The ACE "adapted-Vee" [75] recognizes that sufficient public safety and emergency response enterprise detail is available to immediately identify opportunities for delivery of improved capability through improved operational and systems architecture. While requirements development and

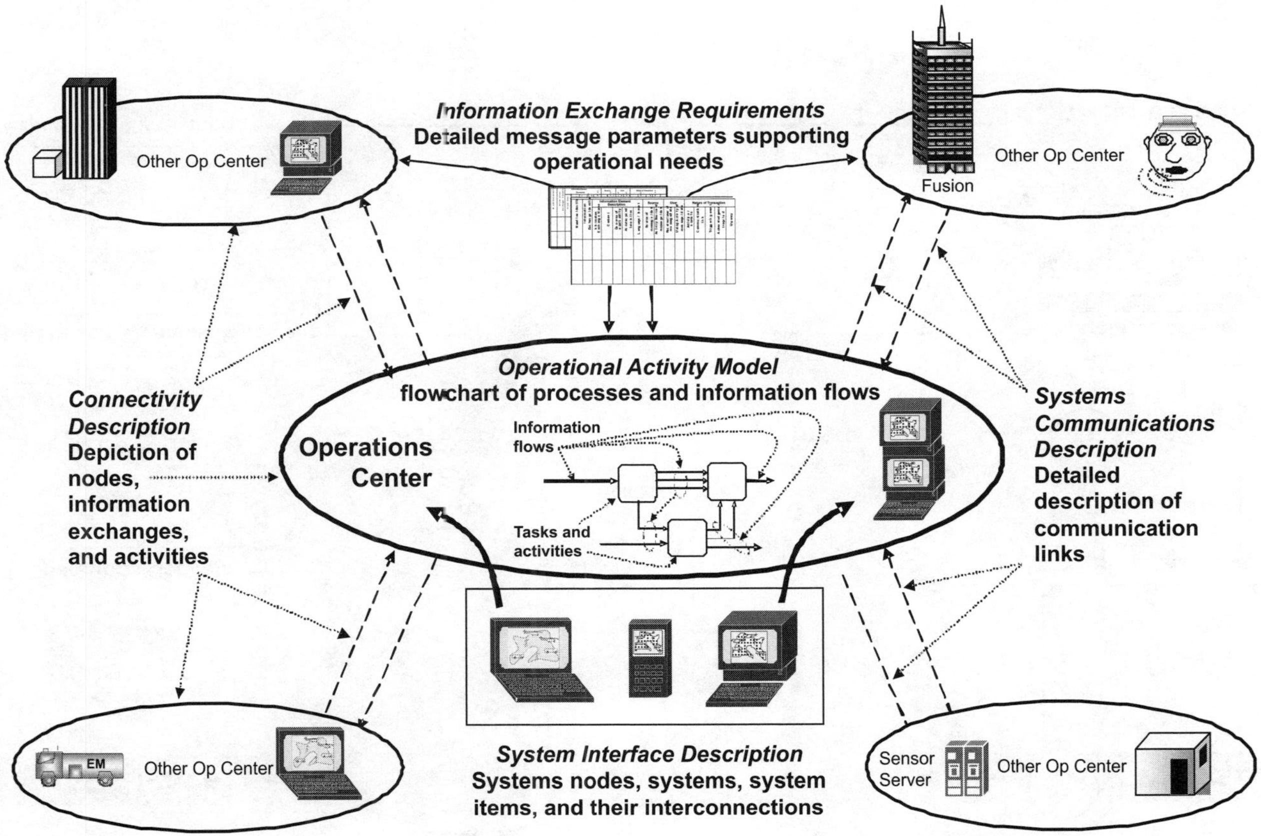

Figure 5.12 Information sharing for OPCEN collaboration.

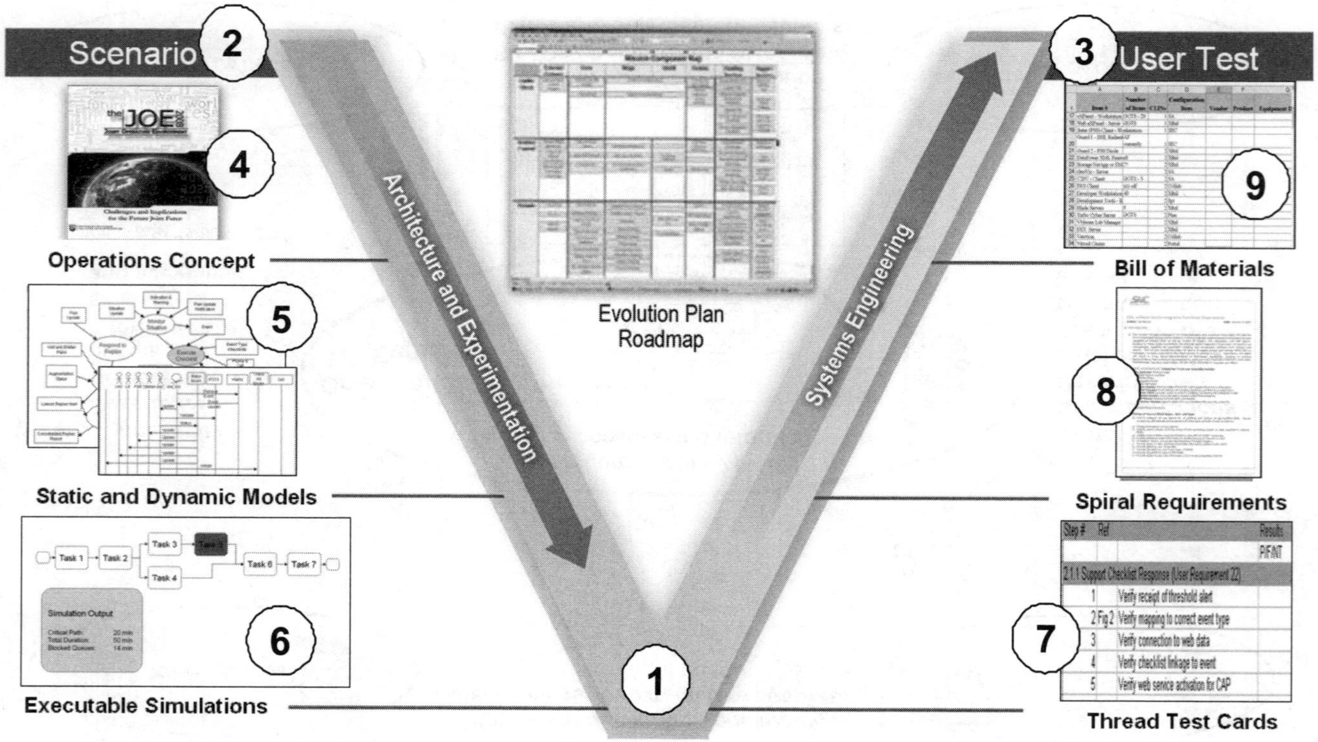

Figure 5.13 Architecture-centric engineering methodology.

analysis are critical to systems engineering, the requirements are not central to begin the systems engineering process as implied by the traditional systems engineering "V." Thus, the ACE adapted-Vee presumes a continuing evolutionary spiral modernization approach that is tracked and refreshed against a living evolution plan roadmap. Furthermore, this approach permits improvements for continuously functioning OPCENs.

2. Threat scenarios that describe the most important capabilities of an OPCEN provide the basis to define the enterprise components, or "piece parts" of the public safety and emergency response enterprise, that contribute or detract from positive OPCEN performance.
3. Focusing on what constitutes a successful improvement spiral, backed up by testing to confirm success, performance measures associated with improved user capability can be better understood. By tailoring real-world emergency response scenarios to include these user capabilities, the basis for virtual scenario "game play" is provided for detailed modeling and simulation.
4. Officially sanctioned CONOPs, doctrine, and derived standard operating procedures provide the basis to identify the piece parts that constitute the public safety and emergency response enterprise and how they integrate or interwork to deliver capability. These piece parts are called *enterprise static objects* and are matched with use-case scenarios to understand which static objects are important in achieving desired capabilities.
5. Dynamic models, including activity models and sequence diagrams, are created and understood by applying doctrine, CONOPs, and standard operating procedures to use-case scenarios—helping to understand the intraworkings of the OPCEN enterprise. By capturing each activity detail, dynamic (time-dependent) models are then abstracted by identifying enterprise static objects (e.g., organization/operator, system, and mission) that are employed in specific threat scenarios. This analysis supports the creation of sequence diagrams with associated tasks that describe each aspect of the scenario.
6. Once models are created, they can be used to create an executable architecture composed of simulations. Simulations are important because they can be used to evaluate enterprise capability relative to predicted measures of performance. Sequence diagrams are then used to create associated simulations loaded with timing and resource detail. Once a simulation is created, the behavior of people, organizations, and tools working together to respond to the emergency is captured quantitatively (as data) through simulation runs. In this way, both benchmark as-is simulations can be compared with various improvement options tested by the simulation. This approach provides the basis for comparative analysis of the results of multiple simulation runs, thus providing insights about the cost and time associated with alternatives for achieving the desired capability (see the business case analysis practice in Section 4.6). When the cost of various improvement options is evaluated using simulations that demonstrate improved performance, a return on investment can be calculated. This return-on-investment data is than

able to support stakeholder review and spiral procurement and sustainment decisions in the OPCEN.

7. The use of modern architecting tools to develop integrated OPCEN capabilities allows automatic extraction of traditional systems engineering artifacts into standard word processor format. For example, digital references on electronically captured sequence diagrams provide the basis to automatically generate *thread test cards*. These cards are needed to support the earliest experimentation of integration alternatives—normally a time-consuming systems engineering activity.
8. As stakeholder review decides on a spiral option for implementation, the technical requirements document associated with the spiral is automatically generated by taking data from both the integrated architecture and requirements management system. This automatic synthesis of architecture and requirements data into word processor format documents provides a jump start for traditional systems engineering implementation activities.
9. As integration and installation planning are advanced, associated bills of materials for the hardware and software needed for the particular OPCEN improvement is also able to be automatically generated from an integrated architecture tool, again streamlining an important but time-consuming systems engineering activity.

5.7.3 Modeling and Simulation

Using the UML 2.0 modeling standard, the *as-is* OV-5 Operational Activity model views (that depict as-is capabilities, operational activities, relationships among activities, inputs, and outputs, etc.) are used to define and expand the as-is and to-be architecture. These views include organization, operator, and mission objects from the as-is AV-2 view of the as-is interoperability enterprise architecture. The AV-2 Integrated Dictionary [76] is an architecture data repository with definitions of all terms used in all interoperability enterprise architecture work products.

In addition to the as-is activity model, other dynamic models include:

- OV-6c: Operational event-trace description—One of three products used to describe operational activity that traces actions in a scenario or sequence of events [77].
- SV-10c: Systems event-trace description—One of three products used to describe system or service functionality, and identifies system/service-specific refinements of critical sequences of events described in the operational view [78].

A mission thread is the overarching sequence of events and corresponding activities or tasks planned in response to a specified threat scenario. The sequence diagrams are then populated with objects from the AV-2 library and others added if needed. The sequence diagrams both provide:

- Understanding of the as-is and to-be operational and data flows associated with satisfying key mission and interoperability capabilities;

- Foundation for developing system architecture.

The SV-4 Systems Functionality Description and SV-2 Systems Communications Description models, respectively, can also be derived from these OV-5s as shown in Figure 5.14.

The elements of Figure 5.14 are labeled to simplify its explanation as follows:

1. Leveraging ACE-generated artifacts described in Figure 5.13, steps 2, 4, 5, 6 and Figure 5.15, steps 1-5, *to-be* systems engineering supporting the execution and implementation of the stakeholder review-selected spiral is supported with integrated architecture tools. *Breadth* aspects supporting follow-on engineering stem from work-to-date done with the static models (integrated dictionary referred to as AV-2 in the DoDAF). The *depth* activities associated with dynamic models support follow-on systems engineering associated with capability improvement linked to improved performance measures.
2. Dynamic modeling sequence diagrams are expanded with specific-system component selections traded for implementation in to-be architecture.
3. A system-of-systems model provides a system component perspective for the entire enterprise. Although this is a large, complex model, its creation and maintenance is simplified by modern architecting tools that keep track of the individual system components along with their system attributes and associations to other system components. That way, when multiple systems components are "dragged and dropped" onto a common architecting tool palette, all the related associations and attributes are inherited, or brought into this new view.

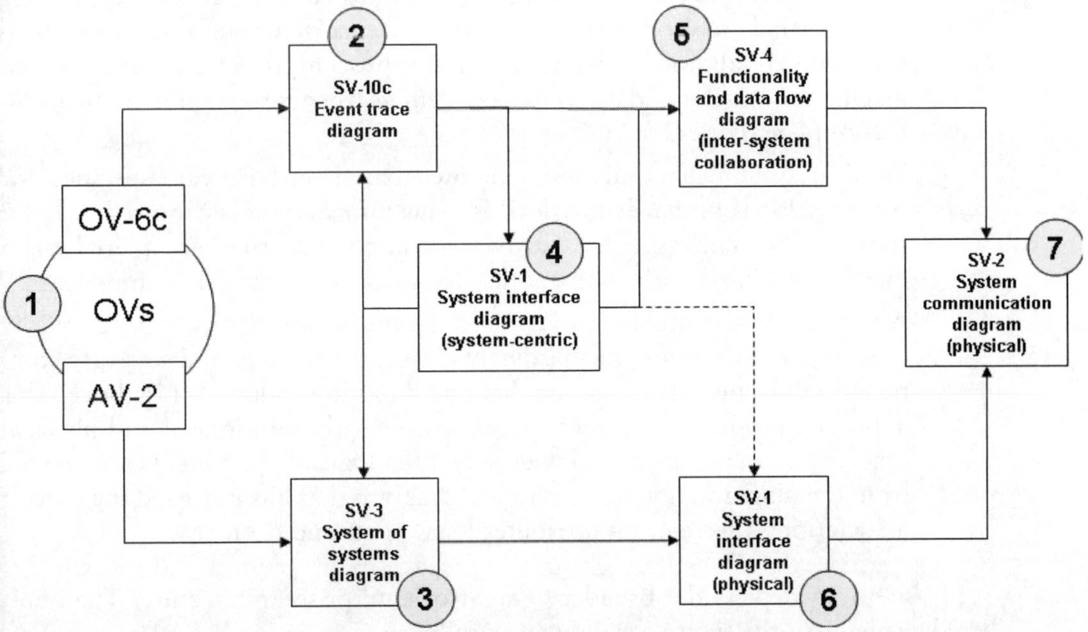

Figure 5.14 Derivation of additional architecture views.

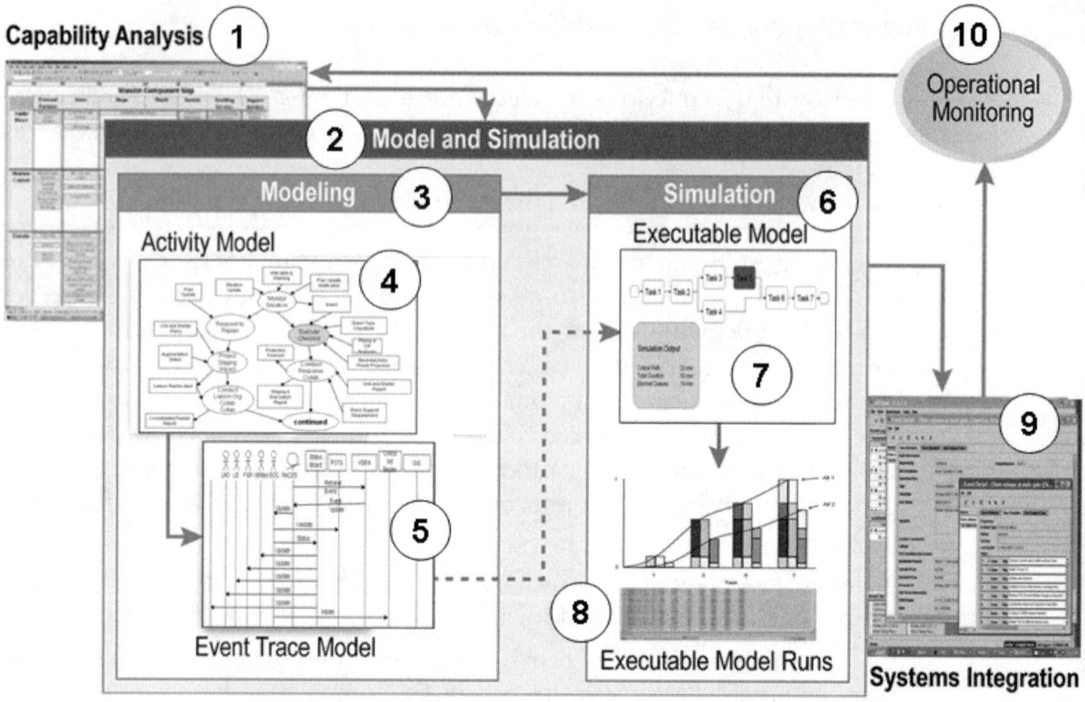

Figure 5.15 Architecture simulation environment.

4. Static-model system-component diagrams are further refined from an improved understanding of system component usage in context of further abstracted sequence diagrams realized in more detailed use-case scenario definitions.
5. Functionality and data-flow diagrams are automatically generated by using an integrated architecture tool to invert sequence diagrams into communications and collaboration diagrams. This approach provides a quick-look at machine-to-machine data flow as well as the corresponding "man-in-the-loop" activities.
6. Physical system diagrams are generated based on hardware specified for software loads on hardware devices. This information is captured in AV-2 static model packages for hardware components that are paired on a one-for-one basis with static models for associated software components.
7. A physical systems communication diagram is automatically produced by dragging and dropping the hardware classes from the paired software components onto the same functionality and data-flow palette in (5) above. This approach produces a temporary worksheet with both logical and physical objects automatically associated. When the logical components are deleted from the diagram, it leaves a physical diagram that has inherited the correct associations and relevant attributes (e.g., ports and protocols).

Figure 5.15 depicts the broader context of enterprise architecture. The numbered elements of the figure are defined as follows:

1. Capability analysis provides broad enterprise perspective to identify candidate areas of enterprise dysfunction that could benefit from ACE. This identification is done by viewing major enterprise operating units and subordinate components against key processes for monitor, control, and execution. The performance of these key processes relative to the associated operating components can reveal gaps, overlaps, and overextension when applied to a high-level scenario, such as national planning scenarios.
2. The results of capability analysis then supports the start of modeling and simulation activity. Enterprise business components revealed as dysfunctional relative to one or more high-level scenarios are able to be linked with modeling use cases. This is achieved by digital linkage between scenario activities contained in capabilities analysis and use cases contained in modeling.
3. Modeling focuses on enterprise system-of-systems analysis, both breadth and depth. Breadth is provided by static models that capture the organization/operator objects, systems objects and mission objects (e.g., high-level conceptual information objects) that comprise the enterprise and populate the enterprise integrated dictionary/repository. Depth is provided by dynamic models that relate enterprise objects to sanctioned enterprise doctrine, CONOPs, and standard operating procedures.
4. Activity models provide the initial basis to understand an emergency response or threat-based use-case scenario, depicting process relationships with mission objects, and relating to approved enterprise doctrine, CONOPs, and standard operating procedures.
5. Sequence diagrams provide information flow-level detail for the specific interaction between organization/operator objects, system objects, and mission objects to fully understand an activity model.
6. Simulations take models and create an executable architecture for evaluation against performance measures.
7. Sequence diagrams are used to create associated simulations that are loaded with timing and resource detail.
8. Token values (e.g., automobiles/second, CBRNE sensor alarms/hour, etc.) are loaded into benchmark and option simulations to support comparative analysis in terms of improvement to time and/or resource footprint.
9. Implementation engineering is aided with by-products from simulations such as business process execution language scripts that help automate the integration of system services supporting operations.
10. Enterprise monitoring tools provide an understanding of variance between predicted simulation operations performance and actual real-world distributed OPCEN operations performance. If variance is significant, further capabilities analysis is conducted to possibly identify systemic enterprise deficiency for follow-on engineering spirals.

Figure 5.16 shows the numbered salient points explaining the task elements of the simulation runs for the catastrophic evacuation example in Section 4.5.5.2. These are explained as follows:

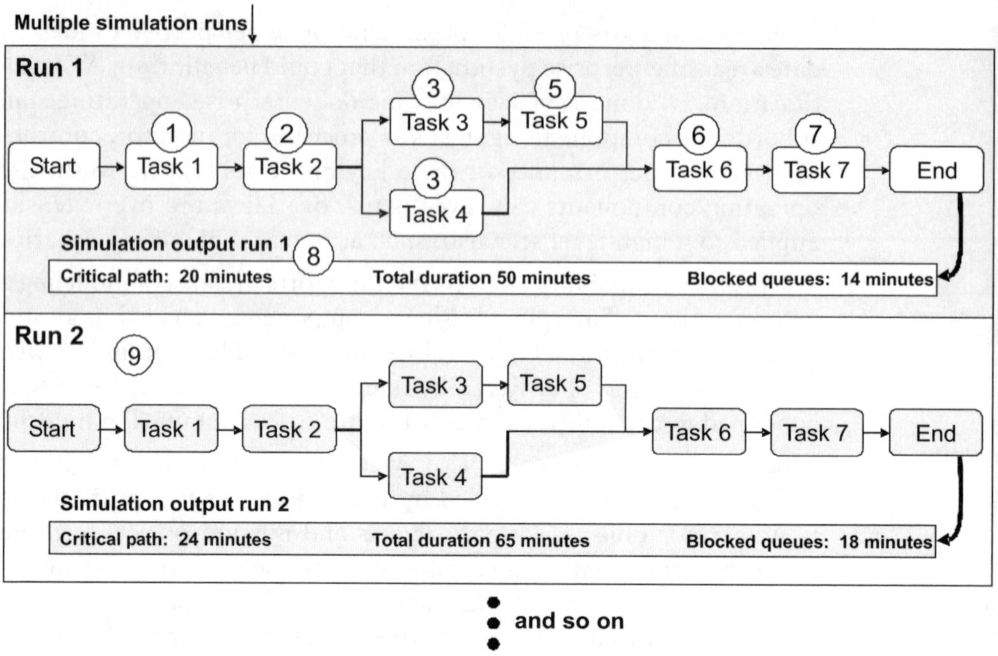

Figure 5.16 Simulation of architecture performance.

1. Monitor traffic flow and resource consumption sensors and portals.
2. Correlate flow/consumption against control limits and generate threshold alert when a control limit is breached.
3. Simulate the flow of traffic and project the location of future backed up traffic, traffic volume and waiting time in backed up traffic.
4. Initiate checklist-linked operator event(s) for multi-OPCEN nodes based on threshold alert and dissemination of the alert to only those operators who need to respond.
5. Query and retrieve asset data (e.g., aid, shelter, augmenting resources) in zone of projected backlog locations—plot both backlog locations and related assets to geospatial visualization backbone.
6. Conduct orchestrated service execution guided by Web service-enabled checklists that supports the automatic workflow of tasks that need to be done in correct order.
7. Join multi-OPCEN node collaboration for execution and after-action.
8. Simulate run based on 10,000 added vehicle flow per major departure route originating within a national capital region.
9. Conduct comparative simulation runs swapping in alternative systems and roles options.

Using the business process modeling notation simulation standard [79], systems event-trace description (SV-10c) sequence diagrams are used to generate simulation task networks that can instrument a simulation with *timing* and *resource-load* information through entry of this data in each individual task node. The timing data is pulled from duration information associated with each business process modeling notation task node while resource load data includes information about the roles, licenses, and so forth, used for each business process modeling notation task node. Using this instrumented simulation, information and action deficiencies such as capability bottlenecks, gaps, and overlaps are empirically derived and measured. This information is used to compare results with benchmark performance metrics to support recommendation for improvement and stakeholder review.

5.7.4 Experimentation

Concurrent with modeling and simulation activity, IT experimentation assets are employed to establish a reference pilot architecture for catastrophic evacuation spanning several governance levels. The reference pilot architecture provides a basis for risk reduction by proving ACE architectural assumptions, verifying vendor trade parameters, updating technology trade studies, improving team learning curve, improving vendor partner collaboration/teaming, providing a service-oriented-architecture-based rapid prototype, and improving fidelity of bill-of-materials and basis-of-estimate costs.

Requirements remain fundamental, but not central, to implementation of the ACE methodology. For this reason, ACE can be used to identify dysfunctional aspects in the as-is public safety and emergency response interoperability enterprise architecture. Once the key problems and opportunities are identified, associated derived requirements then provide the basis for further systems engineering. Architecture modeling helps set up architecture simulations that support bottleneck analyses that focus on process time bottlenecks and areas of excessive resource consumption (e.g., operator touch time). Through architectural modeling and simulation, performance measures are provided for the as-is benchmark and alternative to-be improvement options. These options and their simulation performance results are then available for stakeholder review, thus providing a facts-based decision justification (see business case analysis in Section 4.6) for candidate option selection and implementation.

ACE approaches have been limited by an inability to scale architecture tools usage for large-systems engineering efforts. When graphical architecture tools were all that were available, end products were limited to providing just a basic visual guide for actual systems engineering and development. Today's tools provides an industrial strength architecting tools environment that supports almost seamless production flow of engineering artifacts with such traditional systems engineering functions and test case development, ports and protocols annex, equipment bill of materials, server floor drawings, build plans, functional design diagrams, technical requirements documents, and so forth. Architecting tools that are commonly found in today's engineering houses enable the extraction of various engineering perspectives, such as those described above, into standard word processor documents. These documents often provide the basis of systems engineering deliverables (see Section 4.8). By deriving these documents from an architecture repository that supports multiple version baselines, drawings can be quickly regenerated in response to multiple baseline variations and excursions, and at the same time ensure internal accuracy and integrity.

The ACE methodology [80] is successfully supporting the development of several large-scale, complex system-of-system enterprises. This approach brings together the concepts for transforming vision into improved public safety capabilities, as outlined in this chapter.

5.8 Summary

This chapter has described many fundamental aspects of operational architecture for public safety and emergency response, culminating in an architectural modeling and simulation approach useful for evaluating alternative configurations for best performance. This approach can be used to determine the human and IT/system operations needed among disparate OPCENs in a locale or region to optimally prepare or respond to a homeland security event. Since this approach also allows generation of the bill of materials, we can directly apply the best-practice cost/benefit analysis methodology described in Section 4.6 using alternative architectures and their associated costs from the ACE approach. Thus, ACE permits planners to develop best-value "interoperability by design" integrated OPCEN operational and system architecture, which establishes and sustains the resources, approach, architecture, and performance interoperability planning success factors (see Chapter 1). In this regard, we have presented a fundamental technique used by the World Interoperability Success Plan described in Chapter 7 for mitigating and eliminating the Pearl Harbor–9/11–Katrina deficiencies that define the World Interoperability Failure model.

References

[1] DoD Architecture Framework Working Group, *DoD Architecture Framework, Version 1.0, Volume I: Definitions and Guidelines*, 15 August 2003, p. 1–1, available at http://www.tricare.mil/jmis/download/DoDArchitectureFrameworkVersion1Feb2004.pdf.

[2] Ibid.

[3] Ibid., p. 1–2.

[4] Ibid.

[5] DoD Architecture Framework Working Group, *DoD Architecture Framework, Version 1.5, Volume I: Definitions and Guidelines*, 23 April 2007, p. 1–5, (hereafter referenced as *DODAF 1.5*), available at https://acc.dau.mil/CommunityBrowser.aspx?id=22209.

[6] Ibid.

[7] Ibid., p. 1–6.

[8] Desourdis, R. I., Jr., et al., *Emerging Public Safety Wireless Communication Systems*, Norwood: Artech House, 2001, pp. 9–27.

[9] Clemm, A., "Part I Network Management: An Overview, Chapter 1 Setting the Stage," *Network Management Fundamentals*, Indianapolis, Indiana: Cisco Press, 2006, , p. 8.

[10] Jacobson, C. P., *Quality Measures in a Workflow-Enabled Software Engineering Environment*, draft Ph.D. dissertation, George Mason University, p. 77.

[11] HM Government, *Fire Service Manual, Volume 2, Fire Service Operations*, 3rd Ed., London: The Stationary Office, 2008, p. 18, (referenced hereafter as *Fire Service Manual*).

[12] *Fire Service Manual*, p. 131.

[13] The Institute of Electrical and Electronics Engineers (IEEE), *IEEE Guide for Information Technology, System Definition-Concept of Operations (ConOps) Document*, Software Engineering Standards Committee of the IEEE Computer Society, March 1998, p. 2.

[14] Federal Emergency Management Agency, *National Response Framework*, Department of Homeland Security, January 2008, p. 75, (referenced hereafter as DHS *NRF*), available from http://www.fema.gov/emergency/nrf/.

[15] Federal Emergency Management Agency, *Developing Effective Standard Operating Procedures for Fire and EMS Departments*, FA-197, December 1999, p. 3, (referenced hereinafter as FEMA *SOP*), available from http://www.usfa.dhs.gov/downloads/pdf/publications/fa-197-508.pdf.

[16] Desourdis, et al., Ibid., p. 11.

[17] FEMA *SOP*, pp. 3–4.

[18] Accepted operating practices within an industry or profession.

[19] FEMA *SOP*, p. 8.

[20] Ibid., p. 26.

[21] Ibid., pp. 30–31.

[22] Ibid., pp. 33–42.

[23] Ibid., p. 36.

[24] Ibid., p. 34.

[25] Ibid., p. 35.

[26] Ibid., p. 36.

[27] Ibid., pp. 39–40.

[28] The Public Safety Interoperable Communications (PSIC) grant program in the United States required adherence to a specific annotated strategic plan template.

[29] *FEMA SOP*, p. 37.

[30] Ibid., pp. 38–39.

[31] Ibid., p. 40.

[32] Ibid., p. 41.

[33] *Fire Service Manual*, p. 52.

[34] Ibid.

[35] Local authorities, police (local and British transport police), fire and rescue authorities, ambulance services, national health services (including primary care trusts, strategic health authorities and local health boards), health protection agency (the health protection agency will represent port health authorities if applicable), environment agency, and maritime and coast guard agency.

[36] *Fire Service Manual*, pp. 52–53.

[37] Ibid., pp. 53.

[38] Ibid.

[39] Ibid., p. 54.

[40] Ibid.

[41] Ibid., p. 56.

[42] Ibid., p. 7.

[43] Ibid., pp. 20–21.

[44] Ibid., pp. 23.

[45] Ibid., p. 25.

[46] Federal Emergency Management Agency, *National Incident Management System*, Department of Homeland Security, December 2008, (hereafter referenced as FEMA NIMS), available at http://www.fema.gov/emergency/nims/.

[47] Federal Emergency Management Agency, "ICS Review Materials," *Incident Command System Training*, May 2008, p. 1, (hereafter referenced as FEMA ICS), available at http://training.fema.gov/EMIWeb/IS/ICSResource/assets/reviewMaterials.pdf.

[48] Ibid.

[49] Ibid., p.3.

[50] Ibid., p.4.

[51] FEMA *NIMS*, p. 107.

[52] Ibid., p.109.

[53] Ibid., p. 46.

[54] DHS *NRF*, p. 27.

[55] Ibid.

[56] *NRF*, p. 32.

[57] *NRF*, p. 45.

[58] Derived from Figure 8, DHS *NRF*, p. 63.

[59] Derived from Table 2, DHS *NRF*, pp. 58–59.

[60] DHS *NRF*, p. 66.

[61] Ibid., p. 67.

[62] Ibid., p. 52.

[63] Ibid., p. 68.

[64] Ibid., p. 68.

[65] Ibid.

[66] United States Government, *National Strategy for Information Sharing: Successes and Challenges In Improving Terrorism-Related Information Sharing*, October 2007, (referenced hereafter as USG *NISG*), available at http://www.surfacetransportationisac.org/SupDocs/NSIS_book.pdf.

[67] DHS *NRF*, p. 33.

[68] USG *NISG*, pp. A1–4.

[69] Ibid., pp. A1–5.

[70] DHS *NRF*, p. 48.

[71] FEMA *NIMS*, p. 63.

[72] Architecture-centric engineering has been defined in Jacobson, C., *Quality Measures in a Workflow-Enabled Systems Engineering Environment*, Draft Ph.D. dissertation, George Mason University, p. 19, forthcoming.

[73] Ibid.

[74] Object Management Group (OMG), OMG Unified Modeling Language (OMG UML), Infrastructure, V2.1.2, Needham, Massachusetts, November 2007.

[75] Jacobson, C., Keynote address, "Applied Architecture-Centric Engineering Using UML and DoDAF: A Success Story," *20th Annual Systems & Software Technology Conference (SSTC 2008),* Las Vegas, Nevada, April 29—2 May 2008.

[76] *DODAF 1.5*, pp. 1–10.

[77] Ibid.

[78] Ibid., pp. 1–11.

[79] White, Stephen A., *Introduction to BPMN*, IBM Corporation, no date provided.

[80] An overview of the ACE methodology was presented to the *System and Software Technology Conference–2008* as the keynote address (Jacobson, C., Keynote address, "Applied Architecture-Centric Engineering Using UML and DoDAF: A Success Story," *20th Annual Systems & Software Technology Conference (SSTC 2008)*, Las Vegas, Nevada, April 29–May 2, 2008.

CHAPTER 6

Interoperability System Architecture

A sound interoperability system architecture that supports multiple operation centers (OPCENs) can save lives and billions of dollars. Here we will address how to use the best practices planning and operations design work of the previous chapters and create truly interoperable systems. To promote interoperability across OPCENs, system architecture must facilitate predictable, coordinated action among a diverse and large set of stakeholders. The key stakeholders are first responders and hospital-based first receivers (responders-receivers) who require effective tools so that they can readily share and access information across OPCENs.

System architecture guides detailed system design so that systems meet the strategic operational requirements of OPCENs supporting the responders-receivers. An IT/communication system architecture determines the general hardware and software tools necessary for the reliable and efficient information sharing that is so essential to public safety and emergency response interoperability. More specific technical detail, such as the differences between U.S. and international wireless and land mobile radio standards can be found in a previous work [1], which provides a comprehensive treatment of specific technologies and the associated international implementation considerations.

Individual OPECNs each have their own system architecture to support interoperability needs within the OPCEN. To achieve interoperability within the OPCEN, the various technologies deployed as part of that OPCEN's system architecture must be connected to each other so they work together in a reliable manner. Many of the technologies that must be tied together were not originally designed to interoperate, so the process of tying these systems together within an OPCEN necessitates new and innovative processes. This chapter will provides a framework with which to tackle the even more complex technological challenge of tying together the system architectures of more than one OPCEN.

Successful cutting-edge projects such as the Law Enforcement Justice Information System in southeastern Pennsylvania used the processes set out in this chapter. These projects show that it is possible to effectively tie together technologies across dozens of OPCENs and provide first-responders interoperability among products from competing vendors that were not originally designed to work together.

Chapter 3 identified 25 planning deficiencies that can inhibit the proper planning that is required for a system architecture that supports interoperability. Chapters 4 provided best planning practices to help systems planners avoid those deficiencies and improve operations planning.

In Chapter 5 we described how sound operational architecture practices will yield a clear set of operational policies, procedures, training, and other program elements to improve interoperability.

This chapter provides system architecture views to help policy makers, planners, and technology experts identify the technology strategy they need to enable people in different OPCENs to conduct predictable coordinated multi-OPCEN action. We demonstrate below how IT/communication system architecture can be used to plan ahead the technical capabilities necessary for holistic interoperability among disparate operations centers. The OPCEN system architecture in this chapter offers a framework for individual OPCENs to use so they can assess, adapt, configure, design, and deploy the technologies and systems required to support the desired multi-OPCEN operational architecture.

Throughout this chapter we take a "system-of-systems" approach for working with existing technology in order to achieve near-term holistic interoperability.

Chapter 7 then illustrates how to use these best practices to develop a near-term interoperability architecture that would make possible our World Interoperability Success Plan. Chapter 8 concludes by describing emerging technologies that if used to enhance interoperability will mitigate and eliminate the deficiencies discussed in Chapter 2 that led to disasters such as Pearl Harbor, September 11th, and Hurricane Katrina.

6.1 Interoperability Architecture and IT/Communications Systems

To begin our discussion of the role that system architecture plays in delivering interoperability, we start with a simple picture of how people (receiver-responders) in different locations, supported by the systems of their respective OPCENs, work together. Figure 6.1 shows the relationship among people, process, and technology deployed at multiple OPCENs. This picture illustrates the interoperability challenge: stakeholders need to communicate and act through coordinated, predictable, collaborative processes, using reliable technology to achieve a common public safety and emergency response goal.

Coordinated activity is often needed across many different operations centers. OPCENs can be connected to one another through what we will refer to as an OPCEN backbone. The OPCEN backbone is the set of networks and supportive infrastructure services that interconnect individual OPCENs to each other. Technology and systems must be configured to enable collaborative processes across OPCENs. Anything that prevents or impedes the effectiveness of the desired coordinated action is a barrier to interoperability.

The system views presented later in this chapter provide additional details about the technology and systems represented in Figure 6.1, and discuss opportunities and barriers to interoperability that are inherent in currently available technologies. The scope of this discussion covers all forms of electronically transmittable information. The systems and infrastructure considered here are intended to pro-

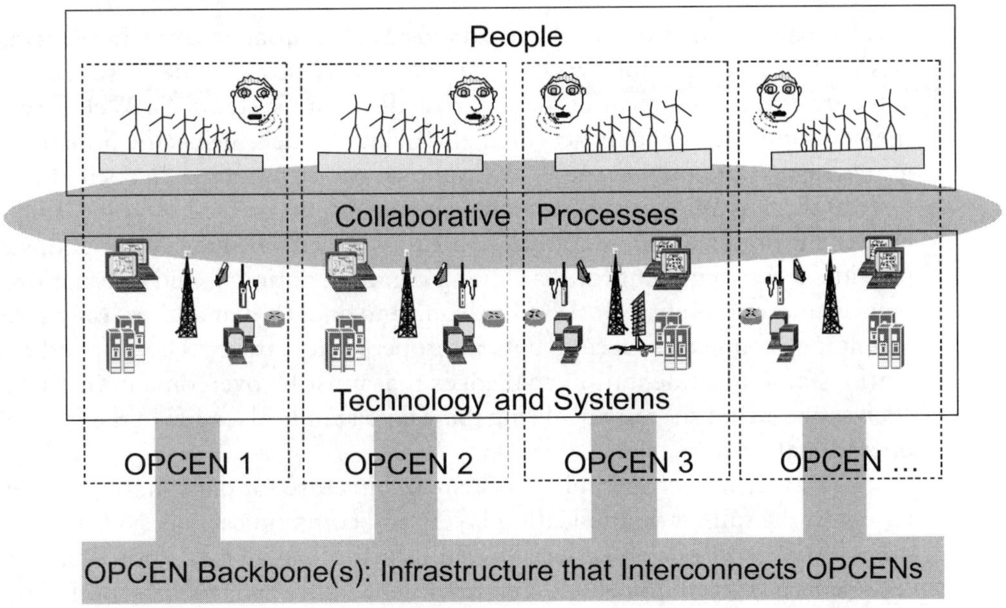

Figure 6.1 Interoperability across OPCENs.

vide the ability to share text, voice, video, images, sounds, sensor data, and any information that may be useful to responder-receivers.

6.1.1 Information Sharing and Application Architecture

A great deal of progress has been made towards promoting system architectures that support simple data exchange according to regional, statewide, national, and international standards. As discussed in Chapter 1, the value, need, and ability to exchange data has led to an expansion of the term "interoperability" in the context of sharing data and voice communications. In the area of data exchange, many domain-specific extensible markup language (XML) schema have emerged, such as emergency data exchange language (EDXL) for the emergency management domain [2] and the global justice extensible data model (GJXDM) for the justice domain [3]. These have been developed in an effort to standardize how data is represented by different systems and to allow the exchange of common data elements based on a shared, agreed upon information format (e.g., XML schema).

In the United States, these domain-specific XML standardization efforts have given rise to the National Information Exchange Model (NIEM) in an effort to coordinate and provide national-level interoperability across multiple information-sharing domains. NIEM is an interagency initiative that leverages established XML schema efforts (such as GJXDM) to provide a coordinated set of interoperable XML standards across domains. Currently the seven NIEM domains are emergency management, immigration, intelligence, infrastructure protection, international trade, justice, and screening. Twenty-five of fifty states are using NIEM, primarily for justice-oriented applications. The NIEM organization is evaluating the criteria for establishing new domains and assessing the capacity to support additional domains [4].

Two powerful initiatives have catalyzed substantial progress in reducing obstacles to data interoperability: the emergence of XML as the de facto technical data standard for communicating data across disparate systems via Web services, and the development of widely accepted XML data models in specific domains. In integrated justice and public safety, for example, this approach delivers real-time information sharing among records management systems offered by competing vendors [5]. The promulgation, adoption, and enhancement of these XML data standards provides a starting point for developing regional, national, and international information interoperability. But XML schema and data format address only some of the technology challenges faced in data interoperability efforts. There are other considerations and implementation challenges that must be overcome in order to deploy Web services that successfully bring these data standards to life in the form of secure operational systems.

To establish interoperability system architecture for data sharing, it is useful to represent the full set of application layers and communications system components that characterize a Web services-based information-sharing approach. Figure 6.2 represents a layered model for Web services [6] and aligns these with the traditional OSI and Internet-layered model [7].

In this illustration, the Web services stack on the right provides a useful picture of the relationship among application, XML schema, and the Web service technologies which provide the underlying messaging. Today's information sharing systems leverage Web services and a growing set of supporting standards and technologies

OSI	Internet	Web Services
Application (Layer 7)	Various (application dependent)	Application
		XML Schema
		Various (e.g. BPEL)
		WSDL MEP
		WS-Addressing WS-Routing
Presentation (Layer 6)		SOAP URI XML Infoset
Session (Layer 5)		HTTP, HTTPS, …
Transport (Layer 4)	TCP or UDP	
Network (Layer 3)	IP	
Data Link (Layer 2)	Ethernet	
Physical (Layer 1)	Various Ethernet media	

XML: eXtensibleMarkup Language
BPEL: Business Process Execution Language
WSDL: Web Services Descriptor Language
MEP: Message Exchange Pattern
WS: Web Service
SOAP: Simple Object Access Protocol
URI: Uniform Resource Identifier
HTTP: Hypertext Transfer Protocol
HTTPS: Hypertext Transfer Protocol Secure
TCP: Transmission Control Protocol
UDP: User Datagram Protocol
IP: Internet Protocol

Figure 6.2 Application architecture for Web services-based information sharing.

to provide application interoperability. While overcoming implementation-specific issues with many of the standards remains the challenge, successful, systems in operation today prove that it is possible to interconnect legacy mainframe systems with disparate client/server applications and tie them in to modern Web-based systems.

6.1.2 Interoperability System Architecture

The number of information systems that are deployed using a service-oriented architecture is increasing; the number of services available within any given OPCEN is growing. Thus it is critical to have a framework for assessing and supporting multiple information-sharing and interoperability initiatives. Many of the products and technologies being deployed today can be configured to interoperate widely with external systems. Where the operational architecture requirements demand it, the technology challenges can usually be overcome. This section will provide a framework for conceptualizing the interoperability architecture across OPCENs. Our new framework will help technology experts, systems engineers, and designers make effective policy choices or coordinate and configure existing products and technologies to meet these requirements across multiple OPCENs. It will also help planners and policy makers visualize how technology will support collaborative actions.

The substantive progress in standardizing XML schema in recent years represents an excellent start in what is termed "information sharing," and there is growing collaborative effort to continue to improve these data exchange standards. The current use of Web services as the standard messaging technology to deliver information among disparate systems enables this data interoperability. Still, for systems to truly interoperate as needed, bridges between systems must function at many levels. This is particularly the case when real-time information sharing is vital, as in the public safety and emergency response area.

Based on lessons learned in several multijurisdictional information-sharing initiatives, as well as the Web services architecture discussed in the previous section, we will develop a general interoperability system architecture.

Figure 6.3 provides a more tactical view of the levels at which interoperability may be built. These are also the layers at which potential barriers to interoperability can impede or prevent information sharing. This interoperability system architecture combines elements of network, platform, system, data, messaging, and application architecture. Interoperability system architecture focuses on the degree to which the user at one operations center (OPCEN 1) is able to share information with and access information from another operations center (OPCEN 2).

Figure 6.3 illustrates and addresses the issues that arise with IT system integration efforts to share information across operation centers. The layers are selected and ordered specifically to provide a user centered view of the systems and technology. The two dotted boxes represent two OPCENs that need to collaborate. The upper three layers that comprise the people components represent the user aspects of interoperability which must be understood and considered in creating the system architecture. Humans ultimately interact with a device of some sort, for example a laptop computer, a mobile data terminal, a radio, a camera or a cell phone. For this reason the device is placed at the top of the systems stack. If the device in the user's

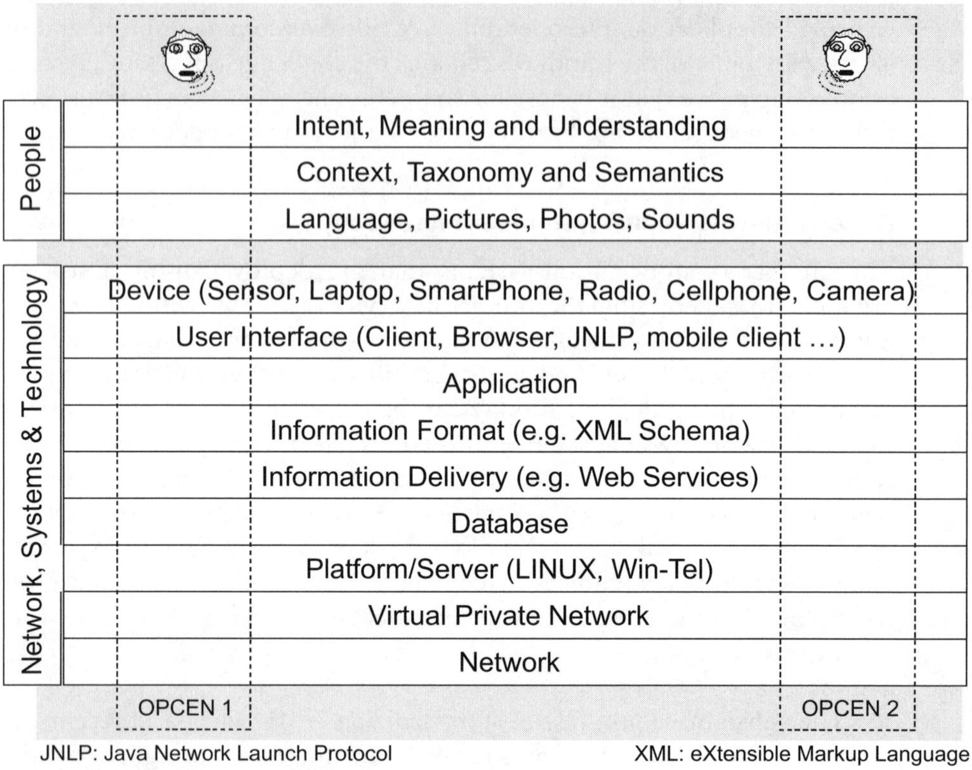

Figure 6.3 Interoperability system architecture.

hands is functioning properly, the user will see or hear words, pictures, sounds, or video images. The lowest "people" layer is the raw representation or expression of information in human terms. This data is experienced within some context (in a particular emergency response situation, or in the context of particular training the person has received). Hence, the middle tier represents the other cues that are observed or expressed in association with the information. How well the user will understand what is presented through the system is an important design consideration, which is why these layers are important in an interoperability system architecture. The top layer in our architecture represents the intent of the person sending the information, the understanding of the person receiving the information, and the (hopefully) shared meaning. While less technical and difficult to specify, this layer of the architecture brings an important set of design criteria that requires understanding the human factor. If the system does not support shared, trusted understanding, then there will ultimately be a failure of interoperability.

The lower nine boxes that comprise the network, systems, and technology components represent nine potential barriers to interoperability as well as nine potential bridges to interoperability. For example, OPCEN 1 and 2 may run different applications that utilize different database technologies and user interfaces. However, if the Web services and XML schema are designed to a common standard, these disparate systems will be able to share information via the Web services and XML schema layer. The days of requiring each OPCEN to run the same application are over. The layers that must interoperate, however, are the people layers.

Considerations of how the human responders-receivers will interpret and understand the information they will access are important design factors. The system design must take into account the intent of responder-receivers when they hit certain keystrokes on a keyboard, or say a particular word on the radio, must be taken into account in system design. These considerations are important, for example, when assessing data content that may be shared from one system to another. While a piece of data may be mapped via a shared schema and made available to OPCEN 1 through a Web services interface from OPCEN 2, if the person receiving the data has a different understanding of the meaning of the data, then interoperability has not been achieved. How people will interpret the information they receive from the systems are important considerations within the system architecture because it affects system design. System architecture may be influenced by usability considerations at the device, the user interface, the schema, or the data content layer.

6.1.3 Information-Sharing System Architecture

Whether assessing and enhancing the interoperability of existing OPCEN IT systems, or planning new systems, the framework of Figure 6.3 can be used to identify barriers as well as bridges to information sharing. If we consider modern information-sharing systems to be ones that utilize Web services and a shared XML schema as the basis for data interoperability, then we can imagine that the two OPCENs need not be interoperable at every layer. They do need to be using the same application, database technology, platform, or even user device. Figure 6.4 shows the two OPCENs interoperable at basically three levels. First, the network and the virtual private network provide secure connectivity between OPCENs over some OPCEN backbone. Second, information delivery and information content (e.g., the Web services and common XML schema) provide the basic ability to move information between OPCENs. Finally, the people must be trained to see, interpret, and understand the information that is being served to them via their local application, connected via Web services with other systems. While Web service technology is the current preferred means of delivering data among disparate systems, we should note there are still challenges associated with Web services in bandwidth constrained networks, and networks with large losses and associated TPC retransmission. Simple Object Access Protocol (SOAP) and Hypertext Transfer Protocol (HTTP) do not always behave well in such lossy networks. Despite these shortcomings, Web services represent the current standard for information delivery between systems, though this is certainly not the only available option.

Information sharing and interoperability require communications among many, not just two, OPCENs. Figure 6.5 shows a generalized system architecture for implementing the interoperability system architecture of Figure 6.4 among many OPCENs. OPCENs A, B, and C each have an interoperability network interface that enables information sharing on an OPCEN backbone that interconnects OPCENs. The dark-shaded OPCEN backbone shown in Figure 6.5 represents the set of network and infrastructure services that interconnect OPCENs.

Each OPCEN in Figure 6.5 may have a diverse set of systems within the systems box, and these details will be further explored in the following sections. We will also

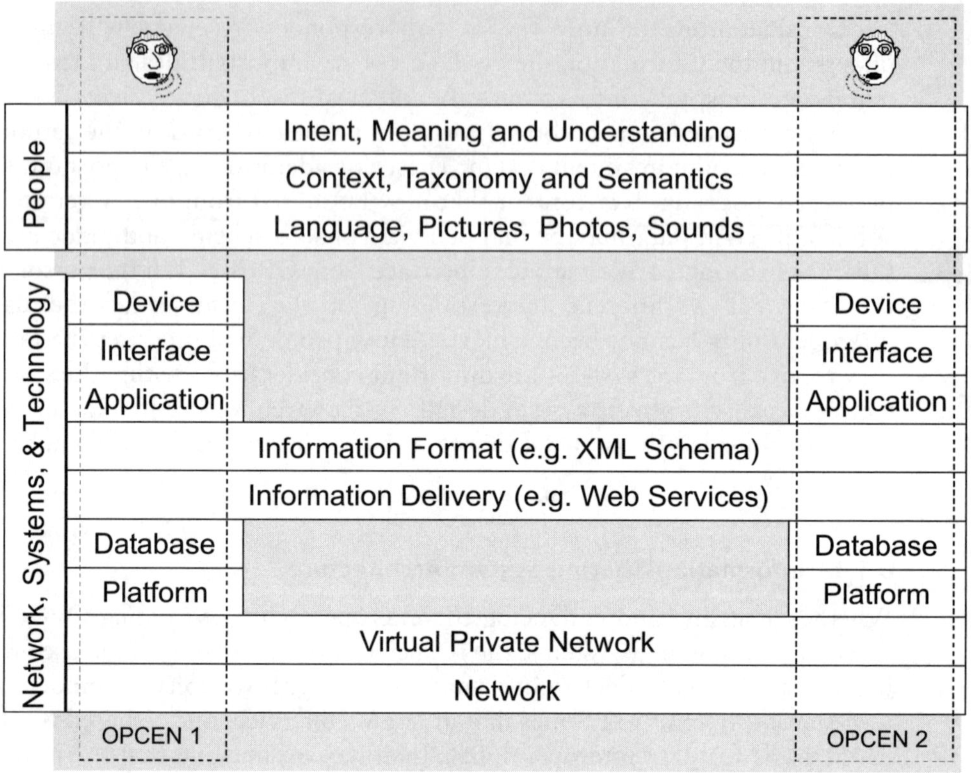

Figure 6.4 Information sharing between OPCENs.

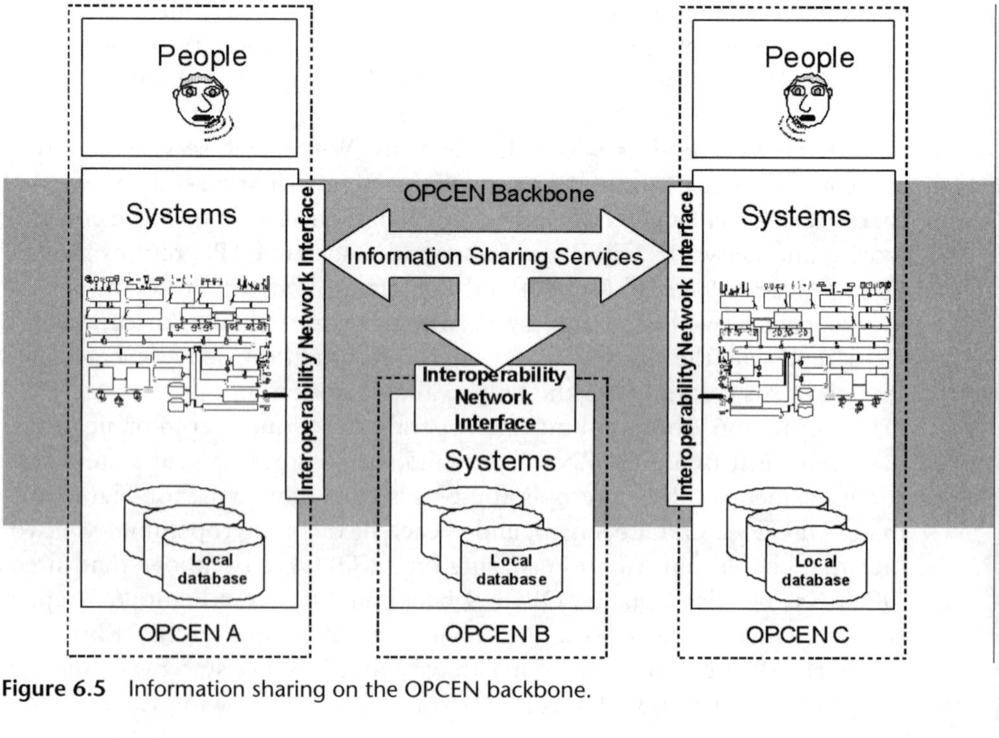

Figure 6.5 Information sharing on the OPCEN backbone.

illustrate the various ways OPCEN backbones can support secure connectivity and information sharing among OPCENs by interconnecting via interoperability network interfaces.

The following sections develop and describe operation center system architecture and provide an extensive set of scenarios that can dramatically improve interoperability through the application of current technologies and the principles and practices already described in Chapters 4 and 5.

6.2 OPCEN System Architecture

Next we consider interoperability among two or more OPCENs and define the technologies needed to implement their collaboration functions for situational awareness and command coordination. For simplicity, we will assume that each OPCEN supports one public safety or emergency response organization (or agency). If multiple agencies are supported by a single OPCEN, we will consider these multiple agencies as a single organization from the system architecture perspective. Each agency may have associated air, land, or water-mobile units *homed* at its OPCEN and employ a variety of radio/wireless means to communicate with those units when in the field. If outside organizations provide mutual aid, or engage in joint or task force operations that use the information resources of a mobile unit's home OPCEN, such organizations will be named "visiting agencies." The visiting agency's OPCEN or mobile units may draw information from—and provide information to—the home agency via its OPCEN and mobile units. Such information sharing will be defined from the best practices set out in Chapter 4 to mitigate and eliminate the Pearl Harbor–9/11–Katrina deficiencies defined in the Interoperability Assessment Checklist.

Figure 6.6 depicts elements of the system architecture that are needed to perform the functions shown in Chapter 5, Figure 5.4. Figure 6.6 shows representative functions needed to support the tasks and activities of the generic public safety and emergency response OPCEN, and certain data flows to support the functions of Chapter 5. Also depicted are the general system-of-systems and associated general technologies that implement these functions. This illustration provides the necessary background for a description of integrated, composite, and federated interoperability architecture. In other words, we will describe the hardware and software that performs existing OPCEN functions so we can then characterize a near-term holistic interoperability architecture.

Here we consider two configurations of OPCENs, one configuration with minimal technology due to limited resources, and a second with state-of-the-art technology from abundant resources. Although the operational architecture for both configurations is identical (see Figure 5.4), and the same functions are attempted, the systems, subsystems, and technologies employed differ significantly and so does the system performance. Given the 25 deficiencies inherent in systems within our World Interoperability Failure model, however, the more advanced technology in the maximal configuration does not necessarily outperform the minimal configuration. Although this maximal technology configuration may allow faster and auto-

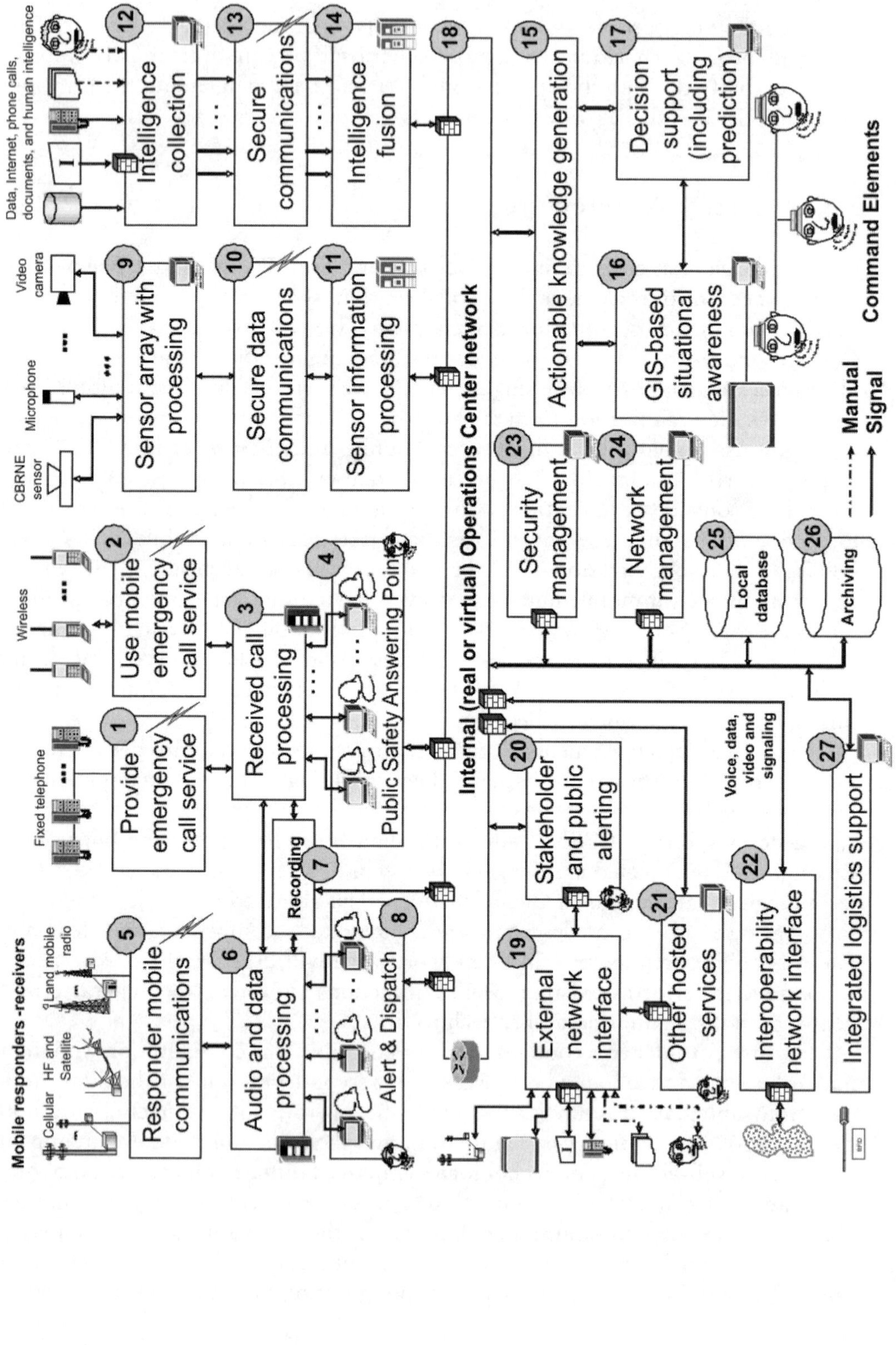

Figure 6.6 Public safety and emergency response OPCEN architecture.

matic information sharing among OPCENs, the 25 deficiencies may restrict or prevent such sharing irrespective of technical capability.

The technologies and systems that typically support the 27 identified OPCEN functions are described for each function as follows:

1. *Emergency call service:* The public switched telephone network (PSTN) is used to process and route calls via telephone company switches to emergency services, including the routing of calls from the public to available call-takers (potentially using automatic call distribution technology), plus call forwarding and retrieval.
2. *Mobile emergency call service:* The processing and routing of mobile (wireless) phone calls from the public for emergency service, which includes routing such calls to available emergency call-takers (potentially using automatic call distribution).
3. *Received call processing:* In addition to the PSAP call-direction switch, this processing includes retrieval of Enhanced ANI/ALI from the service provider (e.g., Enhanced 9-1-1 or E9-1-1 in the United States, 999 in the United Kingdom and British commonwealth countries, and 112 in much of Europe). (112 is also the number used by international mobile networks based on GSM and will require positional information from the cellular operators as in E9-1-1.) Future possibilities include the delivery of images, video, and text—such as immediate message (IM) or short message service (SMS) and telematics data to the PSAP. This function can be supported by computer-aided dispatch (CAD) applications for initial incident data capture. The CAD software may locate the caller's position on a GIS showing mobile units in the vicinity of the caller as well as their availability and status. (In the United States, E9-1-1 Phase I requires that carriers provide the location of the serving cellular tower site, while Phase II requires accuracy to 100 meters. E9-1-1 is available in those PSAPs equipped for the capability and when supported by the cellular carriers.)
4. *Public safety answering point (PSAP):* The telephone audio equipment, including headsets and attached microphones that enable trained emergency call-takers to speak directly to the caller provide an environment conducive to one-to-one voice communications. PSAP call takers can then direct the calls to the appropriate dispatchers or other organizations.
5. *Responder-receiver mobile communications:* These systems could include dedicated or private wireless systems individually owned and operated as a shared resource with other agencies. Examples are Project 25 700 MHz/800 MHz narrowband voice systems and 4.9 GHz broadband data systems. Commercial wireless systems, such as voice cell phone and EVDO (evolution, data only or evolution, data optimize) data systems are also shown. For highly irregular terrain or large geographic areas, satellite or HF radio systems may also support mobile responder-receiver communications.
6. *Audio and data processing:* Audio processing includes trunking and simulcast system control functions such as assignment of remote site assets to radio calls, removal of radio system signaling for extraction of the intelligence, vocoding, and signal voting [8]. For data, this configuration includes re-

moval of radio system signaling for extraction of the intelligence, formatting as required by the radio system, routing of data to the appropriate radio channels, and packet or message switching services [9]. Both audio and data processing may involve encryption and decryption.

7. *Recording:* Analog or digital recorders are used to store all PSAP and mobile voice communications. Minimal configurations of this capability store only PSAP calls for some weeks or months, accessible by standard playback to find specific calls. Maximal configurations will employ all digital recording with immediate playback features as well as long-term archival storage. In addition, data retrieved and collected by OPCEN situational awareness systems can also be recorded, providing a detailed digital history of all information flows supporting OPCEN tasks and activities.
8. *Responder-receiver dispatch:* In minimal configurations, the dispatcher may employ only push-to-talk access to a small number of voice channels, typically a single command channel and a secondary channel for talk-around (i.e., direct mode) or repeater-relayed communications. Dispatcher knowledge of mobile units and the operating geography and environment provide important situational awareness at the OPCEN in support of supervisor command and control. This knowledge can be replaced—or greatly assisted—through the use of CAD systems that provide GIS situational awareness for all mobile assets employing automatic vehicle location (AVL) systems. Conventional and trunked radio control features can be implemented by push-button consoles, specific radio control applications with special-purpose graphical user interface (GUI), or even GIS-based point-and-click access to these functions on the CAD GIS display. In addition, video streams, alarm notifications, sensor array output, and other important data for the tactical environment can complete the dispatcher's tactical situational awareness.
9. *Sensor array with processing:* Multiple sensors in array, such as chemical, biological, radiological, nuclear, or explosive (CBRNE) detectors; microphones (weapon discharges, vehicle presence, spoken words, etc.); RF identification (RFID) device (tracked motion, ceased motion, periodic status transmissions, etc.); video (thermal/visual wavelength motion/exception-triggered, thermal triggered, differentially triggered, etc.); magnetic (metallic devices); and others. This function includes compression of sensor array data—including potential preprocessing for data reduction—to minimize the bandwidth required for wired and wireless transmission of the sensor data.
10. *Secure data communications:* A wide variety of modes and media used to bring processed sensor data to the OPCEN.
11. *Sensor information (including video) processing:* Technologies that can support sensor data processing. These include video analysis, pattern recognition, anomaly detection, timestamp correlation, statistical analysis, exception reporting, and alerting.
12. *Intelligence collection:* Technologies that support intelligence collection, including electronic surveillance, speech recognition, keyword detection, se-

mantic indexing, cryptographic interception, and image, video, and audio collection and recording.

13. *Secure communications:* One or more media are used to communicate the captured data to provide information assurance (IA) between the collection and analysis functions. No data loss, disruption, interception, or decryption is acceptable in this end-to-end transmission.
14. *Intelligence fusion:* The received information content from one or more sources is analyzed collectively to determine capability, intent, timing, and so forth, in the context of other available information (e.g., sensor data, responder-receiver reports, ongoing investigations, PSAP calls, etc.).
15. *Actionable knowledge generation:* Data correlation filters and integration processes are performed in software applications available at OPCEN console positions. Mobile unit locations, recorded emergency calls, all forms of intelligence, processed sensor data, responder-receiver communications, visiting OPCEN situational awareness, and databases and alerts can be fused to provide knowledge of the intended scope of manmade threats (e.g., blast radius), or forecast potential outcome of natural events (e.g., flooding).
16. *GIS-based situational awareness:* Only location-based information can be geocoded, converted actionable information into a format for effective and efficient presentation to command elements. Base maps, orthophotography, and other geospatial technology allow systems to show the location of incidents and OPCEN/other assets in terrain, built-up areas, or within structures in a spatial or geographic format. Using the internal network and data communications, this awareness (arguably modified for mobile or remote users) can be pushed to mobile responders-receivers by dispatchers using mobile applications. This display can be triggered by event-based alerts, such as a door position sensor, heat/smoke sensor, excursion of a mobile unit into identified exclusion areas.
17. *Decision support (including prediction and forecasting):* This function uses actionable knowledge from sensor, intelligence, public, responder-receiver, and external information-sharing networks to simplify and prioritize the decisions-making process for OPCEN command elements. In addition, predictive tools that assess theoretical consequences of hypothetical or real events (e.g., release of a toxic plume) can be used in combination with other data (e.g., prevailing winds, interposed terrain) to predict impact contours pushed to become an overlay to the GIS-based situational awareness display.
18. *Internal network:* This system represents the internal information-sharing network within the OPCEN. It may be implemented in a variety of ways, including:
    - Simple manual means (e.g., "pen and paper" recording with "sneaker net" and telephone message delivery and acknowledgement similar to the December 7th Information Center guarding the Pacific Fleet as described in Chapter 2;
    - Hardwired connection between functions, such as a telephone line or data (network) cable;

- Advanced digital recording and integrated situational awareness provided over service-oriented architecture (SOA) networks. These SOA networks also link through firewalls—including sneaker net—to various external networks for public alerting, responder-receiver alerting, and other external functions.

19. *External network interface:* This interface provides the interfaces to the diverse media both to deliver responder and public alerting, and provide access to external hosted services. A variety of media may be used for these purposes including audio alarms, landline phones, commercial cell systems, paging systems, emergency radio broadcast systems (e.g., AM/FM radio, radio-broadcast television or cable), the Internet and commercial and private broadband networks.
20. *Stakeholder and public alerting:* Alerting often includes the selection of the most appropriate media from those available. A number of vendors offer these applications. Examples of alternative alerting media include reverse 9-1-1, Amber alerts, group wireless calls to responders, sirens, highway reader boards, and mobile test messaging.
21. *Other hosted services:* This interface accesses services provided external to the OPCEN, which could include many of the functions already described in 1 through 20 above, namely PSAP call taking, CAD, database, geospatial imaging libraries, National Oceanic and Atmospheric Administration (NOAA), and weather and Web hosting. In this way, multiple small OPCENs, such as local responder-receiver dispatching (or staging) facilities (e.g., emergency operations centers, information fusion centers, police precincts, fire stations, clinics, public health offices, public works facilities, etc.) can benefit both in terms if interoperability and benefit/cost analysis (see Chapter 4) using these outsourced services.
22. *Interoperability network interface:* This technology provides network interfaces or gateways to disparate voice and data communications systems, records management systems, remote CAD systems, wireless systems, and so forth. This interface is currently a dedicated T1 or phone line, but increasingly it consists of an intranet or the Internet with suitable firewall protection as shown in Figure 6.1.
23. *Network management:* This capability refers to the activities, methods, procedures, and tools that pertain to the operation, administration, maintenance, and provisioning of networked systems [10]. This includes network management systsems, performance management, configuration management, asset management, and exception reporting, alerting, and management.
24. *Security management:* Technologies that protect the availability, confidentiality, and integrity of the associated information systems with the multilayered, end-to-end prevention, detection, and vulnerability management capabilities. It includes systems that provide identity and access management receiving information and event management.
25. *Local database:* This database provides storage and retrieval of data resident to the local OPCEN, which typically includes CAD, records from OPCEN activities, and GIS data. Much data of potential value available in remote

databases such as criminal, license, hydrant, and hazard data as well as building floor plans may be stored in this local database for ease and speed of retrieval, particularly for information relevant to the OPCEN's jurisdiction.

26. *Archive:* This capability represents the repository for data retained on a longer term basis, such as completed incidents, recorded radio traffic, incident photographs, incident video, and so forth, useful in judicial proceedings, training and exercises, operating procedures, and lessons learned. This archive may be a regional or national asset, not only a local archive.
27. *Integrated logistics support:* This capability includes monitoring of the system architecture for status, including impending or actual equipment failures, and provides all sustainment functions including facility/equipment operations and maintenance (O&M), including operator and maintenance training, critical parts ordering/sparing, among others.

6.3 Backbone-Supported Interoperability

Using best interoperability practice, information sharing is maximized among OPCENs using a backbone network to minimize the need to do so using bandwidth-limited wireless systems. Moreover, this sharing can provide improved situational awareness for command elements and dispatchers before they send their responders-receivers into potentially dangerous environments. For this reason, we will first analyze information-sharing system architecture for OPCENs, as depicted in Figure 6.6, over backbone networks in order to perform the corresponding functions. The information shared between OPCENs can be relayed, or appropriated, "over the air" to field units.

Figure 6.7 shows a visiting OPCEN in the center of the figure with potential home OPCENs connected to the same network backbone. Of course, any OPCEN may be considered as the home OPCEN. The potential home OPCENs are shown without detail to simplify viewing, with the highlighted (nongrayed) functional blocks in these OPCEN depictions representing the active functions for a particular mode of information sharing. The backbone, preferably broadband, supports voice, data, and video communications among the collaborating OPCENs. As the figure shows, this backbone could be implemented using fiber, copper, microwave, or satellite communications—perhaps even a combination of broadband wireline and wireless systems. The wireless backbone systems are useful to interconnect to air-, land-, or water-deployable OPCENs. In all cases, information sharing filters are employed according to SOPs such that only preapproved sharing is allowed managed by the network and security management servers highlighted in every OPCEN. This sharing can be threat-scenario dependent.

Clockwise from the upper right of the figure, these information-sharing modes are as follows:

- *Interoperable RF systems*—mobile voice, data, and video communications among home and visiting responder-receivers and other stakeholders and

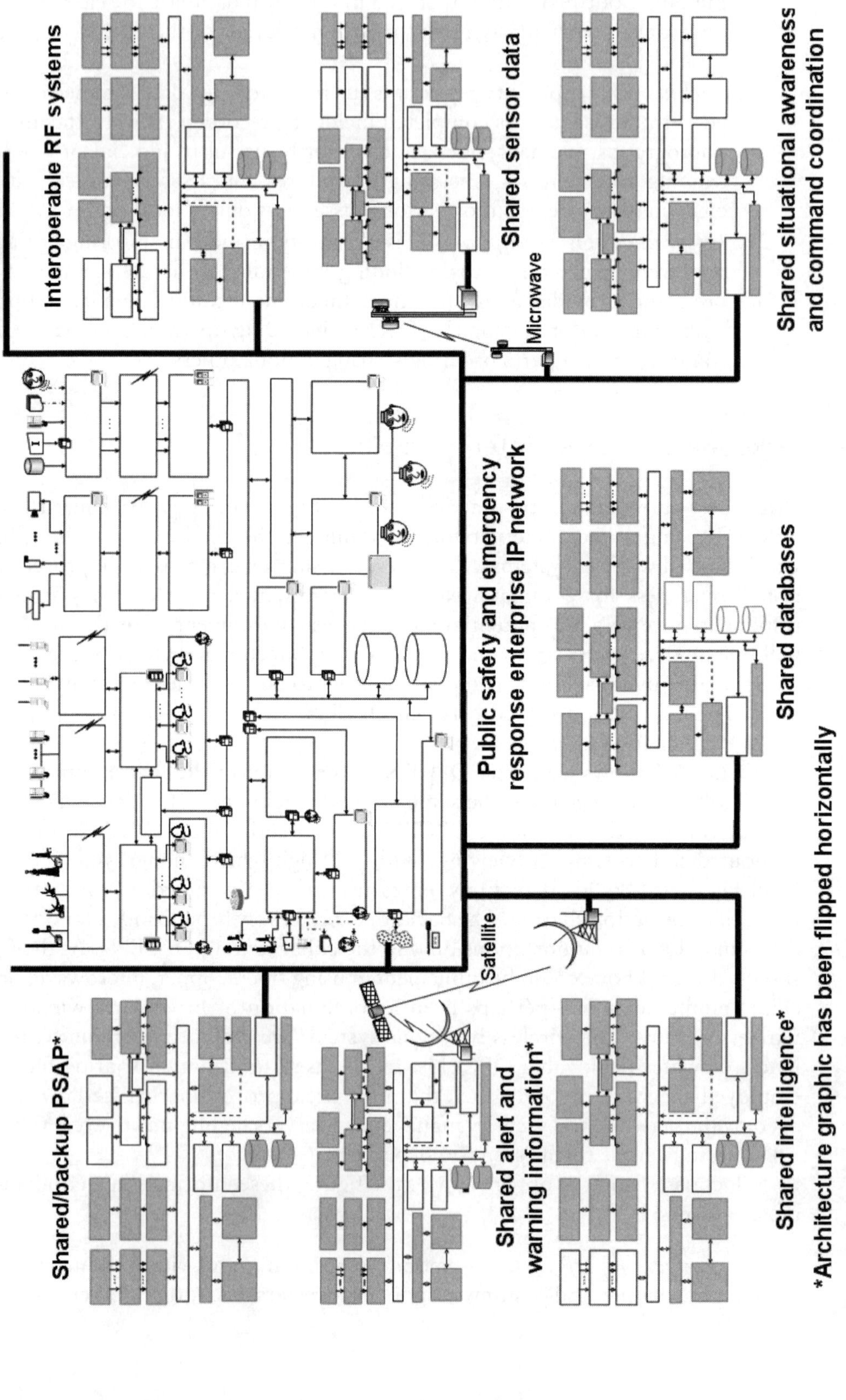

Figure 6.7 OPCEN backbone architecture.

their dispatchers through the OPCEN backbone to support tactical interoperability.

- *Shared sensor data*—provisioning of sensor data among the home and visiting OPCENs to enhance collaborative situational awareness.
- Shared situational awareness and command coordination—command element situational awareness is provided to the visiting and home OPCENs, which can be used to update or improve awareness of all collaborating OPCENs.
- *Shared databases*—the visiting OPCEN gains access to the home OPCEN local or remote databases, and provides access as well to its databases, all under network and security control.
- *Shared intelligence*—raw and fused intelligence under security controls can be shared among visiting and home OPCENs.
- *Shared alert and warning information*—locally, regionally, or nationally generated alerts can be shared among home and visiting OPCENs.
- *Shared/backup PSAP operations*—call-takers at the home and visiting OPCENs can support each other in case of spikes in call traffic or as backup for lost services.

Each of these configurations is detailed in the following sections.

6.3.1 Interoperable RF Systems

6.3.1.1 Gateway Voice Interoperability

Two or more OPCENS can interconnect their respective mobile communication systems through the backbone to enable their responders-receivers and other stakeholders to intercommunicate if they are within coverage of their respective home systems. This configuration is used by the BORder TACtical (BORTAC) system [11] and many other such interconnect interoperability systems. Primarily used for voice communications, at least one of the home or visiting OPCENs must employ a voice communications *gateway*. The use of a gateway between disparate radio systems defines a composite mobile communications architecture.

Gateways are audio interconnect systems that tie together the speaker output and microphone inputs of voice radio systems at block (5) in Figure 6.6 because they all must input or output a voice audio signal. Gateways push spoken audio from the speaker output of one subscriber of one radio system to all subscribers whose radio system microphone inputs are interconnected to that gateway. The subscriber radios that interface the gateway are referred to as donor radios. The gateway may be implemented using two-wire or four-wire audio (telephone lines) or via an IP network (VoIP). The mobile radio systems to be interconnected may operate in the same or different frequency bands and employ diverse technologies. Because audio switch technology preceded that of IP-based gateways, the former is sometimes thought of as first generation and the latter as second generation. These controllable audio switches are integrated into the LMR infrastructure at block (6) in Figure 6.6.

Gateway interconnection between systems is often controlled directly by the system dispatcher or operator. Gateways are implemented as manual patches or remotely controlled software-activated audio switches or IP-based gateways. The specific systems to be interconnected are controlled at the radio *console* used by the dispatcher or OPCEN radio operator using either manual interconnects, push-button, or software-enabled patching capability. These console patches can be implemented between channels or talkgroups on the same radio system or disparate radio systems. For disparate systems, an audio path must be established between the console electronics of the two systems, such as block (18) in Figure 6.6, and the patch requires coordinated action between dispatchers on the interconnected systems. This configuration is shown in the Figure 6.8, in which dispatchers in agency A and B configure their network gateways and console-controlled audio switches to pass the transmitted voice from the subscriber to the cell tower at (a); through the cellular network (b); to the OPCEN at (c); through the audio switch, agency B console, and OPCEN network along (d); pass over the OPCEN backbone to the agency A OPCEN gateway along (e); through the agency A intranet, console and switch infrastructure to the agency A mobile network (f); to the LMR network along (g); to the appropriate tower site along (h); and finally out to the agency A subscriber devices along (i).

If the gateways are IP-based, then they must convert the audio voice stream into IP packets and associated addressing information to allow the exchange of voice between the desired disparate radio systems. This class of products has in common the gateways at the interfaces to the radio systems to be linked as well as servers and clients for processing of the VoIP packets and control of the system. The gateways provide conversion of the protocols of the radio systems to those of the IP-based interoperability system. If the radio system is analog, or if it is digital with an analog interface to the gateway, then protocol *conversion* means conversion from analog to digital, coding the digitized data stream for compression, and insertion of the resulting payload data into IP packets. Again, the use of an IP gateway will support composite interoperability architecture.

If the radio system presents digital IP packets to the gateway, transcoding is required if the radio system does not use the same coding as the interoperability system. In addition, control commands expected by the radio system must be generated. Transcoding and the generation of control signaling for the radio system are not necessarily performed by the gateway. Processing is undertaken to manage the routing of the packet streams to the appropriate gateways for the desired interoperability.

Console patches are widely employed and are relatively low in cost to support. However, there are potential drawbacks to consider. First, patches require dispatcher activity to setup and teardown, delays are experienced in patch coordination and setup, and dispatchers and users must be trained in the coordination and setup process. Patches should be torn down when the need has passed because of the risk of needless and disruptive interference on the interconnected systems. These patch operations are part of established SOPs and accepted by all participating agencies.

All gateways (patches, audio switches, and IP-based gateways) have capacity limitations and coverage issues. Capacity is limited to what can be reasonably supported by the channels assigned to the interconnections. Loading on those channels will increase during major events, thus creating dispatcher coordination and con-

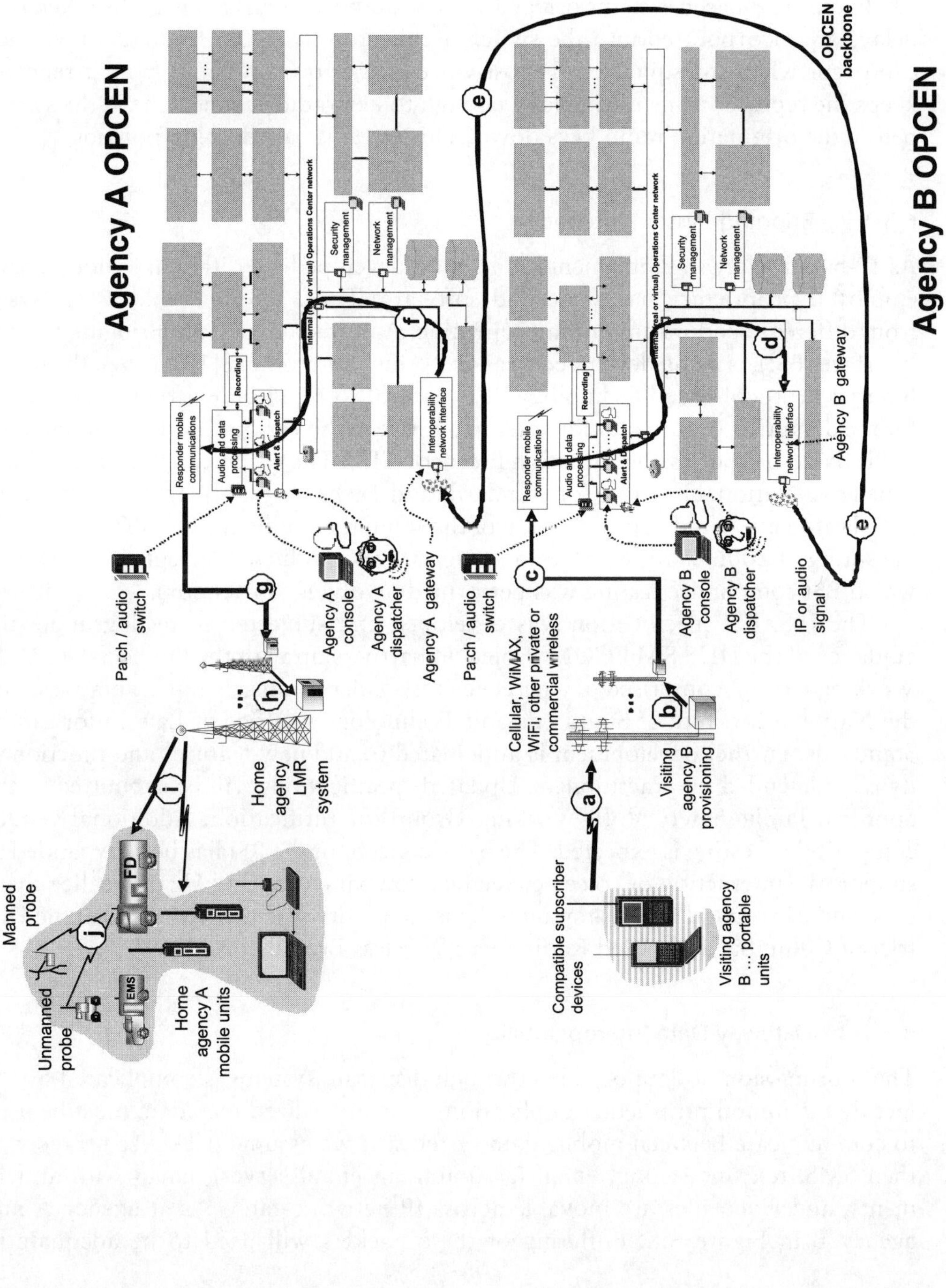

Figure 6.8 Gateway-enabled voice interoperability.

gestion issues. The coverage issue arises when gateways are intended to provide mobile communications for all responders-receivers at an incident, because then all agency radio systems must provide coverage in and around the incident location. There is also an issue of efficient use of spectrum, because each agency gateway interconnection requires an additional operating frequency. In addition, gateways between conventional and trunked systems produce an audio delay and truncation of the initial syllables of a call on the trunked side—risking the infamous "don't shoot" problem.

In repeater-based systems bridged with an audio switch, care must be taken that delays are incorporated into the switch architecture to avoid feedback. Feedback can occur when the squelch tail (post-voice energy transmission) from a repeater keeps the repeater transmitting long enough to key radios connected via the switch before the originating radio keys down. This effect is called "ping-ponging" [12].

6.3.1.2 Bridging System Interface

As IP-based gateway implementations were developed by different vendors, they employed proprietary hardware and software. Thus, interoperable IP gateways from different vendors are not interoperable. A new standard, the Bridging System Interface (BSI), is being developed to address this shortcoming [13]. Given the use of Internet Protocol-based technology, it is natural to build on existing IP standards for the definition of the common interface between the various products. In fact, the BSI is based on the Session Initiation Protocol (SIP). The specification has been published as a National Institute for Standards and Technology (NIST) document, with initial ratification and demonstration of the standard conducted at NIST facilities in the spring of 2006. A roundtable and *plugfest* was conducted in September 2008 at which BSI compliance testing was performed involving six vendors.

The U.S. BSI specification was developed and subjected to testing under the guidance of the DHS SAFECOM Project in partnership with the Public Safety VoIP Working Group, comprised of emergency responders, industry representatives, and the National Institute of Standards and Technology's Office of Law Enforcement Standards. Further development is anticipated to add new features and functionality recommended by practitioners. Updated specifications will be submitted to the approval Public Safety VoIP Working Group for ratification. Additional vendor compatibility testing is expected. The basic concept of the BSI has been expanded to support the interaction of voice between narrowband (i.e., 25 kHz or smaller channels) and broadband wireless systems. This capability was demonstrated in the District of Columbia and called Radio Over Wireless Broadband (ROW-B).

6.3.1.3 Gateway Data Interoperability

The transmission of data or video through disparate systems is complicated by the fact that common proprietary applications or standardized metadata must be used to communicate between mobile data systems. If we assume packetized transport, then SMS, text messaging, email (assuming an email server), email with attachments, and video files are movable across IP network gateways at agency A and agency B in Figure 6.8. Buffering of these packets will need to be adequate to account for differences in wireless transmission rates between agency A and B mobile data systems.

The data stream from one or more mobile units in Figure 6.8 could be sensor data, such as HAZMAT or CBRNE sensor data from a manned or unmanned sensor probe (robot) linked to a vehicle as shown in the figure along path (j). The data stream may include VoIP packets intermixed with sensor data, so voice and sensor-based situational awareness data can be received simultaneously. Operationally, this configuration is valuable for determining whether a responder carrying a HAZMAT probe should be extracted, maintaining voice communications with the responder and data communication with his/her portable device—which may not be visible to the individual in the probed environment. In addition, a dispatcher can be monitoring this voice and data on their GIS-equipped CAD.

6.3.2 Shared Sensor Data

The OPCEN may be configured with its own directly connected sensor array as depicted by blocks (9), (10), and (11) in Figure 6.6. In this case, a visiting OPCEN may want to extract this sensor data to provide CBRNE or HAZMAT risk data to its mobile units or to improve the accuracy and timeliness of its own situational awareness. This configuration is shown in Figure 6.9, in which sensor data is collected at the agency A sensor array and transmitted over paths (a); processed and pushed over the agency A intranet to the interoperability gateway over path (b); then over the OPCEN backbone (c) to the visiting agency B; connected to a microwave hop over path (d); then continuing after the microwave hop on OPCEN wireline backbone (e); then to the visiting OPCEN's gateway and over its intranet through to the mobile data communications system interface via path (f); then to wireless infrastructure via path (g), then to the appropriate tower site via path (h); and then on to the agency B subscriber devices via path (i). The processed sensor data pulled from agency B mobile units provides them (and, although not shown, their OPCEN) with improved situational awareness—particularly for environmental conditions which may be detrimental to agency B responders. Of course, the processed sensor data format must be understandable to the mobile subscriber devices, so translation software may be required in agency B OPCEN infrastructure.

6.3.3 Shared Situational Awareness and Command Coordination

Holistic interoperability requires that command elements share common awareness of the situation and are able to coordinate commands to their respective responders-receivers as well as sharing information with other stakeholders. This situational awareness involves filtered sharing of all aspects of situational awareness, which may vary among a home OPCEN and each visiting OPCEN. In this context, a visiting OPCEN is an OPCEN seeking information from the home OPCEN where the information is produced, retained, or otherwise available with appropriate security precautions. For example, hurricane or tsunami occurrence information—with standardized GIS presentation—would be valuable predictive information to share with all potentially affected OPCENS, including likely supporting as well as directly affected agencies.

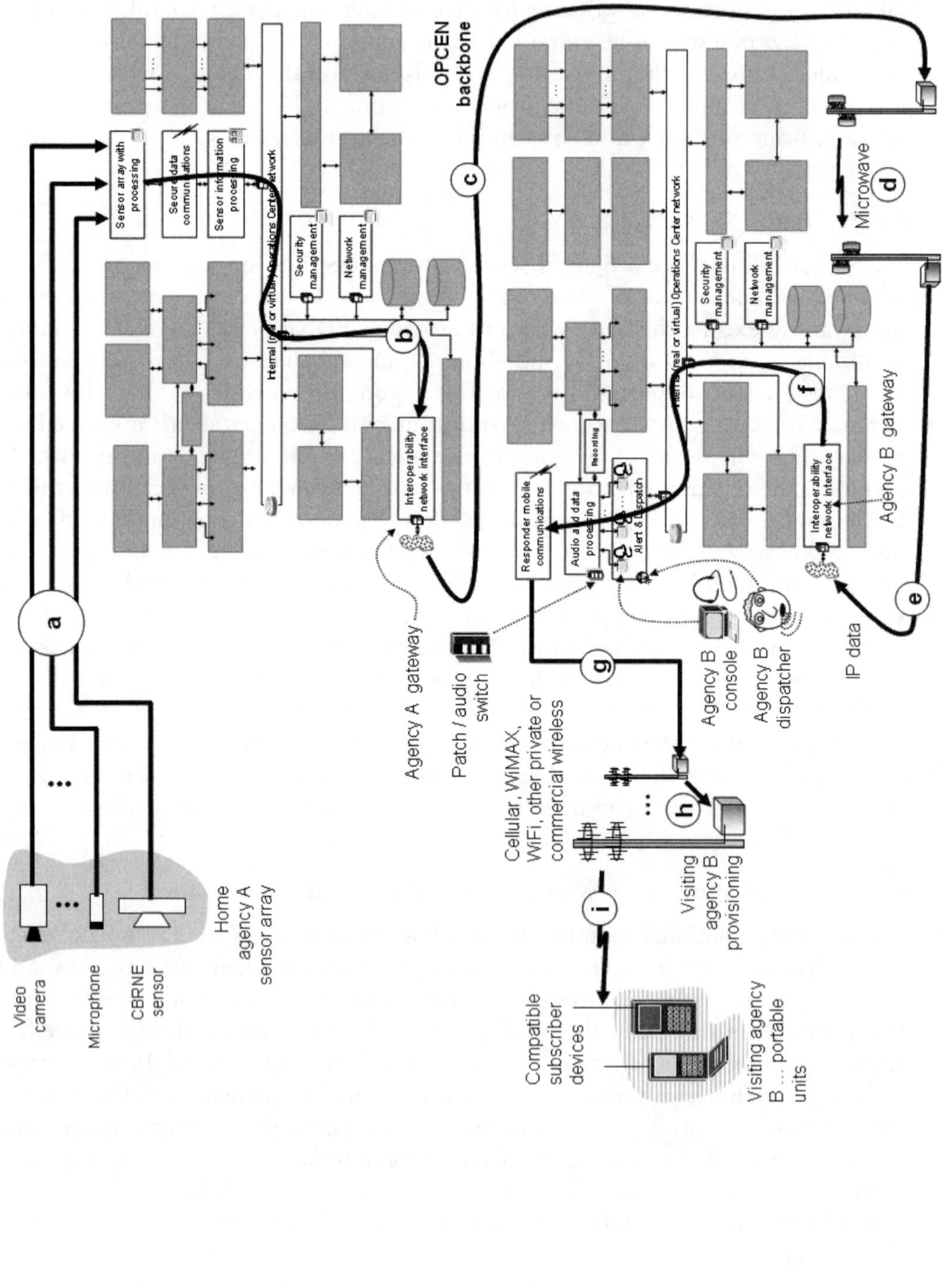

Figure 6.9 Gateway-enabled shared (processed) sensor data.

The complexities of sharing advanced situational awareness involve the selection of a common metadata for an integrated or federated situational awareness/coordinated command architecture, or suitable translations for a composite architecture where the situational awareness applications are different. If common situational awareness applications, such as Incident Management software in EOC OPCENs, are employed, then the resulting Internet-backbone-based shared awareness and coordinated command forms an integrated architecture. If a unique translation between such tools is performed, than the resulting composite architecture would nevertheless support interoperability. Ideally, if proprietary situational awareness and command coordination products can be developed with a common nonproprietary interface, then the resulting federated architectures provide the benefit of competition among vendors, but not at the cost of public safety from failed holistic interoperability.

Figure 6.10 shows bidirectionally shared situational awareness and command coordination. Agency A pushes security-filtered situational awareness and command coordination information over paths (a) and (b) over through its interoperability gateway; then over the OPCEN backbone network path (c) to the agency B interoperability interface; then over the agency B intranet paths (d) and (e) to its knowledge management, situational awareness, and decision-support applications. Again, security as well as network management servers maintain appropriate information filters among collaborating OPCENs, so only information allowed for sharing under the current threat scenario is accessible. This accessibility varies for each home-visitor pair of OPCENs.

6.3.4 Shared Databases

An important shared-information capability of a home OPCEN is providing visitor access to local and remote archived databases. This requires the proper security precautions and is particularly important for law enforcement investigations that might require access to view license plate registrants, building floor plans, utility conduits, HAZMAT precautions or other data. In this scenario, personnel from the mobile units of the visiting agency require access to data that is held by the home agency. The network management function and database servers provide access to the requested information through use of a federated query. Following proper security policy, the requested information is delivered from the databases over the home agency A intranet to the interoperability gateway over path (a); then using the OPCEN backbone over path (b) to the visiting agency B interoperability gateway; then on through the dispatcher monitoring and mobile radio system interface via path (c); next over path (d) to the mobile data infrastructure; then path (e) to the appropriate mobile data tower; and finally over the air path (f) to the requesting subscriber device. See Figure 6.11.

6.3.5 Shared Intelligence

As was shown in the 25 Pearl Harbor–9/11–Katrina deficiencies, intelligence sharing and appropriate interpretation were among the greatest failures of leadership. To correct these deficiencies, and employing a means of rating believability while

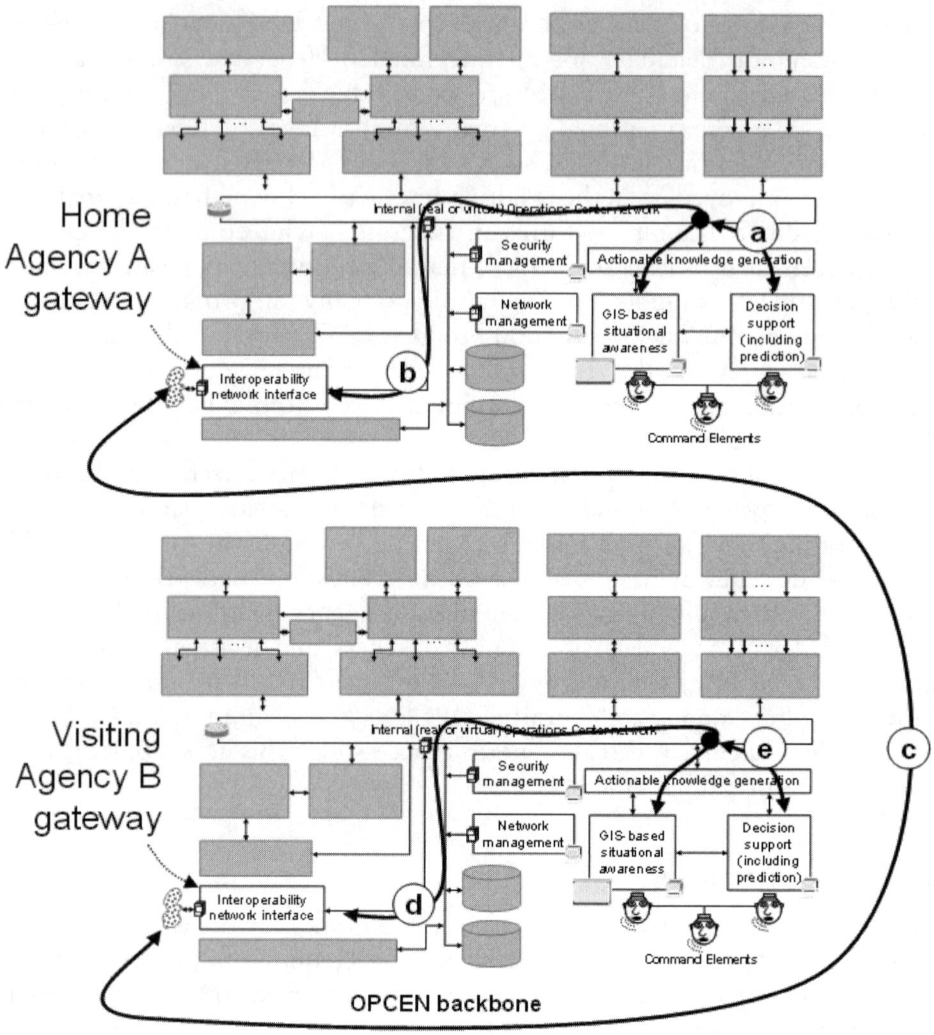

Figure 6.10 Gateway-enabled shared situational awareness and command coordination.

preserving sources and means, actionable (and other) intelligence information will be shared among OPCENs. This intelligence has many possible sources, and only those depicted in Figure 6.4 are shown in Figure 6.12. In all cases, this intelligence is fused with other available information.

The figure shows that after a suitably verified and authorized request, intelligence information—both raw data and fused results—are pulled by the authorized visiting OPCEN, or pushed to that OPCEN if deemed important to the home OPCEN. The information flows from the diverse sources along path (a), follows the analysis fusion process and in both filtered raw form as well as fused, passes along the home OPCEN intranet path (b) to the interoperability gateway; then it passes through secure transmission over the OPCEN backbone to the visiting OPCEN via path (c); through path (d) to be fused by the visiting OPCEN fusion function and delivered to the situational awareness block; and then finally to from path (d) to (e) at the decision-support block.

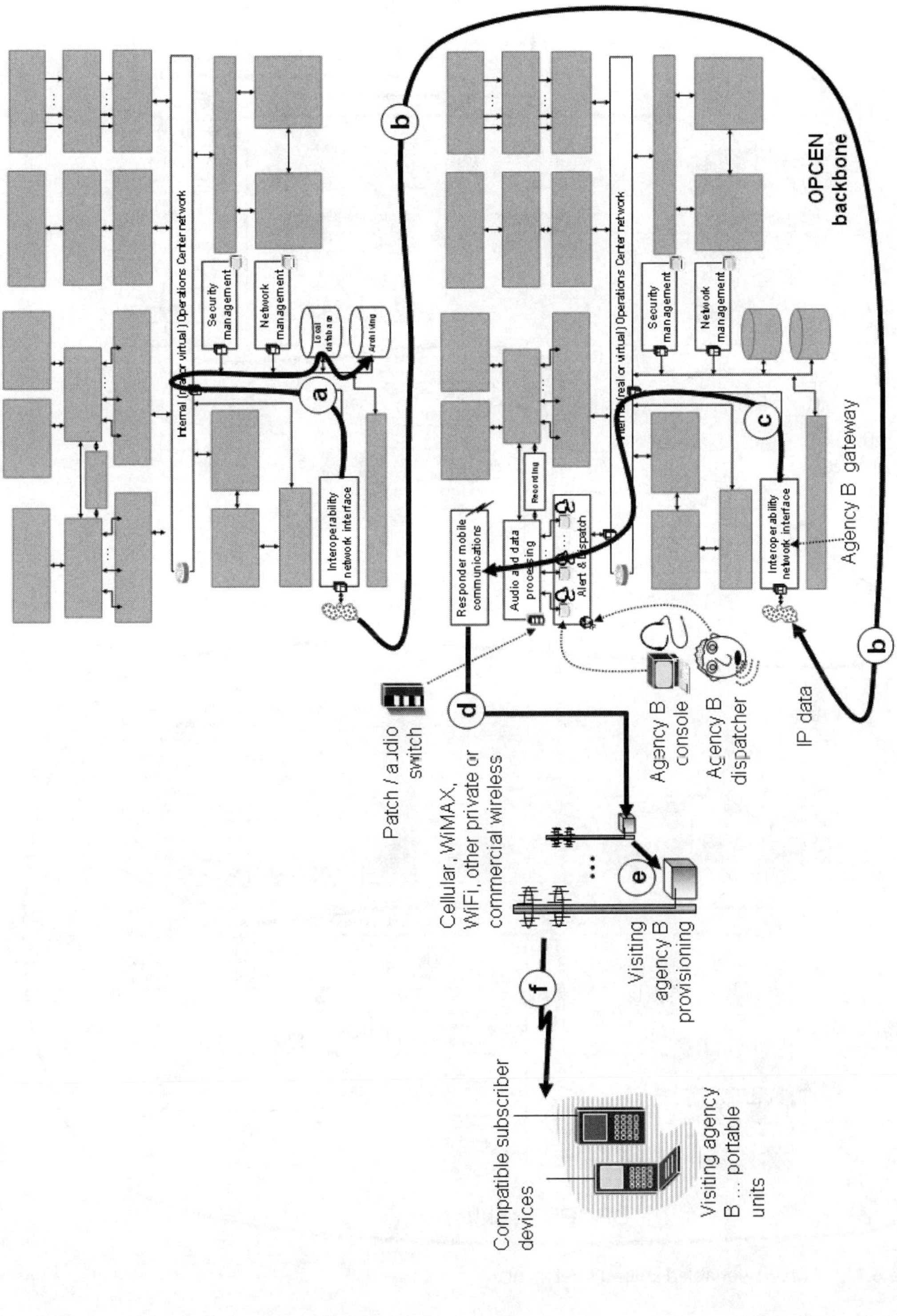

Figure 6.11 Gateway-enabled shared databases.

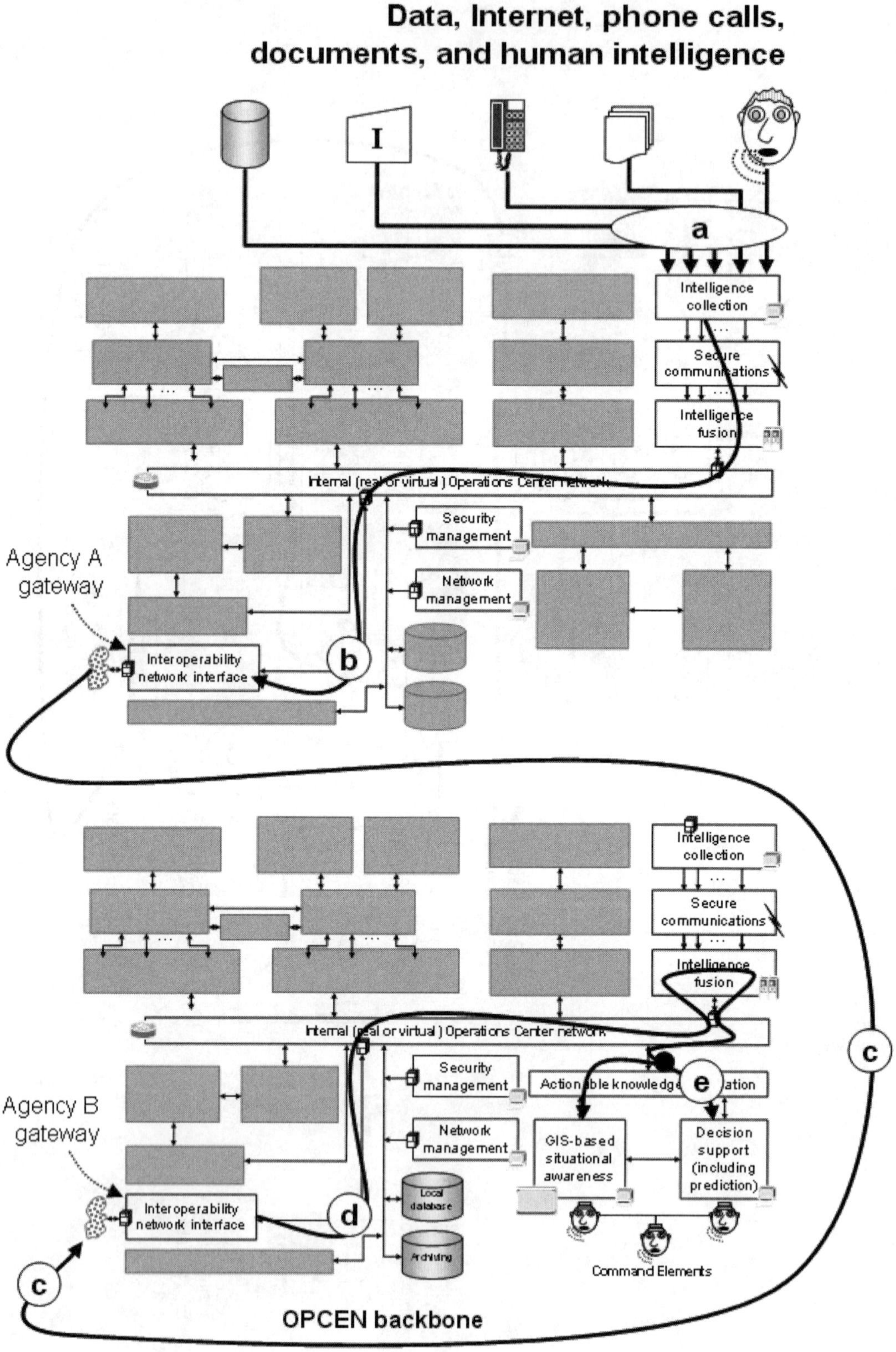

Figure 6.12 Gateway-enabled shared intelligence.

6.3.6 Shared Alert and Warning Information

OPCENs generate or receive alert information from other OPCENS and often 7×24 cable news programs, among other sources. If a common alerting protocol is adopted, it should be generated and upon warning passed to all other collaborating OPCENs over the backbone and seek acknowledgment of receipt. In addition, this alert may include the designation of preplanned tasks and activities coordinated among OPCENs and their mobile units. These alerts may be the result of intelligence regarding a manmade threat or weather service identification of severe weather in advance of its arrival in the region. Again, verification of receipt and identification of intended collaborative action plans to address this planned threat are identified and executed. All relevant agencies are so notified and verification acknowledgements required at the originating (assumed home) OPCEN.

Figure 6.13 shows an alert and warning coming from mobile responders-receivers associated with visiting OPCEN B over path (a) to the appropriate cellular tower; then to the cellular system interface over path (b); from the cellular system to OPCEN B interface over path (c); through the OPCEN mobile communications infrastructure and dispatcher position (monitoring) over path (d) across the OPCEN intranet to the interoperability interface; from the external side of the interface over path (e) to the satellite portion of the OPCEN backbone; over the satellite path (f); across the remainder of the backbone path (g) to the home OPCEN's interoperability interface; over the home OPCEN's intranet via path (h) to the decision-support function for development of the official alert and warning message, and then to the stakeholder output block; and finally transmission of the alert and warning message out over various media using path (i). The alert and warning messages can also be pushed over several media, including the OPCEN backbone, so all collaborating OPCENs receive alerts and warnings.

6.3.7 Shared/Backup Public Safety Answering Point

The public is often the first to detect an incident, if not to provide a warning based on suspicious activity before an incident, or witness approaching natural threats (levee breaks in the absence of sensors, fire, etc.). In such moment-to-moment cases, the PSAP is the OPCEN's vital incident warning and detection capability. Thus, an immediate backup capability is necessary for every PSAP on the OPCEN backbone. In addition, major incidents often lead to many emergency calls, rapidly adding to call-taker and PSTN circuit congestion. The backup PSAP can then serve to offload home PSAP congestion onto one or more visiting PSAPS. The two PSAP case is depicted at Figure 6.14, in which the home PSAP at OPCEN A is backed up by the PSAP at OPCEN B.

The figure shows the emergency callers over wireline or wireless on bidirectional paths (a) and (b), respectively; bypassing the failed or congested home OPCEN PSAP to the home interoperability interface on path (c), and then over the OPCEN backbone path (d) to the visitor OPCEN interoperability gateway; finally then to and from the visiting OPCEN's PSAP to provide backup or congestion relief call taking. The connections are shown bidirectional because we assume here that the visiting OPCEN PSAP is actually dispatching the home OPCEN's responders-receivers. Of course, the implication is that an integrated, composite, or feder-

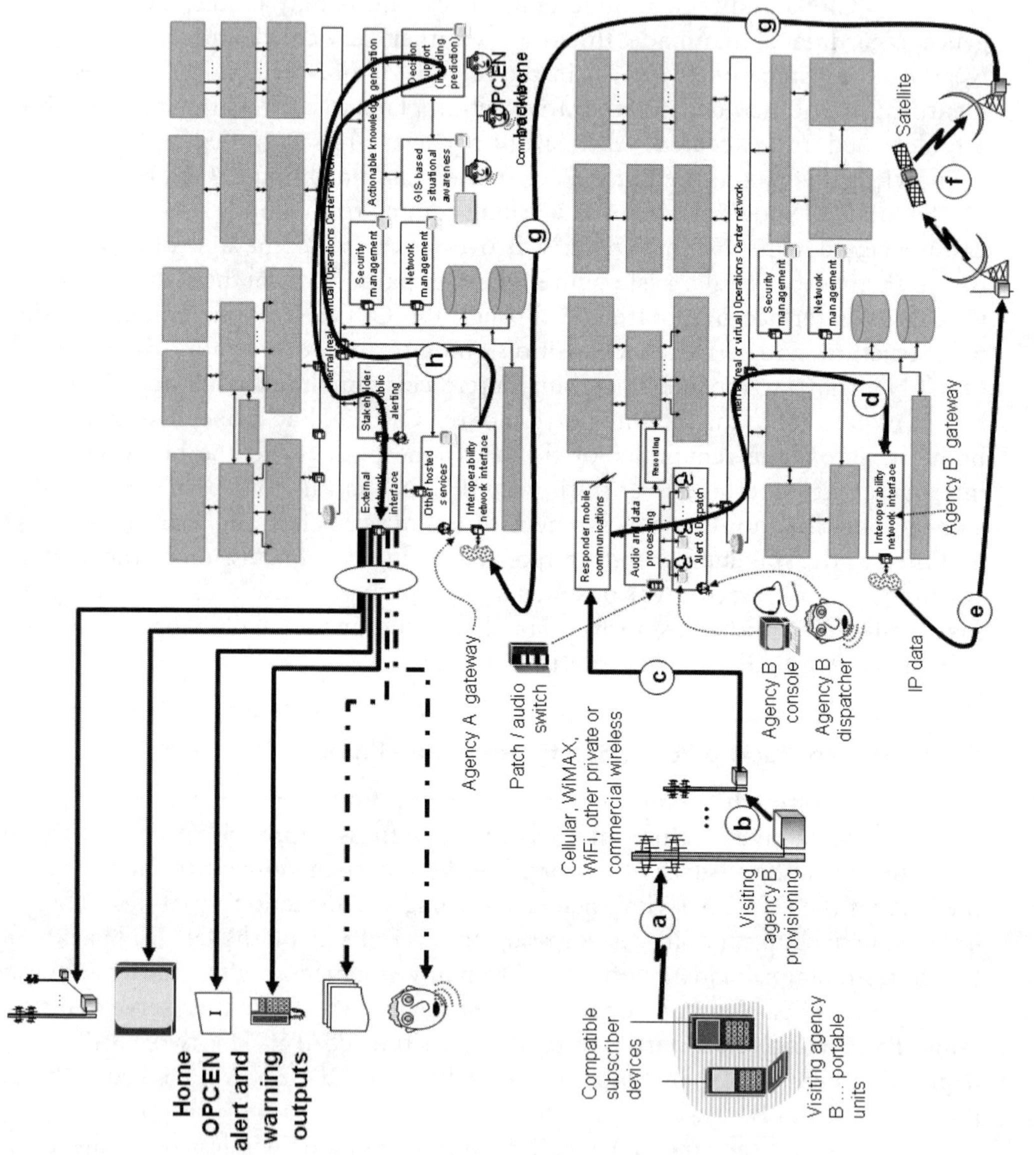

Figure 6.13 Gateway-enabled alert and warning transmission.

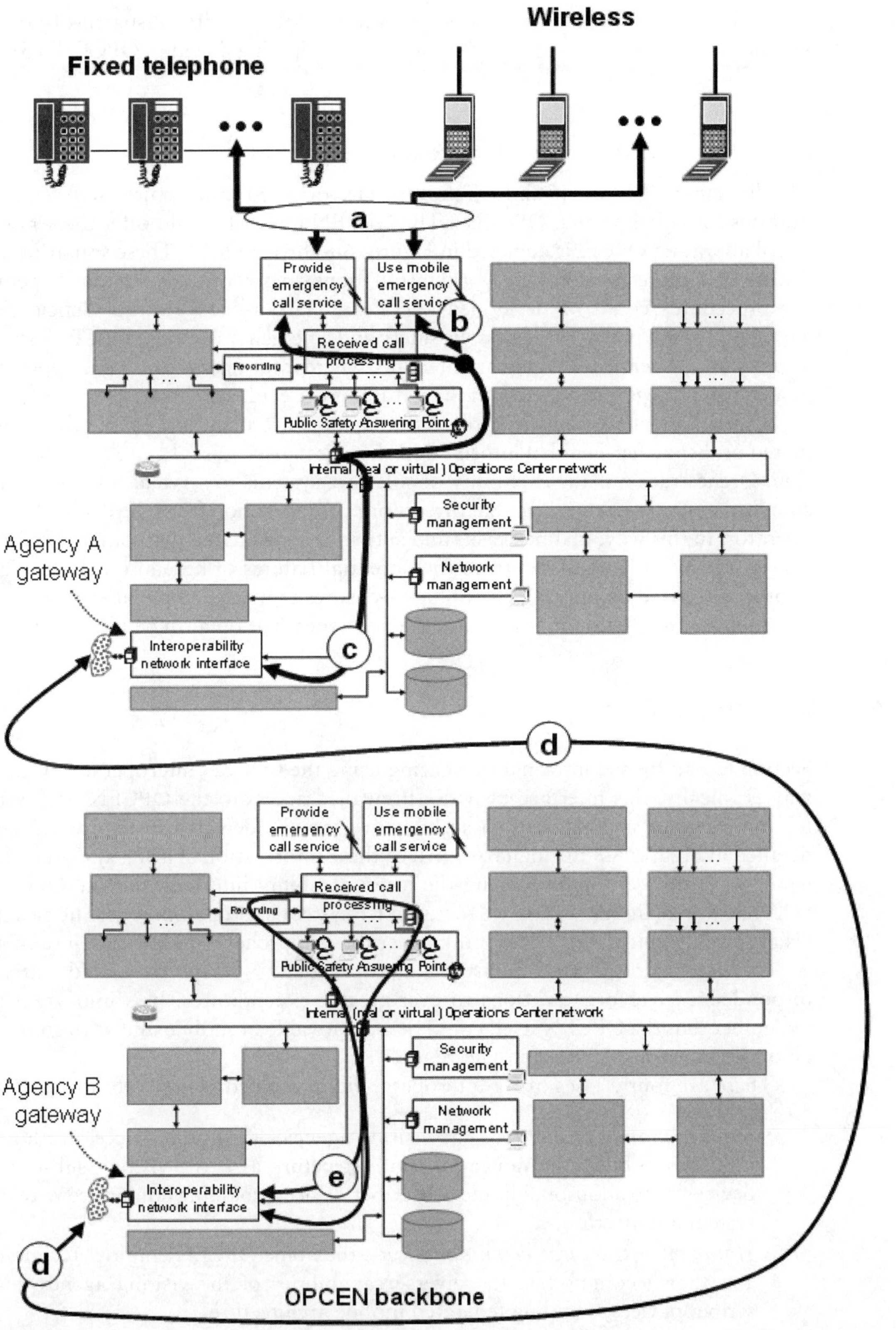

Figure 6.14 Gateway-enabled shared/backup PSAP.

ated dispatching capability exists to support this visitor-dispatch-of-home-mobile-unit activity. Also, SOPs between participating backup OPCEN PSAPs would be required (see Chapter 5).

6.3.8 Service-Oriented Architecture with Orchestration

The diagram in Figures 6.5 and 6.7 show service-oriented architecture (SOA) configurations for collaborating OPCENs. The OPCEN intranet would offer the services to collaborating OPCENs depicted in Figures 6.8 through 6.14. These scenarios are drawn from representative real-world installations. In other words, standard operating procedures could be developed and Pearl Harbor–9/11–Katrina deficiencies eliminated to the point that the public safety and emergency response OPCENs share services on this service-oriented architecture, where information sharing is approved in advance or negotiated based on the nature of the evolving threat.

The complexity of ensuring the right "spigots are open" to enable this automated or dispatcher-enabled information sharing may require the use of orchestration. Orchestration is the capability of software systems to serve as an automated standard operating procedure that directs or prompts action from the dispatcher or operator. In this way, planned tasks and activities coordinated with other OPCENs can be significantly insulated from the potential failures of human initiative. This approach—based on existing technology—ensures that the complexities of achieving holistic interoperability in a crisis are not dependent on a single individual.

6.4 RF Interoperability

Section 6.3 addressed information sharing using the OPCEN interoperability gateway. Typically, this interface provides firewalled access to the OPCEN backbone for the range of service-oriented architecture services depicted in Figure 6.7 and detailed in Figures 6.8 through 6.14. All of these information-sharing system architectures were based on employing the interoperability interface and the OPCEN SOA backbone. In this section we focus on RF system interoperability for home and visitor mobile units. We will not rank order these approaches because such ranking is irrelevant from the holistic interoperability perspective and drives responder-receiver organizations to ever-greater expenditures. It is important to remember that the OPCEN itself could be a deployable or mobile unit with some or all of the functions depicted in Figure 6.6.

There are four classes of RF interoperability presented below as follows:

- *Visiting mobile affiliation*—the visiting agencies are granted access to home OPCEN mobile communications infrastructure as if the visiting subscriber devices were additional home subscriber devices, enabling composite wireless system architecture.
- *Home infrastructure accommodation*—the home OPCEN employs technologies that accommodate the diverse capabilities of the visiting agency subscriber devices, enabling federated mobile architecture.

- *Third-party network or parallel network affiliation*—the home and visitor mobile subscribers all affiliate to an independent or parallel mobile communications capability, forming an integrated mobile communications architecture.
- *Shared systems*—the integrated architecture of a single standards-based or proprietary wide-area LMR system in which all user agencies have intraoperability and programmed interoperability.

Each of these RF interoperability approaches is described below. Note that home and visiting OPCENs may employ both RF and gateway-enabled RF interoperability simultaneously depending on the number and extent of events or incidents and according to their standard operating procedures. We have considered shared systems last because they are the most expensive interoperability solution and has been covered in depth in an earlier sister publication [14].

6.4.1 Visiting Mobile Affiliation

Figure 6.15 depicts the many ways in which multiple visiting agencies can access home OPCEN mobile communications capabilities using a variety of existing technologies. In these situations, the visiting agency has made procurements in advance or otherwise become equipped to use the home OPCEN mobile communications infrastructure. Of course, the home OPCEN has accommodated this affiliation by allowing device IP addresses, radio unit IDs, and other identifiers of individual visitor subscriber devices access to the home network. In this case, all home networks are self-owned and operated, including the satellite capabilities (e.g., military, government-owned, or otherwise dedicated-purpose satellites). The figure shows that home agency A has mobile RF subscribers on all of its radio networks, LMR, satellite/HF, and broadband wireless. We will assume that the agency A OPCEN audio and data processing function is capable of allowing integrated voice and data communications across all three of their mobile communication systems for all agency A mobile subscribers.

There are three basic approaches that visiting agency mobile subscribers could use to affiliate with the home OPCEN's RF systems, namely (1) reprogram compatible radios, (2) use a software-defined or multiband radio, and (3) employ mobile gateways.

6.4.1.1 Reprogrammed Compatible Radios

Agency B in the figure (upper-right corner) employs technologically compatible LMR radios operating in the same frequency band as the home LMR infrastructure. Agency B (1) has radio channels (conventional) or talkgroups (trunked) programmed into its compatible radios, (2) develops common voice radio discipline and language based on specific national interoperability specifications, (3) performs a variety of exercises with the home agency responders-receivers, (4) and interacts as much as possible on a moment-to-moment basis over the shared home network, (5) verifies the absence of Pearl Harbor–9/11–Katrina deficiencies (see the Interoperability Assessment Checklist, Chapter 3), and (6) achieves holistic interoperability with the home OPCEN. Agency C seeks to gain access to mobile

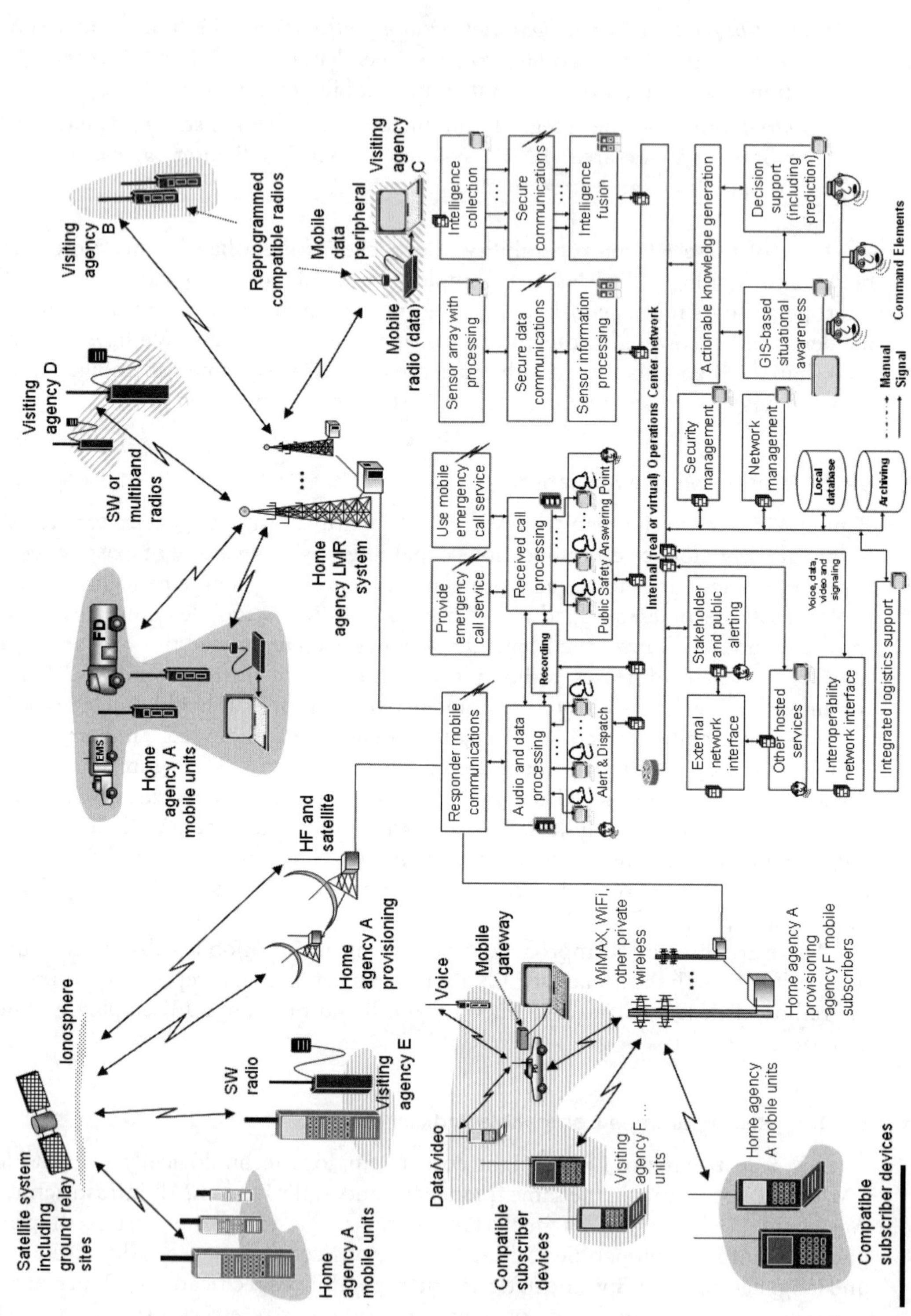

Figure 6.15 Visitor mobile RF affiliation methods.

data via the home OPCEN's integrated voice and data network as shown below the agency B enclave in the figure.

6.4.1.2 Software or Multiband Radio

The agency D enclave—just left of the agency B enclave in the upper right of the figure—uses either software or multiband radios to affiliate to the home OPCEN LMR network. Simply stated, the principle functions of a radio include:

- Processing the analog RF signal;
- Waveform modulation and demodulation;
- Baseband signal processing.

In a multiband radio, two or more separate virtually independent signal paths exist for operation in different frequency bands, often requiring a larger and more expensive radio. In this case, and assuming the proper LMR technology, channels or talkgroups can be programmed to affiliate to the home network.

In a software-defined radio (SDR), the latter two functions are performed in software. An SDR still requires the design of analog RF front-end sections. However, as the technology advances, the boundary moves closer to the antenna. The obvious advantage of an SDR is the ability to modify the radio through the addition/replacement of software, rather than hardware. This replacement capability means that a radio with a well-designed analog front end can be adapted to alternative waveforms or additional waveforms without the need to replace hardware.

SDRs have been in use in the military for many years. They have not seen application in public safety thus far because of cost and form factor (size and weight) considerations. However, a group of products are emerging defined as multiband, multimode radios that are based on adaptation of military SDR technology developed for public safety. Demonstration models of these products are now available and being employed in pilot programs to verify their suitability for public safety and to support refinement of operational concepts. Production models are expected to begin appearing in 2009. When available, these products are expected to approximate current high-end public safety radios in price and size. Now that SDR's are nearing reality for public safety, interest is shifting to the feature sets to be offered, rather than feasibility.

The figure also shows the visiting agency E enclave using common satellite or HF radios as home agency A together with an SDR capable of HF as well as satellite operation. The flexibility of the SDR largely—but not entirely—eliminates the impact of varying technologies on failed RF interoperability. As has been shown, however, achieving RF interoperability is a necessary but not essential condition in order to achieve holistic interoperability among mobile responders-receivers and other public safety and emergency response stakeholders.

6.4.1.3 Mobile Gateways

Visiting agency F in the lower-left portion of the figure not only employs compatible broadband subscriber devices for integrated voice and data RF interoperability with agencies A, B, C, D, and E, but it also employs a *mobile gateway*. The mobile gateway is similar to the infrastructure gateways described in Section 6.3.1, except that it is operated most often as a "hands-free" device in a ground, air, or water-mobile platform—searching for planned signals. For voice, a mobile gateway is simply an array of audio interconnect circuits permitting multiple voice radios to share the same reversed audio (i.e., all microphones connected to all speakers) from all radios. In its simplest form, a *vehicular repeater* [15] is a mobile gateway intended to permit such cross-frequency-band or cross-band voice communication from a visiting portable radio user within line-of-sight of the mobile platform (or in it) and the home OPCEN infrastructure, most often used (but not shown) with LMR systems.

For broadband data (and VoIP voice) applications, the mobile gateway serves as a "best-effort router" to find and maintain the best performing packet data communications path for visitor mobile data communications with the home OPCEN. In general, middleware at the home OPCEN coordinates communications with the mobile gateway such that the mobile and OPCEN applications continue to operate independently of RF dropouts as the mobile platform moves within the RF environment. Although not shown, this mobile gateway can employ satellite as well as terrestrial mobile radios—offering a near ubiquitous geographic best-effort RF coverage for the mobile platform. This device can be useful in minimizing the need for LMR and other terrestrial infrastructure in wide-area (remote) systems in which satellite capabilities will remain lower cost compared with wide-area LMR life cycle acquisition and sustainment costs.

6.4.2 Home Infrastructure Accommodation

The second RF interoperability approach is based on the home agency (A) planning their own systems to accommodate visiting agency mobile units. There are four basic approaches to accommodate visiting units, namely (1) distribute (or swap) radios from an available cache, (2) deploy mobile gateways, (3) employ tower-site gateways, and (4) employ tower-site SDRs or multiband radios. These four approaches to accommodate visiting mobile units are depicted in Figure 6.16.

6.4.2.1 Radio Swap and Cache

The first of the voice elements in the SAFECOM interoperability continuum (see Figure 1.1) is swapping or otherwise provisioning radios to visiting agency mobile units. This provisioning presumes that caches of mobile or portable radios are prepositioned or otherwise available such as those carried in mobile command centers. Prepositioned caches provide stores of radios in locations where they are accessible for the event or emergency in question and must be maintained with fully charged batteries (or have available vehicular power connections), and are preprogrammed to operate directly on home radio channels or talkgroups. Of course, the added radios will impact channel congestion or talkgroup delay if the home system operates close to effective capacity.

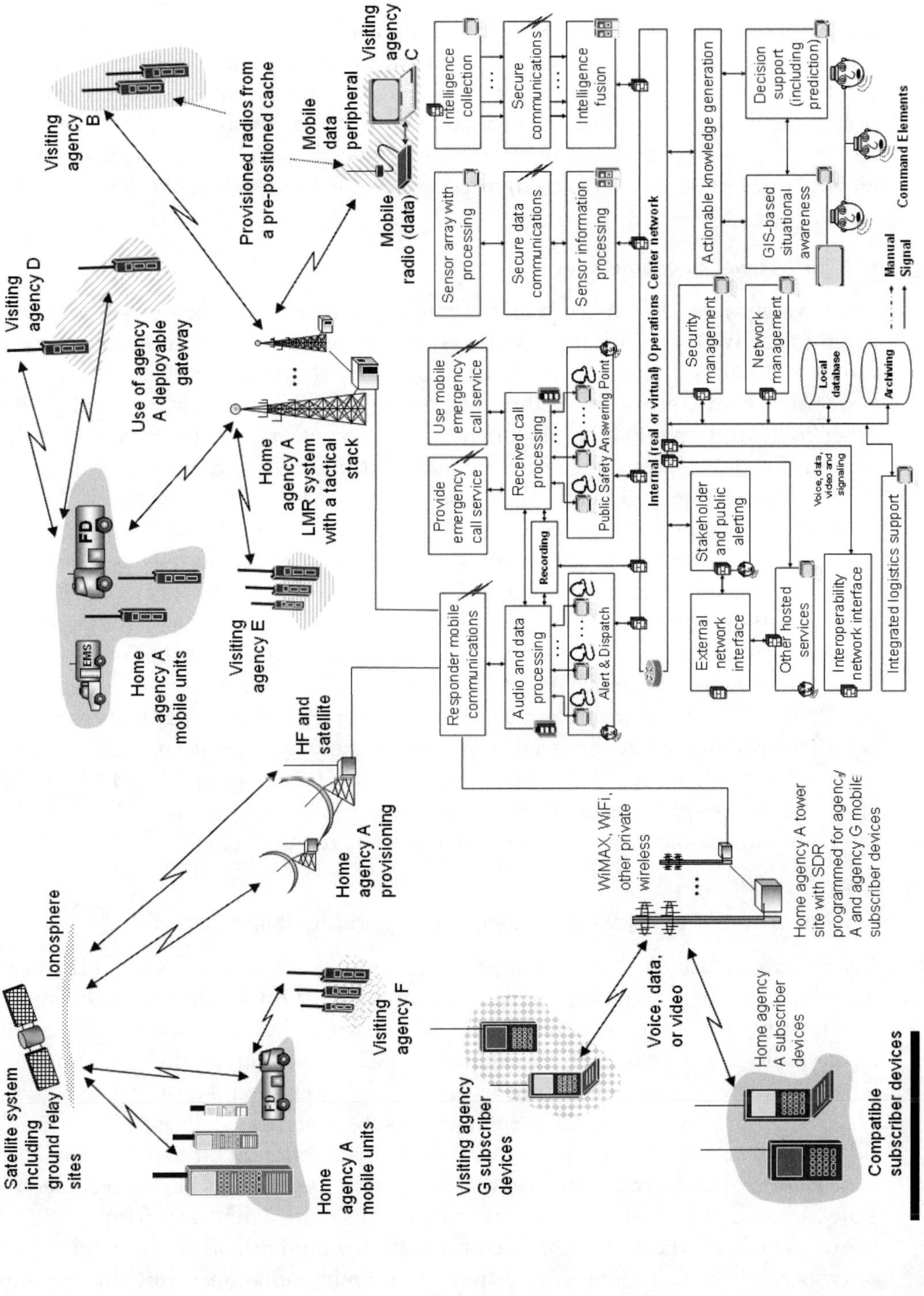

Figure 6.16 Home infrastructure accommodation methods.

6.4.2.2 Deployable Gateways

The figure shows deployable gateways providing voice communications via a deployed agency A mobile gateway (see Section 6.4.1.3) for agency D visitors so they may use their own radio technology and spectrum and still communicate with the agency A's OPCEN and its mobile units while operating as if they were using their own home system. Agency A has also deployed a mobile gateway so agency F can have remote voice communications via agency A's HF or satellite communication systems. The only difference in the deployable gateway from Section 6.4.1.3 and use of gateways in this subsection is the ownership. In this case, agency A owns and operates these gateways configured to accommodate agency F.

6.4.2.3 Tower Site Gateways

The coverage issue associated with the gateways discussed in Section 6.3.1.1 is mitigated for conventional (nontrunked) systems by placing one or more conventional radios of different frequency bands and a remotely controlled gateway at each radio tower site, with control (say, IP-based control) at the home OPCEN and perhaps other collaborating OPCENs [16]. If an incident or event occurs within the coverage contour for the highest transmitted frequency at the radio site, then coverage for all other radios will be contained within that highest frequency coverage area. Thus, all visiting mobile units can employ the tower site as a cross-band repeater or for mobile communications with their own agency E home OPCEN (see the agency E enclave in the figure just left of the agency A LMR towers). This home OPCEN can then use its interoperability interface to include the visitor's OPCEN and dispatchers/operators in the communications (see Section 6.3.11). The dispatchers at the home OPCEN can remotely control which radios are interconnected, providing the benefits of on-demand RF interoperability without the coverage gaps of disparate system interconnect through the backbone interoperability gateway. The use of multiple radios and gateway at a tower site has been dubbed the TACtical STACk (TACSTACK) [17]. Use of this approach assumes that proper RF levels are maintained and levels of intermodulation and other forms of interference are mitigated.

6.4.2.4 Tower Site Software-Defined Radios and Multiband Radios

Visitor agency use of SDR described in Section 6.4.1.2 was focused on the use of portable or mobile devices to affiliate with the home OPCEN mobile radio system. In this configuration, consideration should be given to installing the multirole SDRs at the radio tower sites. This configuration has the advantage that the expense of fewer numbers of base station radios at a much lower capital cost can be achieved rather than provisioning a large number of subscriber units. These SDR base stations would be able to interoperate with a variety of different types of mobile radios. Base station SDR software for commercial mobile carriers has been developed that generates, receives, and processes existing commercial signal waveforms [18]. The use of an SDR for the TACSTACK configuration would provide the flexibility in RF interoperability, including supporting all mobile units operating in the same frequency band as well as disparate bands.

6.4.3 Third-Party Network or Parallel Network Affiliation

There is a third class of RF interoperability techniques in which both home and visiting mobile units affiliate to a third-party capability parallel mobile communication system, including commercial systems. The manifestations of this class of RF interoperability approaches includes (1) shared channels, such as implementation of designated mutual aid or interoperability channels on the home OPCEN LMR infrastructure; (2) deployment of a parallel LMR network explicitly for multiagency use; (3) use of terrestrial and satellite commercial network not (in general) owned or operated by the collaborating agencies. These RF interoperability approaches are depicted in Figure 6.17.

6.4.3.1 Shared Channels

For agencies in the same band with overlapping coverage, the sharing of channels can be a simple, low-cost approach to interoperability, depending on the implementation. In addition to being equipped to operate in the same band, the agencies must be otherwise over-the-air compatible, such as using the same modulation. In RF congested areas, channels may be most efficiently used when shared with other jurisdictions. Typically, shared channels are selected—by agreement between agencies—from the pool of available, licensable channels in a given band or from channels specifically set-aside for interoperability. The expression "mutual aid" is frequently applied to both channel designations. However, for clarity, we refer to the former as *mutual aid channels* and the latter as *interoperability channels*, respectively. The upper right of the figure shows agencies A, B, C, and D all using shared mutual aid or interoperability channels

Mutual Aid Channels Interagency-planned mutual aid channels are licensed for the exclusive use of the participating agencies and are commonly organized through county, regional, or state planning organizations (or managing authorities) that exercise general supervision and disciplinary control. Guidelines for using mutual aid channels are set by the managing authority for the system and may include moment-to-moment and day-to-day operations as well as major events or incidents. Large multijurisdictional mutual aid systems can be supported through the use of multiple LMR base station tower sites. Some of these systems include multiband operation through combination with gateways (see Section 6.3.1.1). Use of gateways, however, introduces the associated coverage issues and reduced spectrum efficiency, such as the need for multiple channels (one in each of the bands interfaced). Complexity and cost depend on the number of base station sites and, if supported via gateways, the number of bands interconnected. Thus, cost can vary from very low to fairly expensive.

Interoperability Channels Interoperability channels are not intended for everyday operations; they are reserved for events such as natural disasters or terrorist activities. In the United States, dedicated interoperability channels are not licensed to any particular agency, but are a general resource for use to support multiagency interoperability. The U.S. Federal Communications Commission (FCC) has set aside national interoperability channels in four public safety bands as follows:

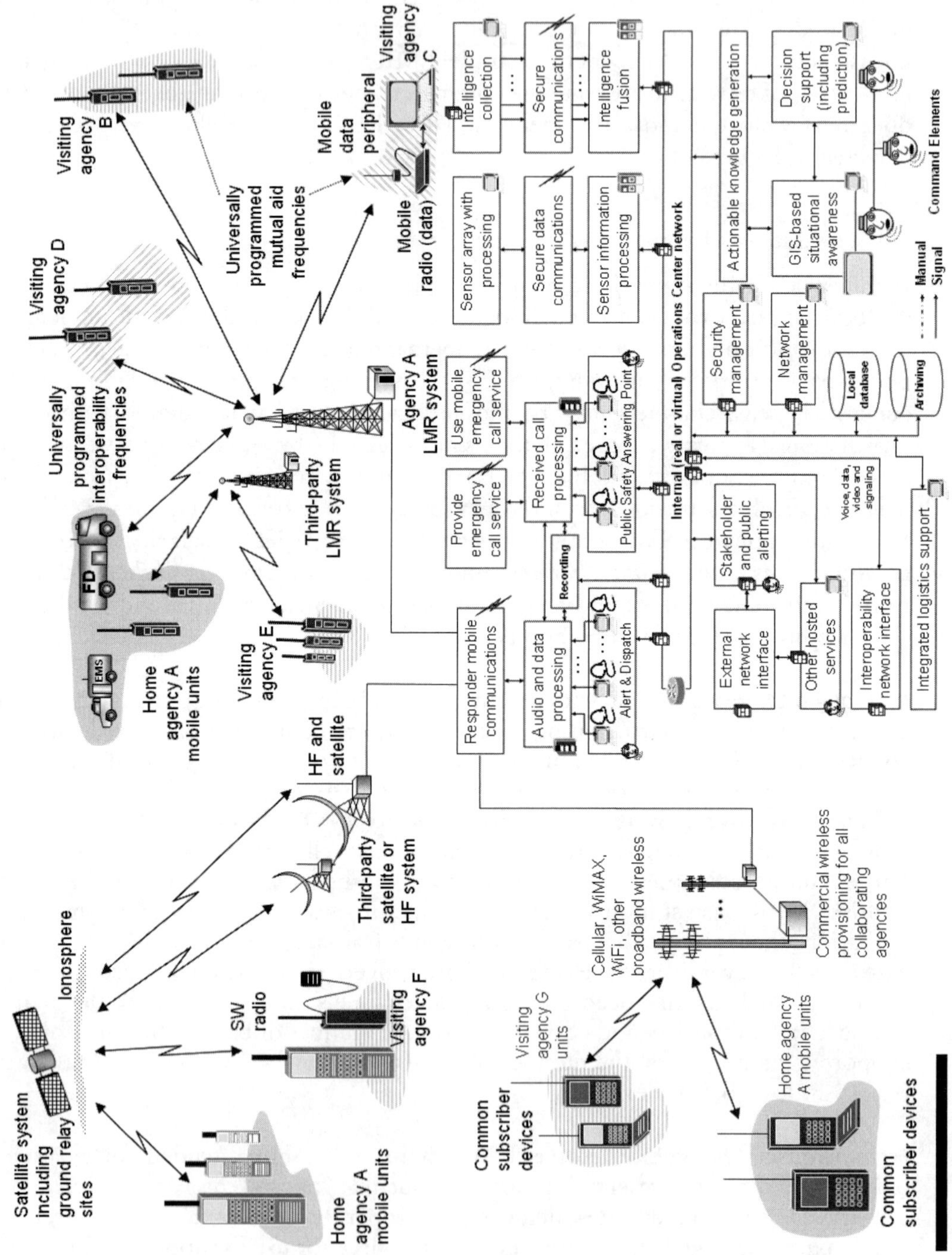

Figure 6.17 Third-party or parallel infrastructure affiliation methods.

- VHF High Band (five frequencies),
- UHF (ten frequency pairs),
- 800 MHz NPSTC (five frequency pairs), and
- 700 MHz Narrowband (64 frequency pairs).

> [FCC] national [interoperability] channels are pre-designated so that all agencies and responders may prepare in advance for operations that may be out of their normal region of concern and with other agencies that are not in their normal sphere of operations [19].

The UHF, 800-MHz, and 700-MHz allocations have both calling and tactical channels. The calling channels are used to contact a dispatcher and request the use of a tactical channel. As soon as the tactical channel is granted, the calling channel should be released.

In addition, the U.S. National Telecommunications and Information Administration (NTIA) has specified forty federal government frequencies that can be used by non-federal government public safety entities for communications involving coordination and cooperation with federal government agencies. Eligibility requirements are found in the Code of Federal Regulations Paragraph 90.20 and rules governing the use of these channels are prescribed in the NTIA manual [20]. Interoperability channels are a general nationally allocated spectrum resource for this purpose and may not be licensed for any other application.

Dedicated interoperability channels are meant for conventional analog radio operation (lowest common denominator), with the exception of the 700-MHz band. Radios operating on the 700-MHz narrowband interoperability channels must be capable of using the digital Project 25 standard [21]. Each of the 700-MHz interoperability channels is assigned a designated use from the following list: calling channel, general public safety, law enforcement, fire, EMS, mobile repeater, mobile data, and other public service. The use of mutual aid systems is spelled out in the operational instructions provided by the managing authority.

6.4.3.2 Parallel Land Mobile Radio Systems

Wide-area land mobile radio systems have been built to provide RF interoperability among different jurisdictions and services, where radio base-repeater channels are programmed into the portable and mobile radios of participating (and contributing) agencies. Parallel systems have been designed with the resulting coverage dependent upon whether a sectorized or wide-area channel is chosen. The wide-area channels provide a calling and alerting function, while the sectorized coverage supports more localized tactical operations. In this way, home and visitor subscribers (e.g., agencies A and E in the figure) can use local jurisdictions channels for their singular agency operations, while having a field capability (without dispatcher intervention) to rapidly switch to interoperability channels—following usage standard operating procedures—for mutual aid requests. Used on a scheduled basis, these parallel systems provide joint operations and task force communications for limited area as well as wide-area operations.

6.4.3.3 Commercial Networks

Agencies A and F in the figure are using commercial satellite or HF services. As shown in the figure, commercial terrestrial and satellite networks are third-party systems that support all mobile units within coverage independent of the home or visitor status of an agency. If the commercial system has a national footprint, particularly in built-up areas, than commercial-grade RF coverage (well below public safety best practices) is available to all responders-receivers and other stakeholders irrespective of governance level, discipline, or jurisdiction. Commercial systems provide a straightforward means of achieving ad hoc wireless coverage, provide a backup system to land mobile radio, and can be included in gateway-based solutions, but they are not controlled (e.g., operated and maintained) by responder-receiver organizations.

6.4.4 Shared Systems

Shared systems used by multiple agencies with programmed interoperability arguably provide the best performing interoperability solution for participating OPCENs and their mobile units. We consider three shared-system configurations as follows: gateways linking proprietary LMR systems, proprietary LMR systems, and standards-based land mobile radio systems.

6.4.4.1 Proprietary Gateways

Since all land mobile radio systems must produce an audio signal, the audio gateway approach is a natural interoperability device whether or not IP enabled. To this point, therefore, we have only considered audio interconnect gateways linking disparate proprietary radio systems and radios from different vendors, as well as incompatible systems from the same vendor. This incompatibility is caused because the control signals from LMR tower sites to the corresponding OPCEN mobile communications infrastructure (block 5 in Figure 6.6) employ proprietary digital technology. These control signals tell the trunked land mobile radio infrastructure how to process a subscriber radio call, such as locating other subscribers in the same trunked talkgroup and notifying them of an incoming call.

If the land mobile radio systems to be interconnected through a gateway are provided by the same vendor, and these systems are suitably compatible, the vendor can provide a digital land mobile radio gateway that handles these proprietary control signals. In this way, dispatchers or operators controlling this gateway have extended functionality in configuring patches between visiting and home agency land mobile radio systems. This increased functionality would nevertheless have to be enabled and authorized through appropriate SOPs by the owners of the independent systems.

6.4.4.2 Proprietary Land Mobile Radio Systems

A detailed description of proprietary and evolving standards-based land mobile radio systems is provided in a sister publication and will not be repeated in this book. Since use of a single system enables RF interoperability among all system subscribers, then it is the remaining Pearl Harbor–9/11–Katrina deficiencies (see Chapters 2 and 3) that will prevent holistic interoperability, and not the land mobile radio system. However, the interoperability gateway and RF interoperability techniques described above will be required to link other visiting mobile subscribers to this proprietary system.

6.4.4.3 Standards-Based Land Mobile Radio Systems

Project 25. Project 25 (P25) is an evolving digital land mobile radio standard in the United States [22] and is accepted as the standard to meet for public safety digital voice and data radio systems in North America and other countries. Deployment has spread through much of the world with the notable exception of Europe and Africa. The process of developing the P25 standards has been long (underway since 1989) and is sometimes contentious. The divergent interests of the industry suppliers that provide the equipment have been the major impediment to progress. Standards development is conducted under the auspices of a committee within the Telecommunications Industry Association's (TIA) Standards and Technology Department. The American National Standards Institute (ANSI) accredits the TIA.

The Project 25 standards define RF-subsystems (RFSSs), which are basically free-standing radio systems adhering to the P25 standards. According to the standard [23],

> The RF Sub-systems are the building blocks for wide area system construction and will connect with any other configuration of equipment or RF Sub-system. The RF Sub-system can be made from any collection of site equipment (single station / site or multiple station / site) whose only requirement is that the equipment supports the Common Air Interface [see below], and contains all the necessary control logic to support the intersystem interfaces and call-processing.

Figure 6.18 [24] shows the interfaces addressed in the P25 suite of standards, the majority of which are with the RFSS.

These interfaces are:

- The common air interface (CAI);
- The inter-RF subsystem interface (ISSI);
- The console subsystem interface (CSSI);
- The subscriber data peripheral interface and data interface;
- The network management interface;
- The telephone interconnect interface.

There are also suites of standards for services (i.e., encryption services and trunked services) and equipment (i.e., transceivers and audio interfaces). Initial attention was focused on the CAI and this is where the set of standards is most complete. The CAI includes specification of channel access methodology, modulation and vocoder technology, data rate and format. Under the heading of System Interfaces, work is underway on the ISSI discussed below. The CSSI is expected to build on the ISSI standards. As has been noted [25]:

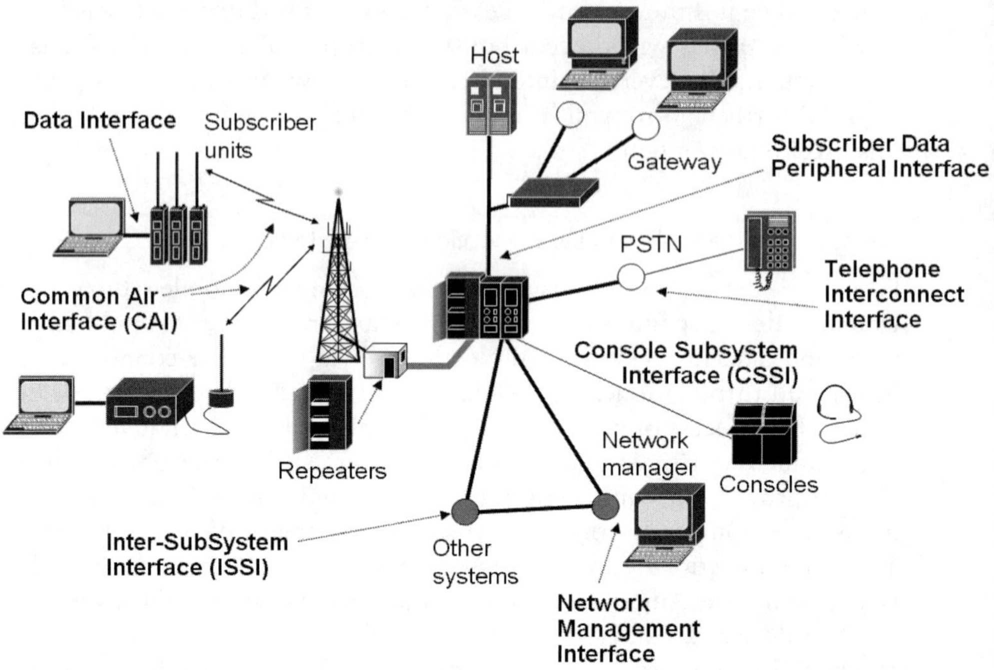

Figure 6.18 Project 25 standard interfaces.

> because of the constant change in the nature of technology, standards are, in truth, never complete. Standards development is a living process. This is a practical result since there will be design improvements in the protocols, which will require changes to the protocol specifications . . . Thus, P25 standards will not remain fixed but will evolve with time and circumstances.

The complete suite of documents covers full system functionality. One can expect incompatibilities because the P25 standards specify interface requirements, but not implementation. Thus, there will be differences in manufacturing designs that can result in unexpected compatibility problems. Thus, a program is being planned to certify compliance with P25 standards and compatibility among the products of the P25 vendors. This program is referred to as the Project 25 Compliance Assessment Program (CAP) and the initial focus has been on the CAI.

Terrestrial trunked radio. LMR systems based on the terrestrial trunked radio (TETRA) standard [26] have been deployed in about 100 countries in the world (65 are outside Europe), but not in North America. About 47% of TETRA systems were implemented for public safety users. The remaining users are a combination of transport, utilities, commercial, oil and gas, and others. The important interfaces in the TETRA standard are shown in Figure 6.19.

These interfaces, like those of the Project 25 standard, are necessary to develop integrated information-sharing architecture for public safety and emergency response.

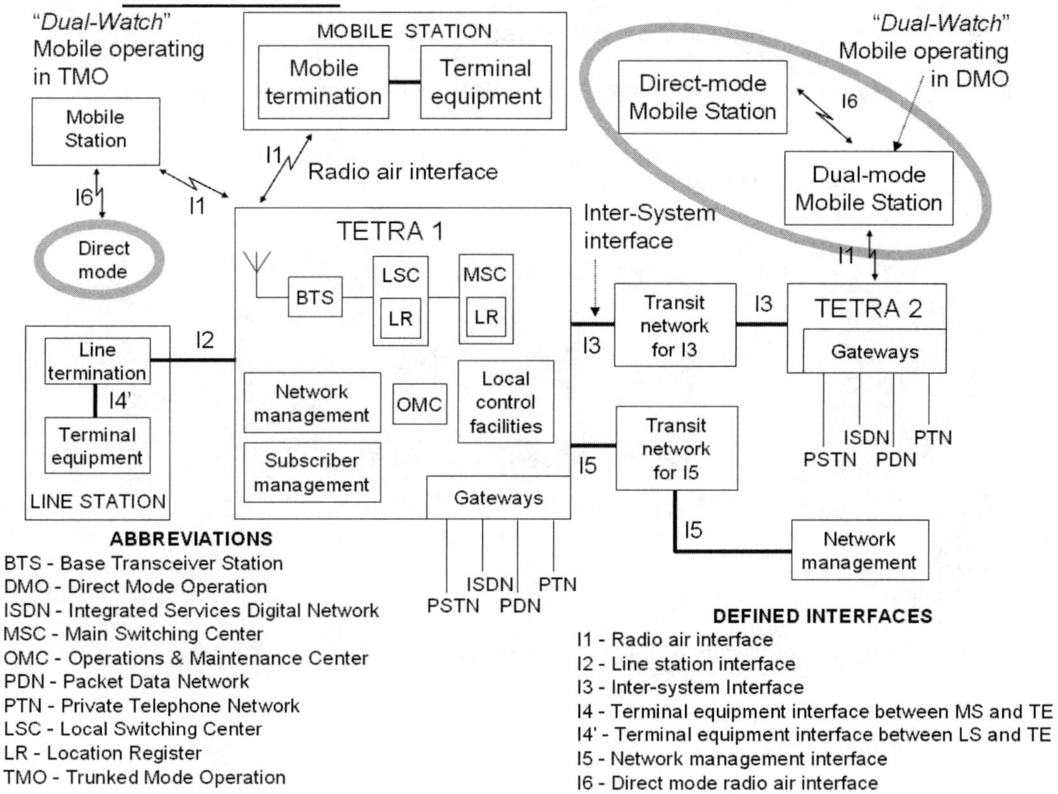

Figure 6.19 TETRA standard interfaces.

6.5 Summary

System engineers will need to develop a system of interoperability architecture to meet the requirements and challenges of providing effective information sharing in public safety, public health, healthcare, and many other areas. This chapter provided tools to visualize and specify interoperability requirements for complex systems that will be capable of providing data interoperability on a broad scale. Ultimately, information sharing must be incorporated system-wide and must extend beyond intelligent fusion collection endpoints and interoperability interfaces. Access to any information source of interest should be enabled from a wide range of user applications through the use of a distinct service layer. Figure 6.20 shows how such a distinct service layer can be conceptualized to allow any application to reach information stored across multiple data sources located at local or remote OPCENs. The definition of such a distinct service layer is a key step in designing data interoperability features of OPCEN systems so that they will function as within a service oriented architecture.

In order to support the operational requirements that good interoperability planning will identify and document, system architects for individual operations centers (OPCENs) will need to satisfy requirements that involve interconnecting with other OPCENs and supporting responders-receivers' information needs regardless of their affiliation or geographic location. This growing need for

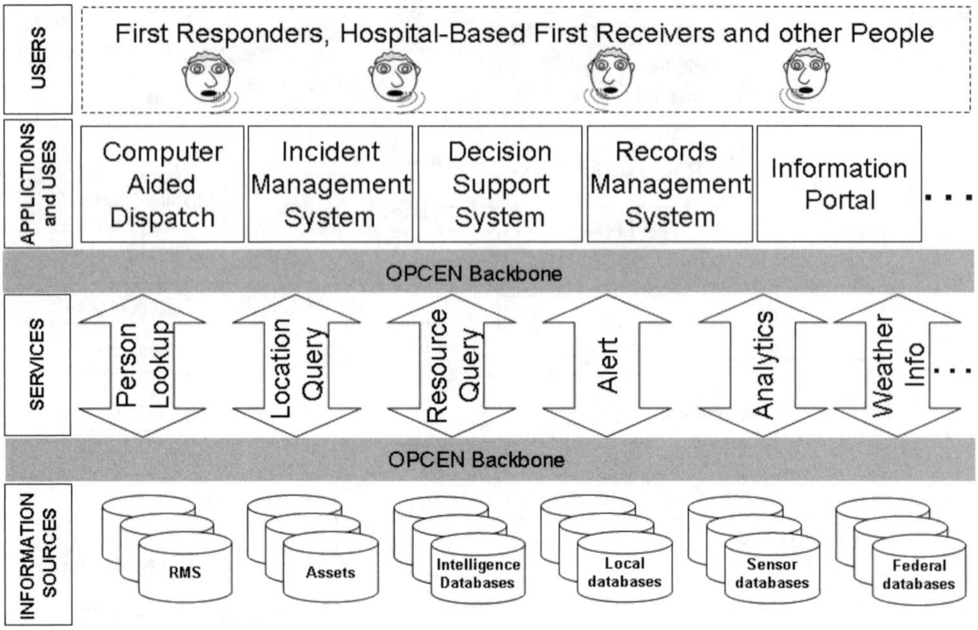

Figure 6.20 Distinct service layer to enable information sharing.

multi-OPCEN interoperability requires a system-of-systems view for delivering the needed technical communications capabilities to these diverse responders-receivers. This chapter provided the interoperability system architecture views and approaches that will ensure that best-practices planning and operational architectures can be implemented. This will assist those seeking to conceptualize and create systems that deliver information and communications capabilities needed to support true interoperability as defined in this book.

References

[1] Desourdis, R., I., Jr., et al., "Land Mobile Radio Systems," Chapter 5 of *Emerging Public Safety Wireless Communication Systems*, Norwood, MA: Artech House, November 2001.

[2] Raymond, M., S. Webb, and P. Aymond, "Emergency Data Exchange Language (EDXL) Distribution Element, v. 1.0, OASIS Standard EDXL-DE v1.0, 1 May 2006. http://docs.oasis-open.org/emergency/EDXL-DE/V1.0.

[3] Roy, D., and J. Murphy,, "National Exchange Model Program: Overview and Status," October 2008. http://www.niem.gov/pdf/NIEM_ESC_Brief_10102008a.pdf.

[4] Ibid.

[5] Rosamilia, P., "Creating Multi-Vendor Interoperability Among Law Enforcement and Transportation Security System," *IEEE 2006 Conference on Technologies for Homeland Security*, Cambridge, MA.

[6] Jeckle, Mario, and Erik Wilde, "Identical Principles, Higher Layers: Modeling Web Services as Protocol Stack", *XML Europe 2004*, Amsterdam, April 2004, http://www.idealliance.org/papers/dx_xmle04/papers/03-05-04/03-05-04.html.

[7] Zimmermann, Hubert, "OSI Reference Model - The ISO Model of Architecture for Open Systems Interconnection," *IEEE Transactions on Communication*, 28(4), April 1980, 425-432, http://www.comsoc.org/livepubs/50_journals/pdf/RightsManagement_eid=136833.pdf.

[8] Desourdis, R., I., Jr., et al., pp. 249-372.

[9] Ibid., pp. 415-506.

[10] Clemm, A., *Network Management Fundamentals*, Indianapolis, Indiana: Cisco Press, 2006. [Au, need location]

[11] Desourdis, R., I., Jr., et al., pp. 536-538.

[12] "Guide to Radio Communications Interoperability Strategies and Products," AGILE Report No. TE-02-02, National Center–Northeast (NLECTC-NE), April 1, 2003.

[13] Mitchell, R., C. Eckel, J. Mathis, "Implementation Profiles for Interoperable Bridging Systems Interfaces," NIST/OLES VoIP Roundtable, September 2008.

[14] Desourdis, R., I., Jr., et al., Chapter 5.

[15] Desourdis, R., I., Jr., et al., p. 271.

[16] Maryland Governor's Office of Homeland Security, "Public Safety Communications Interoperability in Maryland," Report of the Interoperability Project Team, February 2005.

[17] Ibid.

[18] Steinheider, J., "Software Defined Radio Comes of Age," *Mobile Radio Technology*, February 2003. http://www.vanu.com/wp-content/resources/intro/software-defined_radio_comes_of_age.pdf.

[19] Federal Communications Commission, "Public Safety and Homeland Security Bureau," *Tech Topics*, http://fcc.gov/pshs/techtopics/.

[20] U.S. Department of Commerce, National Telecommunications & Information Administration, "Procedures for Federal Radio Frequency Management," January 2008 edition. September 2008 revision. http://www.ntia.doc.gov/osmhome/redbook/Manual.pdf.

[21] Desourdis, R., I., Jr., et al., pp. 278-316.

[22] Ibid.

[23] "Overview of APCO 25 Standards Suite and its Relationship to e-Government Interoperability Framework," *Added Value Applications*, September 19, 2006.

[24] Desourdis, R., I., Jr., et al., Figure 5.9.

[25] "Overview of APCO 25 Standards Suite and its Relationship to e-Government Interoperability Framework," *Added Value Applications*, September 19, 2006.

[26] Desourdis, R. I., Jr., et al., pp. 319–371.

Selected Bibliography

Williams, Matt, "Citywide Wireless IP Network Launched in New York," govtech.com, April 15, 2008. Available at: http://www.govtech.com/gt/286778.

"NYC: Interoperability? Who Needs It?" dailywireless.org, July 1, 2008. Available at: http://www.dailywireless.org/2008/07/01/nycwin-interoperability-who-needs-it/.

Morgenstern, Henry, "NYCWiN Interoperable Communications, A Report of the New York City Wireless Network Innovations," *The Counter Terrorist*, September/October 2008.

Luna, Lynnette, "IPWireless Back Pushing TD-CDMA and Public Safety is a Target Market," *Urgent Communications*, January 12, 2009. Available at: http://urgentcomm.com/networks_and_systems/news/ipwireless-pushing-td-cdma-0114/.

Desourdis, Robert, et al., *Emerging Public Safety Wireless Communication Systems*, Norwood, MA: Artech House, 2001.

Eutelsat Satellites, see http://www.eutelsat.org.

Astra Satellites, see http://www.ses-astra.com.

Intelsat Satellites, see http://www.intelsat.com.

Evans, B. G., *Satellite Communication Systems*, Peter Peregrinus, 1991.

Maral, G., and M. Bousquet, *Satellite Communications Systems*, New York: John Wiley, 1993.

www.iridium.com.

www.globalstar.com.

www.orbcomm.com.

Law Enforcement Tech Guide for Communications Interoperability: A Guide for Interagency Communications Projects, U.S. Department of Justice, Office of Community Oriented Policing Services, 2006.

Public Safety Wireless Interoperability National Strategy (WINS), A Set of Fifteen Papers on Interoperability Methods Published by the Public Safety Wireless Network (PSWN) in 2001.

Federal Communications Commission, "Public Safety and Homeland Security Bureau," *Tech Topics,* http://fcc.gov/pshs/techtopics/.

National Interoperability Field Operations Guide (NIFOG), U.S. Department of Homeland Security, Office of Emergency Communications, Version 1.0, September 2007.

NTIA Manual of Regulations and Procedures for Federal Radio Frequency Management, www.ntia.doc.gov/osmhome/redbook/redbook.html.

Project 25 Document Suite Reference, P25DSR, Telecommunications Industry Association, Mobile and Personal Private Radio Standards Committee (TIA TR-8), November 3, 2008. Available at www.its.bldrdoc.gov/resources/p25/p25DocSelection.pdf.

Trunked Solutions, Proprietary Trunked Systems, Public Safety Wireless Interoperability National Strategy (WINS), August 20, 2001.

Merritt, Rick, "Forum's Call to Arms on Cognitive Radio," *EETimes,* April 17, 2006.

Powell, John, "A Pubic Safety Regulatory Perspective on Cognitive and Software-Defined Radio," *Mission Critical Communications,* August 2004.

"FCC Opens Proceedings on Smart Radios," Federal Communications Commission News Release, December 17, 2003, Washington, D.C.

"Wait 'Til' Next Year," *Urgent Communications Today,* December 4, 2008. Available at http://urgentcomm.com/policy_and_law/mag/public_safety_crunchtime_1201/.

"Policy-Based Spectrum Access Control for Public Safety Cognitive Radio Systems," NPSTC Software-Defined Radio Working Group, Interoperability Committee Meeting, Seattle, WA, The Shared Spectrum Company, September 17, 2008.

Jackson, Donny, "Cognitive Radio Heads for Finish," *Urgent Communications,* November 1, 2008.

"Human Behavior Inspired Cognitive Radio Network Design," *IEEE Communications Magazine,* December 2008.

Overview of the APCO 25 Standards Suite and its Relationship to e-Government Interoperability Framework, Added Value Applications, September 19, 2006.

Wendelken, Sandra, "Motorola and ETSI Address TETRA in North America," *Radio Resource Magazine Online,* December 9, 2009. Available at: http://www.mccmag.com/onlyonline.cfm?OnlyOnlineID=67.

National Strategy for Information Sharing: Success and Challenges in Improving Terrorism-Related Information Sharing, Office of the Whitehouse, October 2007, http://www.whitehouse.gov/nsc/infosharing/index.html.

Evans, B. G., *Satellite Communications Systems,* Peter Peregrinus, 1991.

CHAPTER 7

Achieving Interoperability

For responder-receivers and other stakeholders to perform tasks effectively and safely, they must be able to accurately understand and exchange information on a moment-to-moment basis. This moment-to-moment information sharing by diverse sources is vitally important because of the significant nonlinear impact of manmade CBRNE and natural climate issues that threaten the public and economy. One misunderstood "yes" or "no" answer could result in a catastrophe that might otherwise have been prevented. To illustrate the importance of each word, consider the questions stakeholders may ask to which they then receive one-word answers, such as; *Did you evacuate the low-lying areas? Is that person on the watch list? Did the attacking aircraft come from the north?*

A Katrina responder-receiver states [1]:

> "[i]n any operation, particularly in a chaotic environment, there needs to be a balance between 'going outside the system' and following a plan and a procedure." ...[he] suggested "there needs to be a well-disciplined, systematic approach based on a solid plan that is sufficiently flexible for a variety of situations. Experienced personnel know where the pitfalls are and can make decisions where flexibility is required. Doctrine, policies, training, and exercises should be developed that meet the needs of a trained and ready workforce."

In many scenarios, a simple "yes" or "no" answer could affect thousands of lives. In the event of something like a biological attack, a simple answer could affect millions of lives. Interoperability planning must be sensitive to all details. Effective information sharing is undermined by the 25 Pearl Harbor, 9/11, and Katrina deficiencies we have identified and remedied in this book. Most of the deficiencies outlined in Chapter 2 were deficiencies of leadership. All deficiencies were used as the basis for the Interoperability Assessment Checklist in Chapter 3. This chapter shows how the best practices described in Chapter 4 are used to address these deficiencies. In this regard, the technology recommended is dictated by known and possible deficiencies. That said, adequate technology is necessary to resolve deficiencies, but human factors must be addressed as well.

A heuristic representation of the classic public safety and emergency response interoperability is shown in Figure 7.1 [2]. In the figure, the horizontal access is duration of an event or incident (from hours to weeks), and the vertical scales repre-

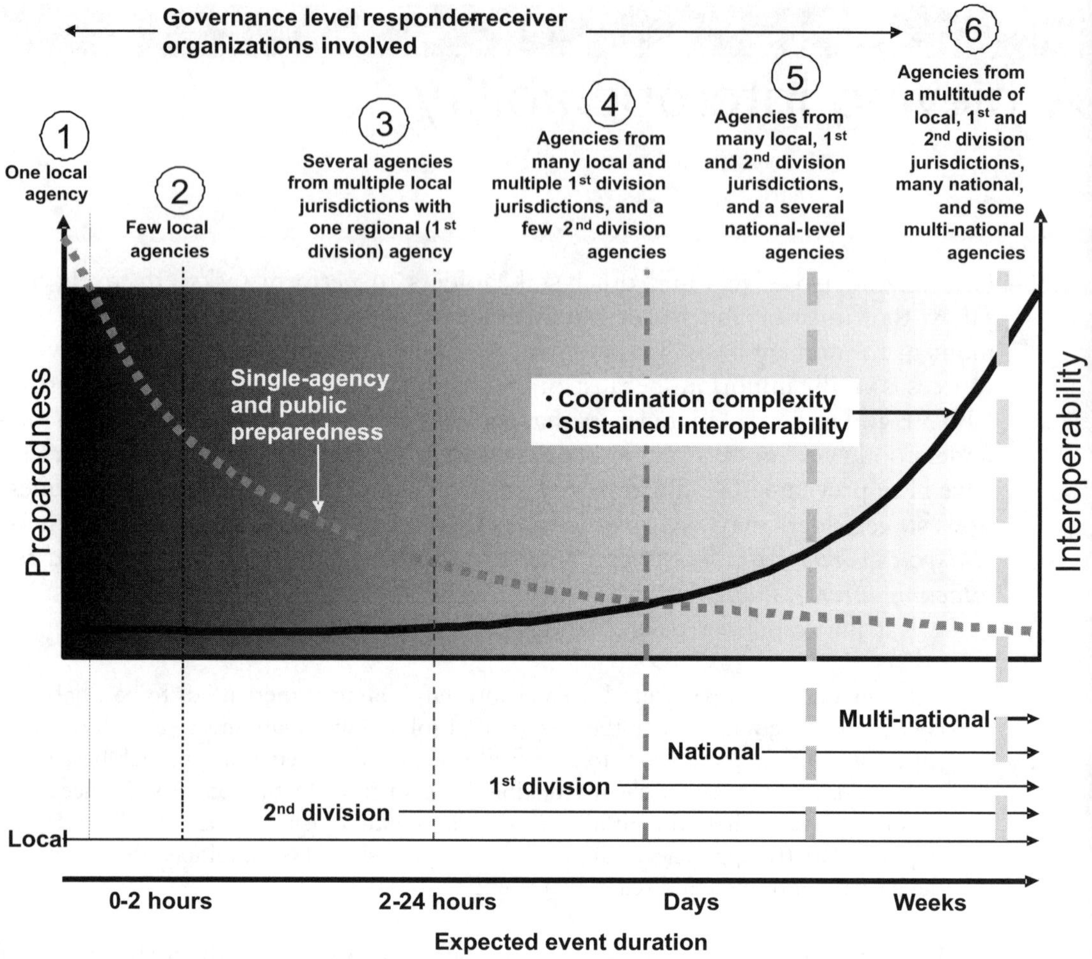

Figure 7.1 Heuristic response continuum model.

sent preparedness (left scale) and interoperability (right scale). In this context, preparedness reflects the potential capability to meet homeland security needs, while interoperability is the required level of shared understanding for predictable collaborative action during the event or incident. The two curves plotted in the figure include arrows showing the appropriate vertical scale.

Above the time scale in the figure is shown the levels of governance that must interoperate, that is, achieve *shared trusted understanding* that drives their *predictable collaborative actions* throughout the event or incident. The numbered circles are used to identify different points for discussion along this *response-scenario continuum*. For example, the preparedness curve shows local agencies at point (1) are well prepared for short-duration incidents or events typically of duration less than two hours, usually covering a limited geographic area within their local jurisdiction.

At point (2) in the figure, a small number of local agencies from the same jurisdiction are involved as we move along the continuum, but the response remains at most a few hours in duration and the area is still limited to a single local jurisdiction. Note that the preparedness curve falls away quickly for involved agencies as we move to the right and require increasing interoperability. At point (3) in the figure, an agency form the next-higher governance level (2nd level in this example, such as a county level) plus several local agencies from different jurisdictions are involved, but interoperability requirements are still low compared to their level of preparedness to support this interoperability. For the local and regional agencies, years of proximity during regional events, personnel relationships and mobility, professional groups spanning jurisdictions, and other social functions have led to some level of familiarity and trust among the associated responder-receivers.

As the incident increases in duration, there is a correlation with increased affected area and numbers of responding agencies. Individuals begin to come together who have had little or no common training or social interaction, and the number of involved disciplines—all speaking a different vernacular—has increased significantly. Thus at point (4) in the figure, a 2nd-level agency (state or province) is involved, and preparedness and interoperability curves are at the same height, meaning each agency is *on the margin* of performing collaboratively with other organizations. Ultimately, at points (5) and then (6) in the figure, the need for interoperability far exceeds the preparedness of the involved agencies. There is limited situational awareness, limited standard vernacular, and minimal common understanding—but the nonlinear sensitivity of information and potential impact looms as an ever-larger threat to public safety in these large-scale scenarios.

The problem at points (5) and (6) in the figure is that the specific details of information sharing for scenarios like those in the figure are not defined. The high-level frameworks available today do not provide the specific information engineering requirements needed to ensure interoperability no matter the scope of the incident. This chapter will describe how the multinational and national public safety and emergency response information-sharing architecture should be developed following the processes described in Chapters 4 and 5, and will lay out a step-by-step approach for achieving the necessary holistic interoperability using best practices. In this regard, this process must include planning for the integrated interoperability elements of governance, resources, approach, architecture, performance, and acceptance. It can be employed in parallel with stop-gap measures to improve interoperability while proper long-term planning is adopted and proceeds. Short of "sound-bite solutions," a dual top-down and bottom-up process using these practices is defined in what follows.

7.1 Summary of the Interoperability Problem

The cause of the interoperability problem is at the heart of human nature, namely to do what is in one's, one's organization's, *perceived* best interest and keep its information to itself, when in fact information sharing greatly improves the preparedness and capabilities of all stakeholders. Because of this deep misperception and its consequences, best practices are rarely integrated into a coherent plan to achieve interoperability, often not even for an organization's internal operability. Too many times "just buy something" has been the quick-sought solution, with little or no time spent following best practices to develop the most needed capabilities—because there is "never time to do it right, but always time to do it over."

In our worst-case World Interoperability Failure model, the many leaders, planners, bureaucracies, vendors, and consultants employed at all governance levels continue to maintain their parallel and noninteroperable IT/communications projects as they seek control of the maximum available resources irrespective of the best interests of public safety. The as-is IT/communication systems and operations shown as (1) in Figure 7.2 are in place and have not been designed (or intended) to share information. Their de-facto interoperability approach is to await disaster (2), then rapidly rush to "buy something," identified as path (X) in the figure. In this tempting rush-to-buy, leaders appear to have acted swiftly, but this course of action is contrary to best practices and often ignores the entreaties of more experienced (but lower ranked) planners at their governance level. These more experienced planners may advocate to use best practices by taking observations from actual incidents and events (3), determine what changes to operating procedures, policies, and training are required (4), and then from that specify the technical IT/communications systems needed to support these improved operational policies and procedures (5). Unfortunately, however, the national funding (grant) approaches often promote the rapid, unplanned, and largely unjustified rush-to-buy partial Band-Aid solutions as their criteria are overly simplified.

As shown in the figure, proper planning for the to-be IT/communication systems and to-be operations after the incident, (4) through (6), is bypassed in the just-buy-something approach in path (X). For example, grant processes in our World Interoperability Failure model support acquisition alone, don't mandate (or mention) best practices as defined in Chapter 4, don't require compliance with specific information-sharing architecture, and accept rapid minimally planned response. Unless multinational and national governments pursue a more unified plan as presented below, information stovepipes will be preserved at the cost of sig-

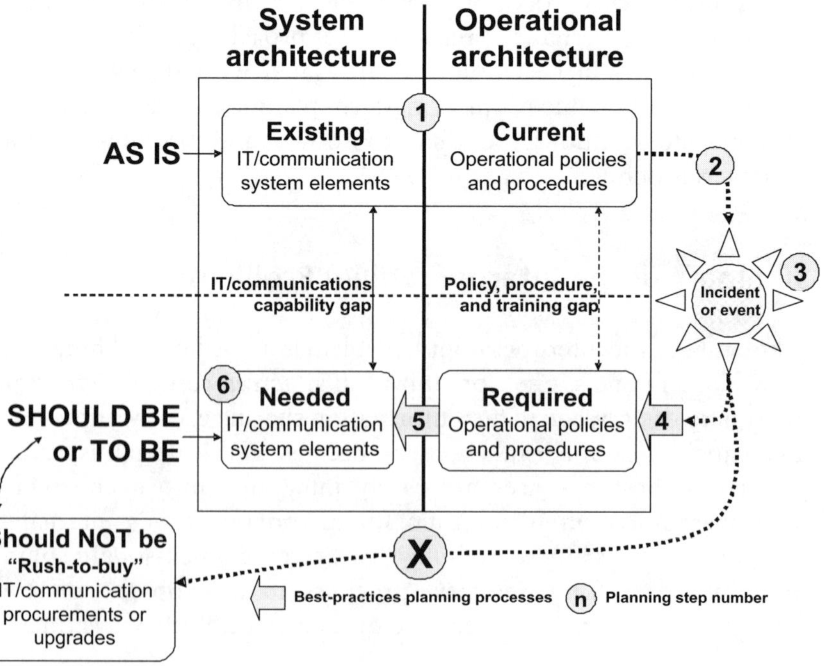

Figure 7.2 The as-is/to-be progression diagram.

nificant duplicative resources. In our model, neither the public safety nor its treasure should continue to be squandered in this way.

7.1.1 Failure of Existing Frameworks

Existing national frameworks, such as those described in Chapter 5 for the United States and the United Kingdom, do not provide a specific threat-scenario-based information-sharing architecture for the detailed integration of all governance levels and the responder-receiver/stakeholder organizations. In other words, these frameworks do not provide nationally standardized information architecture and best-practices-based planning tools to support their respective subnational governance layers. These frameworks do not provide governance layers below the national level with a unique standardized information-sharing architecture in order to "plug in" in a predictable, expandable (geographically), and extensible (added media, e.g., video, and content) manner for regional as well as national and perhaps multinational interoperability.

The failure to provide such an information-sharing architecture prevents the holistic interoperability needed to best support our responders-receivers and all other stakeholders in public safety and emergency response. Instead, these frameworks stop short of this necessity, and instead expresses that other governance layers are free to fill in the broad frameworks provided as they choose. This approach does *not* employ best practices available to help these local and regional organizations achieve interoperability through the proper development of information-sharing requirements. These requirements must be determined properly, defined explicitly, and be developed and provided collaboratively through all governance levels in a given geographic region according to a usable standardized template. Instead, the high-level framework approach ensures the maintenance of hundreds to thousands of information stovepipes that effectively establish and sustain the Pearl Harbor–9/11–Katrina deficiencies as design criteria, despite the oft-repeated consequences.

Arguably, the reason for such failure is that the techniques and best practices necessary to achieve interoperability are not simple, and unless they can be simply stated to many leaders in our World Interoperability Failure model, they will be ignored. Moreover, best practices to establish this information-sharing architecture are unknown, unused, or otherwise forsaken. At the highest governance levels, the sound bite that we will "go to the moon within the decade" has immediate understanding and value, whereas, "building an information-sharing architecture for all responders-receivers and other stakeholders to achieve holistic interoperability" means little or nothing, the value lost in a sea of words. For this reason, a better statement would be the more succinct "public safety through community interoperability." In this way, we address information sharing among all stakeholders and mitigate or eliminate the Pearl Harbor–9/11–Katrina deficiencies.

7.1.2 Failure to Promote Interoperability

In specifying the roles and responsibilities of state governments in the *Katrina Report*, the authors state that [3]:

> State documents detail the specific options and steps available to the chief executive, including an analysis of gubernatorial prerogatives, including: First and foremost, the governor must understand and accept the fact that he/she is the primary person responsible for response and crisis management within his/her state. All citizens look to their governor as the person ultimately responsible. That is not to take away from the local responsibility of mayors, city councils, and county commissions, but, in truth and fact, "the buck stops at the governor." Secondly, although the governor must be the leader of his/her state, the governor must also be prepared to delegate. This statement may seem rather simplistic since every governor in the United States is confronted with so many governmental and administrative decisions, on a daily basis, that they obviously need to be able to delegate. On the other hand, in the case of an emergency catastrophe situation, the number of issues that arise are exponentially greater than ordinary day-to-day issues of government, they are unusual, sometimes technical in nature, they require instantaneous decisions, as opposed to general governmental issues which commonly allow for consideration and even collaboration among advisors and affected entities. In these regards, in order to delegate, it is extremely important that the governor has surrounded himself/herself with an outstanding group of cabinet officials who are not only qualified but who are both qualified and capable of responding in emergency situations. This is most particularly true of the adjutant general of the state's National Guard, the director of the state's Department of Homeland Security, and the office of the director of the state's office of Emergency Management. Obviously each of these positions is a key appointment for every governor, but when confronted with a catastrophic emergency, the importance of the quality and qualifications of the persons holding these positions becomes extraordinarily important. Thirdly, an emergency operations center and a communications system which are capable of and designed to operate under emergency conditions become a key element of the governor's ability to communicate, manage, and lead through the crisis. Finally, there must be pre-planning ("emergency operations plan") that sets out clearly policies, procedures, and responsibilities that will be required to meet all known emergency catastrophe situations. These must be coordinated with local emergency management officials and local government officials.

It is insightful to acknowledge proper leadership and preplanning as described in this chapter, but this planning must start with effective national-level guidance, which has not occurred. National governments have not properly enabled holistic interoperability by demonstrating the major advantages of information sharing as a minimal approach, or tied grant funding and spectrum licensing to specific and detailed information-sharing requirements. This failure has only a few causes in our World Interoperability Failure model, such as:

- Leaders and planners do not know how to specify such a detailed information-sharing *open* architecture, and are afraid of the cost consequences of doing so.
- Bureaucracies, consultants and vendors do not want open architecture and approaches because of the impact on their otherwise controlled resources.
- Planners and their consultants do not know how to integrate best practices or develop mandatory measures to do so, such as organizational development or enterprise architecture, despite the repeated consequences of failing to employ them.

- Leaders and planners await the "perfect" solution, hoping that industry and the public safety and emergency response community will agree one day—perhaps in the distant future—on a single information-sharing architecture.
- Leaders and planners believe the problem is too big for them to solve and are timid about making sweeping change despite the promise of improvement and despite its obvious criticality to basic government purpose.

All of these reasons are strongly correlated with one or more of the Pearl Harbor–9/11–Katrina deficiencies in the Interoperability Assessment Checklist of Chapter 3.

Overall, ensuring the public safety is a fundamental government responsibility that requires holistic interoperability, and it cannot be achieved if those who benefit most—the responders-receivers and other stakeholders—are not shown, or helped to understand, the benefits of doing so. Instead, they are handed large, broad framework documents that "leave it to them" to determine how they will operate and what equipment they will use, but do not provide them with best practices in developing those operations or selecting that equipment. As shown in Chapter 4, these practices are available and have, in fact, been used by organizations seeking true progress worldwide, but largely not in the public safety and emergency response community as defined in our worst-case World Interoperability Failure model.

Promotion of the value of information-sharing and interoperability is evident from the many lessons learned that are embodied in the Pearl Harbor–9/11–Katrina deficiencies, and from the simple application of the Interoperability Assessment Checklist (see end of Chapter 3). This effort must first understand the specific legal restrictions on the dissemination of criminal justice or healthcare information when associated with an individual's name. Otherwise, embrace the complete availability and benefit from moment-to-moment situational awareness and command coordination.

7.2 Best Practices Determine Interoperability Planning Success Factors

As discussed in Chapter 5 and described in Chapter 6, we wish to develop information sharing among operation centers (OPCENS) at multiple governance levels that span many public safety and emergency response disciplines in response to several chosen threat scenarios. For each level of governance, Figure 7.3 shows the tasks and activities of one or more OPCENs through all stages of homeland security, that is, the repeated OODA loops shown at point (10) required to "prevent, detect, protect, alert, respond, and recover" at point (9) in the figure. In this figure, we assume that the indicated tasks and activities span all participating OPCENs for all specified threat scenarios. The interoperability success planning factors derived in Chapter 1 (i.e., governance, resources, approach, architecture, performance, and acceptance) work together to create the desired OPCEN information sharing for each stage at point (8) in the figure. Each of these planning factors is in turn provided by one or more of the integrated best practices defined in Chapter 4. The goal

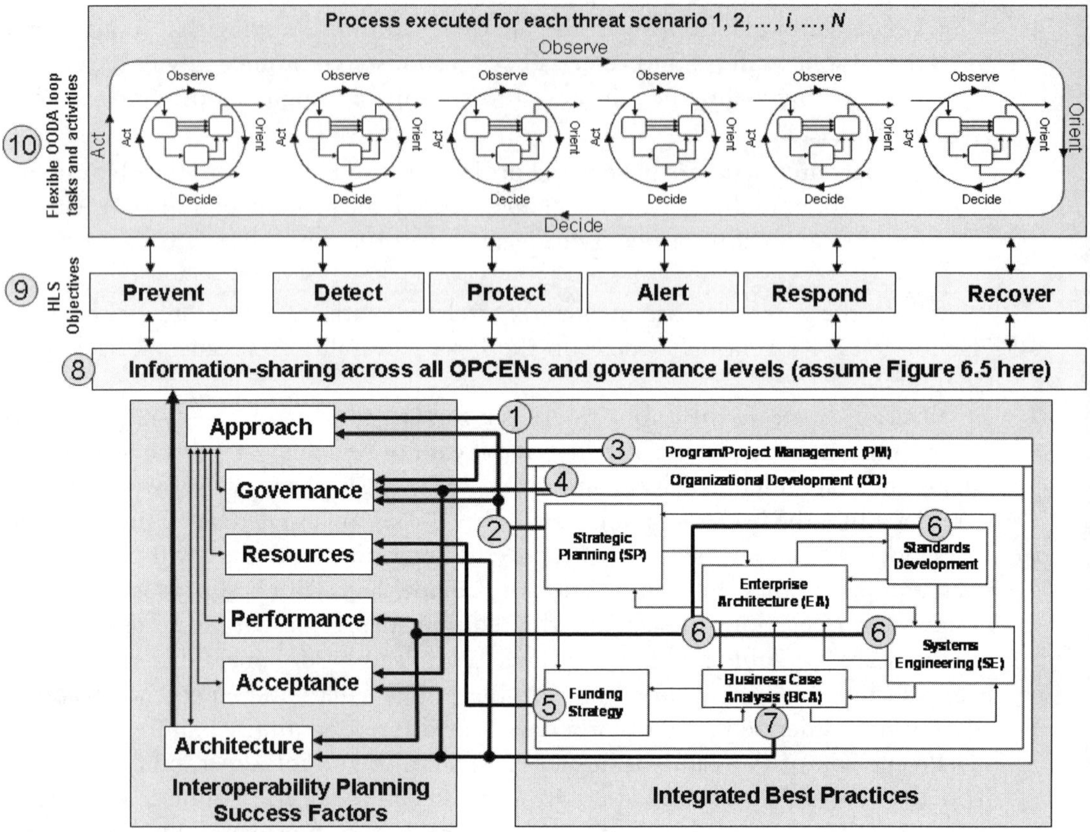

Figure 7.3 Relationship of best practices to interoperability success planning factors.

of this approach is to achieve the best overall public safety response to the threat scenario based on collective OPCEN capabilities and scenario involvement.

Point (1) in the figure shows that the approach planning factor encompasses the use of all best practices as depicted within the associated shaded rectangle. These best practices are program/project management (Section 4.2), organizational development (Section 4.3), strategic planning (Section 4.4), enterprise architecture (Section 4.5), cost/benefit analysis (Section 4.6), funding strategy (Section 4.7), systems engineering (Section 4.8), and standards development (Section 4.9). Point (2) in the figure illustrates where strategic planning defines the overall approach—including specification of the best practices to be used—as well as establishes the vision, mission, goals, and objectives that guide development of governance.

Points (3) and (4) in the figure show that program/project management and organizational development practices determine the governance and acceptance success planning factors, respectively. The program/project management approach determines the rationale and specificity by which the overall program and its individual projects are controlled and results vetted and implemented. The organizational development techniques help build the program and project teams as well as establish "buy in" to the planning process and its results. In this regard, organizational development techniques are also critical in achieving the acceptance planning factor for responder-receivers and other stakeholders as their input is solicited and assimilated into the detailed planning effort.

The best-practice funding strategy at point (5) and business case analysis at point (7) determine the resources planning factor, which is dependent on both the needed and available funding. Other resources, such as RF spectrum and personnel, can all be represented in terms of this financial resource category. The funding strategy must account for the needed resources and plan for both acquisition and sustainment of the ultimate operational and system architecture. This strategy must be focused nationally on achieving specific information-sharing at all governance levels, tying resource availability to completion of specific information-sharing templates as will be defined below.

Point (6) in the figure shows that enterprise architecture, standards development, and systems engineering determine architecture as well as the performance success planning factors. The architecture is also strongly dependent upon the results of business case analysis as implied by point (7) in the figure. Business case analysis formally develops and evaluates alternative architecture and determines the functionality, performance, and other benefits versus the cost and associated risks of each alternative. The alternatives are then ranked and a preferred architecture is selected. Once selected, the operational and system gaps between the as-is (existing) architecture and to-be (highest ranked alternative) architecture are identified. The required improvements to systems (and technology) are addressed from the development of detailed requirements through sustainment planning using best-practice systems engineering at point (6) in the figure. In addition, the necessary concepts of operations and standard operating procedures can be developed or updated (see Chapter 5) to define the OPCEN operational architecture. Maximum use will be made of existing CONOPS, SOPs, systems and technology to minimize costs and training requirements.

Use of best practices ensures that the success planning factors establish holistic (implied) interoperability for the collaborating OPCENs. This information sharing at point (8) in the figure—planned at points (1) through (7)—provides the necessary shared trusted understanding that drives predictable tasks and activities at points (9) and (10) for collaborating agencies. Note that these stages and the necessary information flows are defined for all threat scenarios shown at the top of the figure. These scenarios vary with diversity in critical infrastructure, climatic conditions, and geographic, hydrodynamic and even lithographic features of the OPCENs' jurisdiction. Necessarily, the number and identity of responder-receiver agencies or organizations and other stakeholders involved, as well as their specific stage-dependent tasks and activities, will vary with the nature and scope of each threat scenario. OPCEN interoperability must be designed on the *collective* capabilities; however, as if the OPCENs were working collaboratively towards a common public safety and emergency response goal.

7.3 Application to the World Interoperability Failure Model

This section describes the application of these processes to the World Interoperability Failure model defined in Chapter 3. In this regard, Figure 7.3 showed how to employ best practices to develop the success planning factors needed to achieve holistic interoperability among OPCENs. The figure was applied to the development of information-sharing architecture for the homeland security response spanning all threat scenarios for all participating OPCENs and their associated governance levels. Neither the OPCENs nor the associated governance levels were explicitly depicted in the figure, but they were referenced by the rectangle labeled point (8) in the figure. Point 8 identifies the planning artifacts needed to determine the information-sharing architecture for Figure 5.5 (of Chapter 5) showing an array of collaborating OPCENs. These OPCENS spanned all responder-receiver and stakeholder disciplines and their corresponding jurisdictions and governance levels.

To achieve interoperability across OPCENS, information flows that support tasks and activities of all OPCENs at each governance level must be determined. The involvement of a particular OPCEN and its supported disciplines at a given governance level depends on the threat scenario. For example, a local special event may not involve a national emergency operations center beyond high-level awareness, whereas a major disaster might fully engage the national OPCEN as well as those in the region and locality where the disaster occurred. Thus, we must first develop the relevant threat scenarios for each jurisdiction and the associated or contributing governance level as well as the corresponding information-sharing specifications. Since we don't know which threat scenario will occur, the OPCEN must be capable of performing all scenarios, albeit not necessarily simultaneously. Thus, the OPCENs must have composite, federated, or integrated architecture that achieves holistic (implied) interoperability in addressing any threat scenario.

The various governance levels maintaining responders-receivers and other public safety and emergency response stakeholders from local to multinational are shown under point (1) on the far left side of Figure 7.4. The individual agencies are shown at the local level (three are shown, many more possible), but we assume that (1) agency OPCENs of various disciplines exist at every governance layer, and (2) the agencies OPCENs must be interoperable with OPCENS, responders-receivers, and stakeholders from all other participating governance levels, jurisdictions, and disciplines. These agencies employ individual or consolidated OPCENs that, taken collectively, perform a series of tasks and activities supported by information sharing.

Point (2) in the figure shows international and national conventions for threat scenarios and high-level information-sharing frameworks (for the latter, see the U.S. National Response Framework and U.K. Incident Command System examples in Chapter 5). These conventions are used to plan the form and high-level content of the following documents:

1. *Threat and information-sharing architecture templates*—two sets of templates consisting of (1) threat scenarios defined at the specified governance level (e.g., select threats aligning with the six response-continuum scenarios highlighted in Figure 7.1), and (2) a set of operational and system architecture view templates (see Section 4.5.5) used to define the architecture of the collaborating OPCENs at each governance level spanning the corresponding threat scenarios.
2. *Information-sharing architecture by governance level*—completed operational and system view templates that define the information-sharing archi-

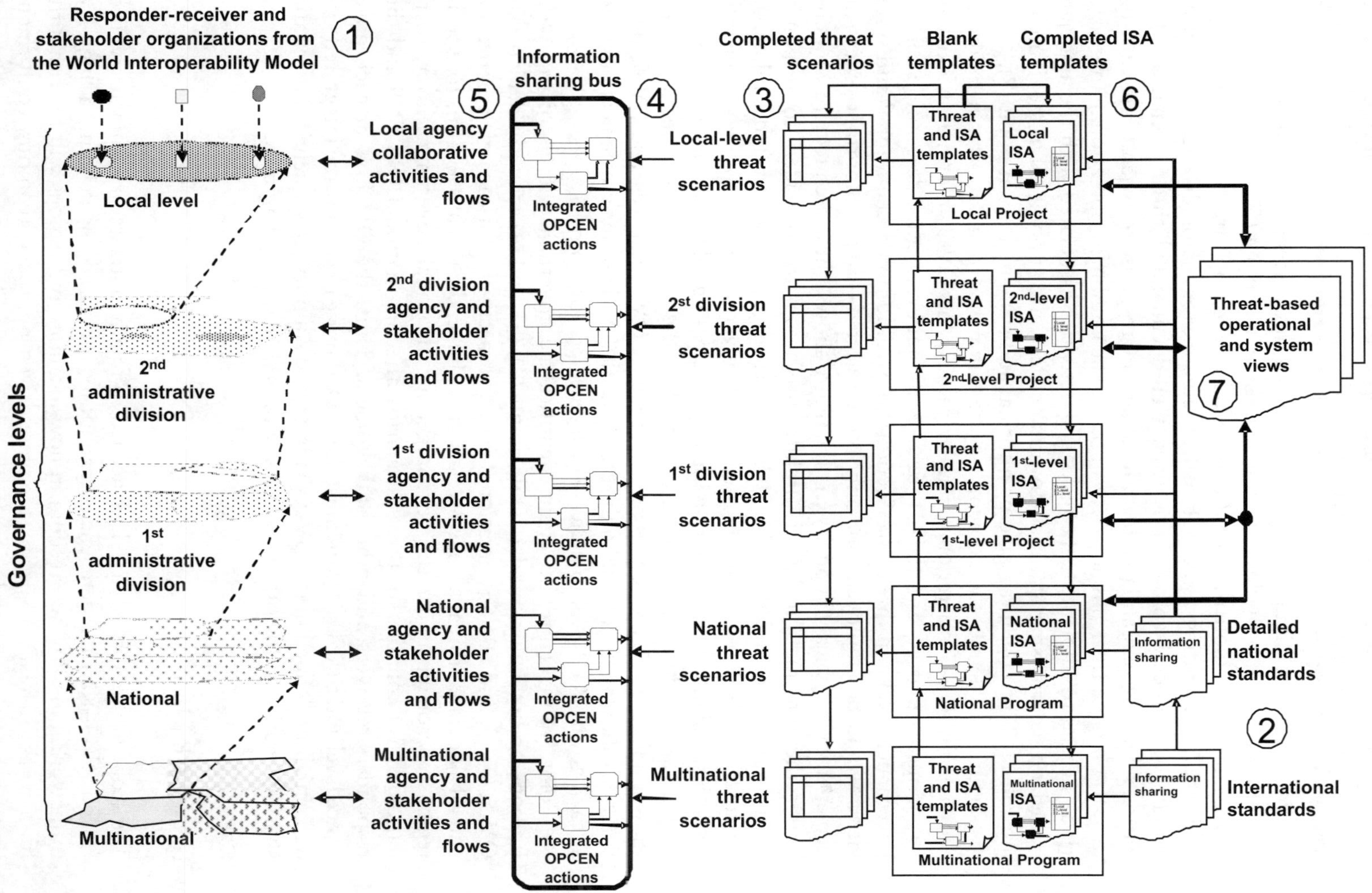

Figure 7.4 Application to the World Interoperability Failure model.

tecture of the collaborating OPCENs at each governance level panning the corresponding threat scenarios.

Point (3) in the figure shows the completed threat scenarios from point (4) that drive the tasks and activities of all OPCENs shown at point (5) at the corresponding governance level. Note that the threat scenarios “roll up” from local to regional to national (and, putting aside political differences for major disasters) to multinational governance levels. In this way, well-documented threat scenarios at the national level are expandable from local agency response to whatever is most prudent to mitigate their public safety impact. The combined operational and system views needed to implement the tasks and activities associated with the associated homeland security stages at each governance level are at point (6) in the figure and are collected into the integrated operational and system views at point (7).

Figure 7.5 details the contents of the completed templates from Figure 7.4 showing the OV-5 operational activity model, the OV-3 information exchange requirements matrix, and the OV-6c operational event-trace description for each governance level and threat (see Section 4.5.5). Additional operational and system views would be employed, but only those selected—albeit important—views are depicted for graphical simplicity. As the figure shows, these architecture views represent the information-sharing architecture templates provided to responder-receiver and other stakeholder OPCENs for each governance level, discipline, and jurisdiction. These templates are used at each governance level to capture the tasks, activities, and information sharing for all participating OPCENs for each threat scenario. These architecture views are created by the enterprise architecture process shown in Figure 7.3. (Note that the OV-5 views, OV-3 matrices, and OV-6c timing diagrams shown are notional only, albeit the OV-5 diagrams are depicted uniquely at each governance level to indicate different operational roles.)

Point (1) in Figure 7.5 again identifies the governance levels of the World Interoperability Failure model, while points (2), (3), and (4) show the depictions of OV-5, OV-3, and OV-6c operational views, respectively, for each governance level and all associated threat scenarios. As stated above, the threat scenarios for specific regions all roll up under the next larger jurisdictional governance layer, so the national-level contains all threat scenarios affecting national OPCENS and their responders-receivers and other stakeholders. Thus, the OPCENs at each governance level and threat scenario must then collaborate to execute the integrated tasks and activities defined in the OV-5 diagram and support the sequential information sharing detailed in the OV-3 matrices with OV-6c timing diagrams.

7.4 The Work Plan

The scope and nature of activities needed to finally mitigate or eliminate the Pearl Harbor–9/11–Katrina deficiencies cannot be obtained by defining a high-level framework and allowing freedom in how localities and regions will fit within that framework. Instead, a specific template-driven approach requiring conformance through spectrum-use and grant-funding leverage with a specific national information-sharing architecture will achieve the required interoperability. Of course, the

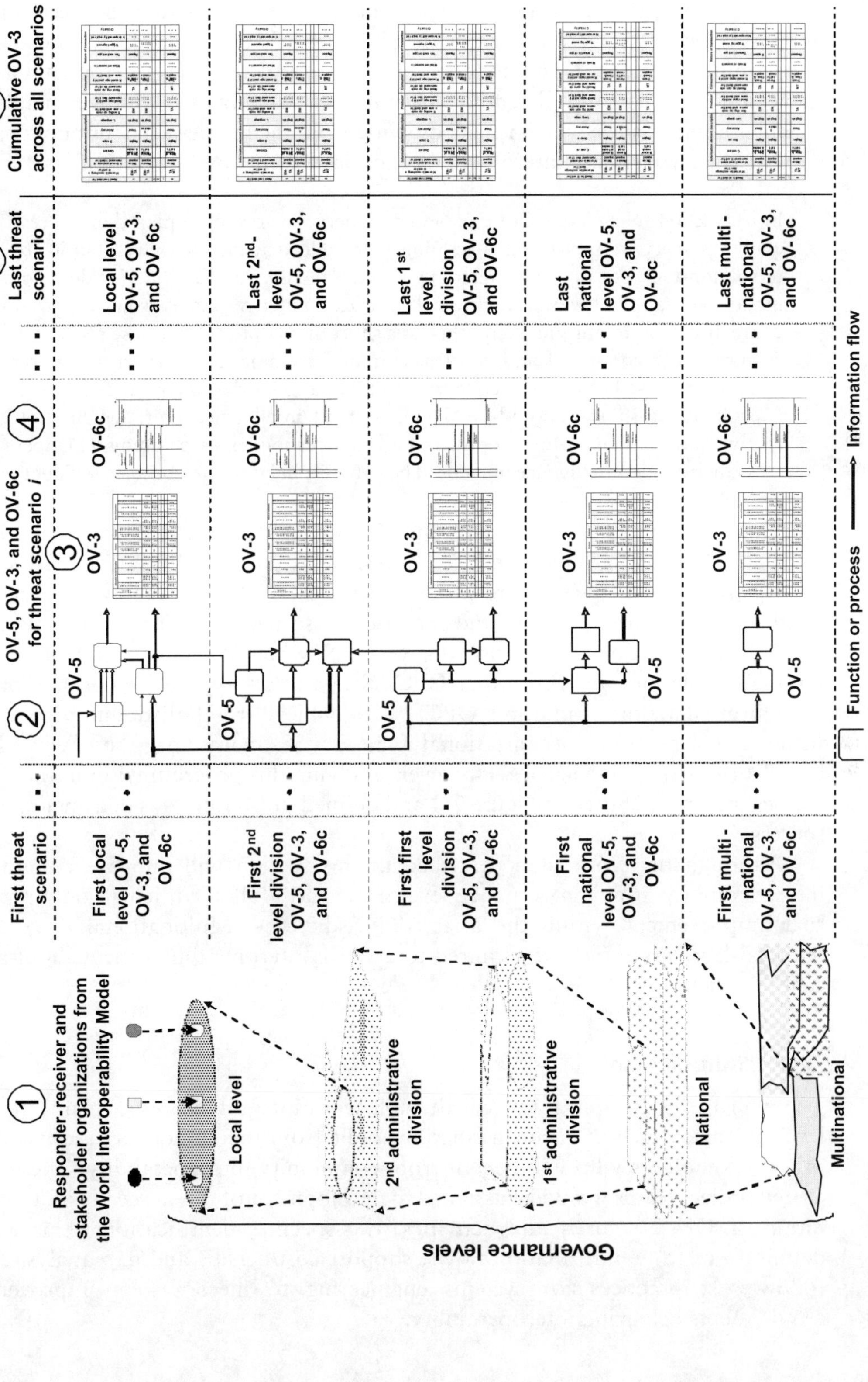

Figure 7.5 Tasks and activities with information xxchange requirements for each threat scenario.

threat scenarios and specific information-sharing content that completes these templates must come from local and regional responder-receivers and other stakeholders under a managed program following best practices. This approach is particularly important given the sensitivity of small amounts of information in executing the homeland security stages. It is analogous to the sensitivity of war-fighter information in a complex battlefield, and no less important as it determines interoperability for homeland security and defense. The military has had to resolve significant interoperability issues, which is why drawing from their experience for public safety and emergency response interoperability displays good governance [4].

> In May 2004, New York City adopted an emergency response plan that expressly contemplates two or more agencies jointly being lead agency when responding to a terrorist attack but does not mandate a comprehensive and unified incident command that can deploy and monitor all first responder resources from one overall command post. In our judgment, this falls short of an optimal response plan, which requires clear command and control, common training, and the trust that such training creates. The experience of the military suggests that integrated into such a coordinated response should be a unified field intelligence unit, which should receive and combine information from all first responders—including 911 operators. Such a field intelligence unit could be valuable in large and complex incidents.

The OV-5 and OV-6c diagrams and OV-3 matrices shown above in Figure 7.5 for each threat and governance level represent the combined tasks, activities, and message flows form one or more of the associated OPCENs. An individual OPCEN would extract from these tasks and activities those actions which it performs alone as well as information flows within that OPCEN, while remaining actions and flows represent collaboration with other OPCENs. In this way, we are considering the collaborative action of multiple OPCENs to achieve true holistic interoperability while avoiding the organizational causes of many of the Pearl Harbor–9/11–Katrina deficiencies. However, applying this program without the use of the best practices shown in Figure 7.3 and defined in Chapter 4 is detrimental to its success.

Although the work plan described in this section will improve community interoperability at any governance level if properly followed, it will only provide such improvement within the affected jurisdictions. Multinational or at least national-level programs are required to achieve interoperability throughout adjacent jurisdictions.

7.4.1 Staffing Plan

The interoperability program will be built from integrated project teams at each level of governance in each jurisdiction, built from the responders-receivers and other stakeholders who will benefit from community interoperability. These integrated project teams would be assigned to development of threat scenarios, the definition of OPCEN tasks and activities, the specific identification of as-is and definition of to-be information flows, support cost/benefit and gap analysis, and follow best practices for systems engineering to meet system improvement requirements achieving interoperability.

7.4.2 Task Plan

The sequence of high-level steps to be performed at the multinational or, more likely, national level is shown in Figure 7.6. The figure shows that the national government must develop a proper program/project management structure following best practices, develop standardized templates for the development of the local-regional-national threat scenarios (see task 3.0) as well as alternative architecture views (task 4.0) composing the information-sharing architecture. The plan shown in the figure establishes a threat scenario continuum from local to national levels, measures existing systems capability to achieve interoperability, and then identifies the minimum system gaps (task 5.0) needed to achieve the desired interoperability through proper systems engineering practice (task 6.0).

The business process modeling notation simulation capability described in Section 5.6 as part of architecture-centric engineering (ACE) provides a computer-based means for evaluating the complexities of architecture performance. Subtasks 4.4 and 5.1 in the figure would include analytic techniques for evaluating performance of alternatives before determining benefits and comparing these benefits to costs and associated risks. In addition to computer simulation, after-action reports for exercises as well as actual events will be used to evolve operational and system implementation to the objective interoperability. In addition, both the as-is

1.0 Establish Program Plan and team	
1.1	Establish Program Management Office (PMO), staff, and PM tools
1.2	Develop best-practices-based Program Management Plan (PMP)
1.3	Establish multidisciplinary Community Interoperability Task Force Executive Committee
1.4	Develop Community Interoperability Strategic Plan using OD techniques to establish vision, mission, goals, and objectives
1.5	Establish criteria for local and regional ISA Integrated Project Teams (IPTs) and work breakdown structures and schedules
1.6	Establish, fund, train, and equip ISA OPCEN IPTs at each governance level from *all* levels and disciplines

2.0 Develop threat and ISA templates	
2.1	Establish Web-based collaborative collection / analysis tools for EA artifacts and SE requirements
2.2	Develop templates for threat scenarios at each governance level
2.3	Develop templates for EA-based information-sharing architecture (ISA), that is, selected as-is and to-be OVs, SVs, and TVs
2.4	Train selected IPT personnel to collect and enter threat scenarios and OPCEN OVs, SVs, and TVs

3.0 Establish threats and as-is ISA	
3.1	Assign IPTs to develop threat scenarios relevant for their local or regional environments
3.2	Assign IPTs to complete *as-is* templates for organizations, OPCEN roles and responsibilities, tasks and activities, and information sharing
3.3	Enter template results into EA tool so as-is OVs and SVs are in tool repository
3.4	Integrate local scenario-based as-is OVs and SVs into regional views
3.5	IPTs conduct local and regional reviews of as-s OVs and SVs all participating OPCENs
3.6	IPTs conduct local and regional reviews of as-s OVs and SVs for all participating OPCENs

4.0 Establish to-be ISA alternatives	
4.1	Establish ISA OVs and TVs defining all interfaces and semantics from best practices and frameworks
4.2	Train each IPT to develop to-be ISA OVs for each threat scenario
4.3	Support each IPT in development of to-be OVs for each threat scenario and governance level
4.4	Determine alternative SVs that achieve best-practice OVs for each threat scenario

5.0 Perform cost/benefit and gap analyses	
5.1	Train IPTs to perform cost/benefit analysis on each SV alternative and select preferred ISA SVs
5.2	Determine gap between preferred to-be ISA and as-is ISA (OVs and preferred SVs)

6.0 ISA Systems Engineering	
6.1	Determine OV and SV enhancement or upgrade requirements
6.2	Perform Systems Engineering to develop and sustain to-be ISA
6.3	Perform appropriate test, verification, and exercise of threat-scenarios and ISA performance

WBS Dictionary

1. PM – Project Management, Section 6.2
2. OD – Organizational Development, Section 6.3
3. OPCEN – Operations Center, Section 5.2
4. EA – Enterprise Architecture, Section 4.4
5. SE – Systems Engineering, Section 4.8
6. OV – Operational View, Section 4.5
7. SV – Systems View, Section 4.5
8. TV – Technical View, Section 4.5

EA – Enterprise architecture
IPT – Integrated project team
ISA – Information-sharing architecture
OD – Organizational development
OV – Operational view
SE – Systems engineering
SV – Systems View

Figure 7.6 World interoperability success task plan.

and postulated to-be information-sharing architecture can undergo iterative evaluation using the Interoperability Assessment Checklist.

7.4.3 Resource Plan

Elimination of the Pearl Harbor–9/11–Katrina deficiencies through proper planning and implementation of community interoperability cannot be achieved without appropriate resources. These resources can be provided in several ways, including controlled grants or allotments. These controls must be applied to ensure that resources are only provided in producing the specific information-sharing artifacts needed to achieve true interoperability. Thus, the completion of specific templates described in Section 7.3 by all grant-funded agencies and governance levels must be a precondition for receiving the requested resources. If resources are not tied to results, interoperability planning will invariably fail to achieve its goals.

7.5 Success Planning Factors

Each of the success planning factors must be considered in development of a multinational or national program to achieve interoperability in public safety and emergency response.

7.5.1 Governance

An executive organizational body will be convened to control the information-sharing architecture at each governance level with proper representation from all responder-receiver and other interoperability stakeholders within the local community, not just the principal public safety and emergency response organizations of police, fire, and EMS (see Table 3.1). The number of such bodies at any governance level will be based on geographic and demographic boundaries, utilizing existing bodies as core (albeit expanded if necessary) to the maximum practicable extent. The charter of this organization, developed using organizational development tools, will be standardized by the national government and delivered as a template to complete as needed by the designated local or regional body. This charter will mandate the use of a specific common approach for threat scenario and architecture view development throughout the associated geographic region, dictated by the national templates, but provides the greatest leeway for local or regional government traditions. Proper voting procedures should be followed if deemed necessary.

The governance body at the local and regional level (i.e., below the national level):

- Provides the oversight—and OPCEN site/staff access—for the local project team collecting threat scenario and as-is information-sharing architecture data.
- Verifies that the project team is following the prescribed templates for collected information and supports this collection where possible.

- Validates and votes on results of the information collection, threat scenario development, and information-sharing architecture views.
- Support the development of to-be alternatives for information sharing in their local or regional jurisdiction.
- Provides the political, public, and press interface to the progress and accomplishments of the project team, insulating them from extraneous effort.

7.5.2 Resources

Each local or regional program will employ resource staff capable of identifying available grants and other sources of funds to support the project planning and implementation activities. The resources staff will monitor local or regional implementation requirements, support the generation of cost estimates, and then use these estimates as resource requirements which must be met. All resource staff will report to the national program, providing requirements and managing the acquisition of resources for sustainment as well as acquisition. This staff would develop the funding strategy (see Section 4.7) that assigns funding so the national program office can coordinate among resource requirements based on the priority of threat scenarios and the documented needs. By tying these resources to implementation of open systems architecture, increased competition for better performance at lower costs increases the capabilities available for the same resources.

7.5.3 Approach

The best practices in Chapter 4 and the guidance provided in this chapter constitute the approach success planning factor. As stated above, failure to employ best practices as shown in Figure 7.2 (e.g., program/project management, organizational development, strategic planning, enterprise architecture, business case analysis, funding strategy, systems engineering, and standards development) risks the imbedding of the Pearl Harbor–9/11–Katrina deficiencies into the information-sharing architecture. Thus, both the template-based to-be design approach described in this book and the associated best-practice processes must be used to achieve interoperability.

7.5.4 Architecture

The realization of specific information sharing through integration of operational, system and technical architecture—developed from the application of best-practice enterprise architecture and systems engineering to fill identified operational capability gaps—is an innovation in public safety and emergency response. As in the World Interoperability Failure model, bureaucracies, consultants, and vendors pursue minimal to no change in the status quo, which maintains the Pearl Harbor–9/11–Katrina deficiencies. Instead, bringing best practices to acquisition of only the new architecture that is required, maximizing use of existing proven capabilities and in-use systems and technology, will most efficiently serve responders-receivers and stakeholders.

A second innovation required to improve interoperability is to delineate specific information-sharing architecture at all levels of governance. Instead, multinational and national architecture that provides specific open information-sharing interfaces and templates at all local and regional levels will foster interoperability, rather then continued support for stovepipes enforced by following a high-level framework. These open interfaces will also increase competition for interoperability solutions and lower costs of available capabilities.

7.5.5 Performance

Architecture performance for information sharing can be measured in three ways, including use of the Interoperability Assessment Checklist, the BPMN simulation, and after-action reports. These performance measurement approaches must be used, rather than ignored, to maintain the status quo and identify failures continuing as Pearl Harbor–9/11–Katrina deficiencies. These performance measures provide direct feedback to the systems engineering and other practices to establish a formal improvement capability at the local and all regional levels.

7.5.6 Acceptance

Arguably a critically important element of achieving interoperability is to gain the acceptance of responders-receivers and other stakeholders to appreciate the value of information-sharing to improve the fundamental performance of public safety and emergency response for the public. The benefit improves situational awareness and command coordination for all OPCENs and their associated responders-receivers and other stakeholders, increasing the safety and effectiveness of these public servants. Providing the proof of benefit from improved community interoperability through captured examples, and correlating these benefits with the application of best practices to interoperability planning, will achieve this acceptance.

7.6 Contributing to the World Interoperability Success Plan

It will take a coordinated effort throughout the global emergency response community and across governance organizations to affect the scope of change described here in order to significantly mitigate the 25 deficiencies discussed throughout this book. Leaders and planners at multinational organizations, national governments, 1st and 2nd level governance levels, and ultimately local levels, with the responder-receiver and other stakeholder organizations at all levels, must have a new birth of commitment to mitigating and eliminating the Pearl Harbor–9/11–Katrina deficiencies once and for all. The benefits of doing so can be tremendous and could provide the responders-receivers and other stakeholders the interoperability they need and deserve to do their jobs. This section provides some brief summary guidance for all such stakeholders in implementing the best practices recommended in this book. Only by implementing these best practices can we begin to deliver true holistic interoperability at all governance levels. True interoperability preserves local control of information and assets by allowing secured collaboration while protecting local interests.

7.6.1 National Governance Level

The existing threat scenarios developed at national levels provide a starting point for the sanctioned development of threat scenarios at each level of governance. These scenarios would then be rolled up to the national and multinational levels. In this way, specific scenarios for the local and regional OPCENs are defined for the development of specific information-sharing architecture. As the threat scenario moves to the right on the horizontal axis of Figure 7.1, more and more agencies are involved in the threat and response scenario.

The national government can provide the funds to develop these threats and roll them up with the corresponding governance-level planners. These scenarios provide the basic planning input to OPCENs at the corresponding governance levels using sponsored organizational development best practice to develop the to-be CONOPS for all agencies involved in the scenario governance level to address each corresponding threat-scenario. The templates for the to-be CONOPS will be developed by national level planners so common terminology and capabilities will be employed nationwide. In addition, templates for standard operating procedures will also be provided. Grant and other funding will be provided to those OPCENs at all governance levels that within national jurisdiction that complete these templates. The funding would sponsor use of organizational development best practice to develop the regional CONOPS for each threat scenario and associated OCEN standard operating procedures. In addition, funding would be made available for an OPCEN to develop its as-is and (based on its standard operating procedures) to-be information-sharing architectures, perform a gap analysis, and execute best-practice systems engineering to create the federated IT/communication architecture for regional interoperability and beyond.

7.6.2 1st and 2nd Governance Levels

The role for the governance levels between the local and national is to roll up the threat scenario templates, meet the appropriate CONOPS and standard operating procedure requirements from the national level, and integrate its OPCENs as appropriate with the corresponding projects at all other governance levels. Each level must develop their own as-is and to-be information-sharing architecture to meet CONOPS and OPCEN standard operating procedure requirements, leading to gap analysis and systems engineering to build the necessary systems architecture.

7.6.3 Local Level

The local level OPCENs define their threat scenarios using national architecture and local realities using the national templates, then provide the completed templates to the relevant 2nd and 1st administrative division OPCENs. In addition, the use provided CONOPs templates in collaboration with adjacent jurisdictions to develop a regional CONOPS for each scenario and then individual OPCEN standard operating procedures mapped to each CONOPS. As at other governance levels, they are sponsored to produce their as-is and to-be information-sharing architectures, perform a gap analysis, and employ systems engineering best practice to meet the resulting to-be systems architecture requirements. While national templates and approaches are needed to facilitate interoperability, the balancing concern of preserving and enhancing local control is also important.

7.6.4 Multinational Level

The multinational level coordinates among nations to define threat scenarios and develop transnational CONOPS, standard operating procedures, as-is and to-be information-sharing architectures, perform gap analyses, and ultimately systems engineering to build the necessary system architecture. This international information-sharing system architecture would involve OPCENs in different countries and governance levels depending on the geographic border regions between collaborating nations.

7.7 Summary

This chapter has provided a World interoperability Success Plan drawing on the planning best practices described in Chapter 4 and the operational architecture development approach described in Chapter 5. In this way, the national government can achieve national interoperability in the least time by providing the local and regional-level planning tools and best practices both needed by leaders and planners at all governance levels nationwide. It can proactively provide the templates and best-practices tools, while leaving the specific requirements development using those tools and best practices to the funding-motivated initiative of these governance levels and associated responders-receivers themselves. In this way, we have defined detailed top-down structure and process with funding needed to achieve holistic interoperability, but driven by bottoms-up proactive control, initiative, and requirements to finally achieve holistic interoperability worldwide.

References

[1] Select Bipartisan Committee to Investigate the Preparation for and Response to Hurricane Katrina, *Failure of Initiative, Final Report of the Select Bipartisan Committee to Investigate the Preparation for and Response to Hurricane Katrina*, U.S. Government Printing Office (Washington, D.C., February15 2006), p. 326, (hereafter referenced as the *Katrina Report*), available at http://www.gpoacess.gov/congress/index.html,

[2] This figure was derived from seminal work in public safety and emergency response interoperability by John M. Contestabile, Director of the Office of Engineering & Emergency Services, working in the Secretary's Office of the Maryland Department of Transportation (MDOT). Among many past and present government duties, Mr. Contestible has been with the department for nearly 30 years, and has been responsible for emergency planning/response/recovery activities for the MDOT since 1996. His interoperability experience includes:

a. Program areas of homeland security and public safety interoperable communications;

b. Chairing the Maryland State Interoperability Executive Committee (SIEC) –Working Group (which developed the Maryland Statewide Plan for Public Safety Voice and Data Communications);
c. Service on national committees for the Department of Homeland Security's SAFECOM Interoperable Communications Advisory Committee;
d. Board member for the Board of the Public Safety Spectrum Trust (which holds the national license for the proposed 700-MHz broadband public safety spectrum in the United States).

[3] *Katrina Report*, p. 47.

[4] National Commission on Terrorist Attacks Upon the United States, The *9/11 Commission Report: Final Report of the National Commission on Terrorist Attacks Upon the United States*, July 2004, p. 322.

CHAPTER 8

Emerging Capabilities

In this chapter we turn our attention to some of the recent and emerging innovations in technology and information sharing capabilities that offer promise to policy makers and technology experts who are dedicated to improving their interoperability plans. As described in our worst-case World Interoperability Failure model, human failures to plan interoperability as revealed in the history of Pearl Harbor, 9/11, and Katrina, continue with us today in public safety and emergency response. The responders-receivers and other stakeholders must continue to compensate with initiative and self-sacrifice for deficient interoperability planning. Our World Interoperability Failure model explains these continued failures in terms of lack of best practices used to develop our public safety and emergency response IT and communication systems. At worst, maintenance of the status quo of bureaucratic stability, consultant employment, and vendor procurements are prioritized over improving interoperability. However, technology is *not* the cause of this failed interoperability; the technology performs as was developed. It is the people responsible for building and maintaining information stovepipes—not the technology in those stovepipes—that must be "reengineered."

8.1 The Emergent Information-Sharing Environment

There are several independent information-sharing initiatives which would be in direct support of the development of information-sharing architecture using the World Interoperability Success Plan. These initiatives span the operational communities of homeland security, public health, emergency management, and environmental security. In this section, we highlight several of these developing and mature activities which may contribute to mitigating and eliminating deficiencies in the World Interoperability Failure model. These exemplars illustrate what operational needs shape behaviors of collaboration, cooperation, and integration.

In this section, we examine information-sharing initiatives including:

- The network of fusion centers for homeland security counter-terrorism intelligence and all-hazards preparedness,
- Global public health networking,
- Geospatial data standards interoperability,

- The common alerting protocol (CAP) used to promote a single format in disseminating information.

8.1.1 The Information-Sharing Environment and the Growth of Fusion Centers

Improving information sharing constitutes a cornerstone of national strategy to protect the public and our institutions and to defeat terrorists and their support networks at home and abroad. As shown in Chapters 2 and 3, the National Commission on Terrorist Attacks Upon the United States (the 9/11 Commission) identified a breakdown in information sharing as a key factor contributing to the failure to prevent the September 11, 2001 attacks [1]. In response to the 9/11 Commission's recommendations, the United States Congress passed and the president signed the Intelligence Reform and Terrorism Prevention Act of 2004. Section 1016 of the law called for the creation of an information-sharing environment and defined it as "an approach that facilitates the sharing of terrorism information." The law required the U.S. president to designate a program manager for the information-sharing environment and establish an Information Sharing Council to advise the president and the program manager.

The Implementation Plan for the information-sharing environment sets forth the following vision [2]:

> A trusted partnership among all levels of government in the United States, the private sector, and our foreign partners, in order to detect, prevent, disrupt, preempt, and mitigate the effects of terrorism against the territory, people, and interests of the United States by the effective and efficient sharing of terrorism and homeland security information.

Creating the information-sharing environment is not about building a massive new information system. The information-sharing environment aligns and leverages existing information-sharing policies, business processes, technologies, systems, and promotes a culture of information sharing through increased collaboration. The information-sharing environment goals are to:

- Facilitate the establishment of a trusted partnership among all levels of government, the private sector, and foreign partners.
- Promote an information-sharing culture among information-sharing environment partners by facilitating the improved sharing of timely, validated, protected, and actionable terrorism information supported by extensive education, training, and awareness programs for information-sharing environment participants.
- Function in a decentralized, distributed, and coordinated manner to the maximum extent possible.
- Develop and deploy incrementally, leveraging existing information-sharing capabilities while also creating new core functions and services.
- Enable the federal government to speak with one voice on terrorism-related matters, and to promote more rapid and effective interchange and

coordination among federal departments and agencies and state, local, and tribal governments, the private sector, and foreign partners, thus ensuring effective multidirectional sharing of information.

- Ensure sharing procedures and policies protect information privacy and civil liberties.

As shown in Chapter 2, one clear lesson of 9/11 in the United States was the need to improve information sharing. The gathering of information about terrorist risks and threats, then ensuring that this information is provided to those entrusted to protect our communities and critical infrastructure, are key to preventing future attacks. In the United States, the *National Strategy for Information Sharing* was developed with the understanding that homeland security information, terrorism information, and law enforcement information related to terrorism can come from multiple sources, all levels of government, private sector organizations, and foreign sources. The *National Strategy for Information Sharing* also sets forth a plan to establish a more integrated information-sharing capability to ensure that those who need information will receive it and those who have that information will share it [3].

In the United States, many state and major urban areas have established information fusion centers to coordinate the gathering, analysis, and dissemination of law enforcement, homeland security, public safety, and terrorism information. As of September 1, 2007, over 58 of these centers were operating or are being established in states and localities across the country. A majority of them operate under national guidelines developed through the Global Justice Information Sharing Initiative and Homeland Security Advisory Council.

A fusion center is defined [4]:

> as collaborative effort of two or more agencies that provide resources, expertise, and/or information to the [fusion] center with the goal of maximizing the ability to detect, prevent, investigate, apprehend, and respond to criminal and terrorist activity. The primary components of a fusion center are situational awareness and warnings that are supported by law enforcement intelligence, derived from the application of the intelligence process, where requirements for actionable information are generated and information is collected, integrated, evaluated, analyzed, and disseminated. Other key components resident in the fusion center include representatives of public safety, homeland security, the private sector, and critical infrastructure communities.

State and major urban area fusion centers are vital assets critical to sharing information related to terrorism and all-hazards preparedness. They serve as primary focal points within the state and local government for the receipt and sharing of terrorism-related and other information. As a part of this strategy, the national government is promoting that state and major urban area fusion centers achieve a baseline level of capability and become interconnected with the national government and each other, thereby creating a national, integrated, network of fusion centers to enable the effective sharing of terrorism-related information. The national government will support the establishment of these centers and help sustain them through grant funding, technical assistance, and training to achieve a baseline level of capability and to help ensure compliance with all applicable privacy laws. However, the use of fusion centers cannot be considered the solution for information sharing in general, as all OPCENs at all governance levels must receive and share information in a network which contains these fusion centers as another node, but not necessarily as a hub.

The national government, in consultation with state, local, and tribal authorities, will compile, document, and disseminate baseline operational standards, the achievement of which will determine whether an individual state or major urban area fusion center is considered to have achieved a baseline level of capability. These baseline operational standards will build on the Global Justice Fusion Center Guidelines. Additionally, the national government will initiate a series of activities to assist state and major urban area fusion centers to adopt and incorporate these baseline operational standards into their business operations. These standards will support the gathering, processing, analysis, and dissemination of terrorism information, homeland security information, and law enforcement information.

State and major urban area fusion centers are *encouraged* to ensure that alert notifications are disseminated as appropriate, to state, local, and tribal authorities, the private sector, and the general public. Specific activities may include:

- Implement a protocol to govern the receipt of federally generated threat, warning, and notification messages.
- Develop and/or maintain a plan (processes, protocols, and communication methodology) to govern the further dissemination of federally generated threat, warning, and notification messages, bulletins and other information products to state, local, and tribal authorities, the private sector and the general public.
- Develop and/or maintain a plan (processes, protocols, and communication methodology) to govern the gathering, processing, and reporting to national entities any actions taken by state, local, and tribal authorities and the private sector in response to federally generated threat, warning, and notification messages.
- Identify and establish a communications platform to support the dissemination of these messages and information products.
- Coordinate with the appropriate national authorities to develop synchronized protocols for sharing information with the private sector.

The *National Criminal Intelligence Sharing Plan* advanced a number of recommendations to improve intelligence-led policing, including "the need to identify an intelligence information sharing capability that can be widely accessed by local, state, tribal, and federal law enforcement and public safety agencies"[5].

Sharing is founded upon trust between the information provider and consumer. Such trust is most often fostered on an interpersonal basis; therefore, law enforcement task forces and other joint work endeavors succeed where colocated, interspersed personnel from different agencies and job types convene for a common purpose. In these instances, sharing can either flourish or falter due to changes in leadership, personality differences, and real or perceived issues.

Trust is fostered and may be further institutionalized by setting standards for participation in the information-sharing process. Personnel vetting procedures are established that range from the most stringent—national security clearances for access to classified information through law enforcement agencies' employment background checks, including criminal history records and indices—to situational.

8.1.2 Integrated Justice Information Systems

The Integrated Justice Information Systems Institute is a nonprofit corporation founded in 2001 as a result of the U.S. Department of Justice interest in raising private sector participation in the advancement of national initiatives affecting justice and public safety at the state, tribal, and local levels. In response, an industry working group, a working consortium of companies seeking to improve the standard of products, services, and relationships across this sector, provided the foundation for the Integrated Justice Information Systems Institute.

Serving as the voice of industry, Integrated Justice Information Systems is designed to unite the private and public sectors to improve mission-critical information sharing for those who protect and serve our communities. Among its strategic goals are: to provide leadership for the industry in information sharing; to be the "voice of industry" in developing policies and programs in information sharing for justice, public safety, and homeland security; and, to enhance collaboration between the public and private sectors.

In detecting, preventing, responding to and investigating crimes, disasters and terrorist acts, the exchange of information among multiple engaged agencies must be timely and accurate and, therefore, highly automated. Most existing computer systems are not designed to facilitate information sharing across disciplines and jurisdictions. Automated information sharing between agencies requires the definition of common standards for linking disparate systems. Local, state, tribal, and federal agencies need to adopt common information-sharing standards to facilitate information sharing

Exchanges using a national standard could help by:

- Allowing a national search on the *modus operandi* used in major cases;
- Forwarding incident and arrest data from police to prosecutors;
- Determining the status of beds, staff, and resources at hospitals to allow emergency OPCENs and emergency medical service personnel to improve resource allocation;
- Sending call data from a 9-1-1 center to multiple dispatch centers;
- Reporting suspicious activities from field officers to investigators at state and national levels and from one fusion center to another;
- Reporting the status of emergency response teams and resources;
- Disseminating Amber and other alerts to press and public;
- Forwarding alarm and sensor information to communication centers;
- Identify units and status of resources in adjacent communities in support of mutual aid agreements;
- Screening persons or cargo entering the country against multiple data sources [6].

8.1.3 Global Justice Information-Sharing Initiative

The U.S. Department of Justice Global Justice Information Sharing Initiative (Global) serves as a federal advisory committee and advises the U.S. attorney general on justice information sharing and integration initiatives. Global was created to support the broad scale exchange of pertinent justice and public safety information. It promotes standards-based electronic information exchange to provide the justice community with timely, accurate, complete, and accessible information in a secure and trusted environment.

On December 16, 2005, the U.S. president issued a Memorandum for the Heads of Executive Departments and Agencies on the Guidelines and Requirements in Support of the Information Sharing Environment that included requirements to develop a common framework for the sharing of information. In compliance with the president's memorandum, the program manager for information-sharing environment has defined the architecture program of the information-sharing environment through three documents:

- Drivers and requirements specification;
- Enterprise architecture framework;
- Profile and architecture implementation strategy.

Compliance with the guidance and requirements outlined in these documents significantly increases information-sharing environment participants' ability to securely connect to, participate within, and share information across the information-sharing environment.

Information-sharing environment participants have been working diligently to make information sharing part of internal policy and business process structure. Some have initiated or completed development of their Information Sharing Enterprise Architecture or their Information Sharing Segment Architectures using the information-sharing environment enterprise infrastructure framework architectural principles and approved information sharing environment standards.

8.1.4 National Information Exchange Model

The National Information Exchange Model (NIEM) is:

- A data standard for enabling information exchanges across multiple communities of interest, using agreed-upon terms, definitions, and formats independent of the way data is stored in individual systems;
- A methodology for achieving consensus among parties to an information exchange on the content of specific exchanges;
- A structured and disciplined approach to data interoperability useful in facilitating the development and implementation of specific exchanges.

NIEM was launched on February 28, 2005. NIEM was initially populated by the work from the highly successful Global Justice Extensible Markup Language Data Model, including data elements, structure, and the governance model. NIEM was expanded in its first release by additional communities of interest, including the intelligence community, disaster management, infrastructure protection, and international trade. Efforts are well underway to implement NIEM in multiple state and local government projects.

The future of NIEM is to:

- Implement exchanges based on NIEM by participating communities of interest involving local, state, tribal, and national agencies and organizations.
- Use NIEM to create national priority exchanges for suspicious activity reporting, person and cargo screening, disaster management, and other critical areas of national significance.
- Expand the participating communities of interest to include health care, transportation, education, and other areas, creating parallel governance structures to include representation in the expansion of the data model.
- Expand further endorsement by local, state, tribal and national agencies.
- Expand training and technical assistance efforts to help participants implement NIEM-based exchanges.

8.1.5 Global Public Health Information Exchange

When the world is collectively at risk, defense becomes a shared responsibility of all nations.

—Dr. Margaret Chan, Director General, World Health Organization; World Health Day 2007.

The diversity of public health surveillance systems employed by various health jurisdictions worldwide can become an impediment to sharing public health information about evolving conditions. Because even nearby local health jurisdictions may utilize different disease surveillance systems, their ability to share timely information may not be effective or efficient because the systems lack interoperable communication tools or present information through unfamiliar user interfaces and presentation formats.

The framework of the newly revised International Health Regulations is a key driver in the effort to strengthen global public health security. Unanimously agreed upon by the World Health Assembly on May 23, 2005, the regulations are the result of experience gained and lessons learned during the past 30 years. This global legal framework includes a commitment from the World Health Organization (WHO) and from each WHO member state to improve capacity for disease prevention, detection, and response. It provides standards for addressing national public health threats that have the potential to become global emergencies. Its success will rely on the capacity and performance of national public health systems, anchored by strong national public health institutes (NPHIs). The new International Association of National Public Health Institutes aims to strengthen and invigorate existing NPHIs, to create new NPHIs where none exist, and to provide funded grants to support NPHI development priorities [7].

Established in 1952, the WHO Global Influenza Surveillance Network monitors, collects, and analyzes the influenza viruses that cause seasonal flu throughout the world. The Global Influenza Surveillance Network is a surveillance system comprised of laboratories and coordinating centers around the world that collect and analyze clinical data and specimens from patients with influenza-like illness during the influenza season [8]. These data are used to recommend vaccine strains for seasonal flu and to identify emerging strains of influenza A virus with the potential to cause a pandemic. Because of the avian influenza pandemic and increasing concerns about the potential of H5N1 virus to cause a human influenza pandemic, a dedicated H5 Reference Laboratory Network was established in 2004 to monitor H5N1 viruses. The WHO Global Influenza Surveillance Network works closely with the United Nation's Food and Agriculture Organization (FAO), the agency responsible for avian influenza virus surveillance [9, 10]. Global Public Health Intelligence Network and the International Society for Infectious Diseases' ProMED-mail are chief sources of reports sent to the WHO. ProMED-mail uses in-country infectious disease experts to improve the credibility of outbreak reports. The experts review preliminary reports and often request additional information about suspected outbreaks. ProMED-mail was the source of the first SARS report and a recent outbreak of yellow fever in Brazil.

The ProMED-mail program also is a model for an affordable surveillance system that may be suitable for countries with limited resources. The basic requirements are modest: a university-based server, a Web site manager, and the personal computers of the subject matter volunteers. Since 1994, national ProMED-mail systems have been established in Brazil, the Netherlands, and South Korea. However, the governments of many countries are reluctant to implement a national version of ProMED-mail, because subscriber-based input may circumvent official reports of the ministry of health in these nations.

The time interval between the onset of an epidemic and its recognition affects the magnitude of the epidemic and, consequently, the magnitude of the response required to contain it. Real-time surveillance theoretically reduces the time interval between outbreak and detection. QFLU is a surveillance program used in the United Kingdom that approximates a real-time (24 hour) clinical surveillance and reporting capability. This timeframe is possible because of the National Health Service electronic record system. QFLU uses analytical tools to monitor the timing, magnitude, and spread of seasonal and/or pandemic influenza. This report, updated daily, contributes to situational awareness and informs decisions about use of drugs, vaccines, and infection control measures.

The U.S. Public Health Information Network (PHIN) is a national initiative to improve the capacity of public health to use and exchange information electronically by promoting the use of standards and defining technical requirements. The standards and technical requirements are determined by best practices related to efficient, effective, and interoperable public health information systems that support both routine public health activities and emergency preparedness and response.

The Centers for Disease Control and Prevention (CDC) serves as the facilitator of the PHIN community. CDC's role in PHIN is [11, 12]:

- *Supporting* the exchange of critical health information between all levels of public health and healthcare;
- *Developing and promulgating* requirements, standards, specifications, and an overall architecture in a collaborative, transparent, and dynamic way;

- *Monitoring* the capability of state and local health departments to exchange information;
- *Advancing* supportive policy;
- *Providing* technical assistance to allow state and local health departments to implement PHIN requirements;
- *Facilitating* communication and information sharing within the PHIN community;
- *Providing* public health agencies with appropriate and timely information to support informed decision making;
- *Harmonizing* PHIN with other federal initiatives.

The overall architecture for PHIN is currently in evolution. The Public Health IT landscape is changing and working to harmonize with national health information technology initiatives. Disciplines, such as enterprise architecture are being increasingly leveraged. Adoption of industry practices, such as service-oriented architecture, is gaining momentum in public health IT [13].

8.1.6 Geospatial Data Standards

Thousands of organizations spend billions of dollars each year producing and using geographic data. Yet, they still do not have the information they need to solve critical problems. There are several aspects to this problem:

- Most organizations need more data than they can afford. Frequently, large amounts of money are spent on basic geographic data, leaving little for applications data and development. Some organizations cannot afford to collect base information at all.
- Organizations often need data outside their jurisdictions or operational areas. They do not collect these data themselves, but other organizations do.
- Data collected by different organizations are often incompatible. The data may cover the same geographic area but use different geographic bases and standards. In addition, information needed to solve cross-jurisdictional problems is often unavailable.

Many of the resources organizations spend on geographic information systems (GIS) go toward duplicating other organizations' data collection efforts. The same geographic data themes for an area are collected again and again, at great expense.

A geospatial data framework improves this situation by leveraging individual geographic data efforts so data can be shared. It provides basic geographic data in a common format and an accessible environment that anyone can use and to which anyone can contribute. In this environment, users can perform cross-jurisdictional and cross-organizational analyses and operations, and organizations can funnel their resources into applications, rather than duplicating data production efforts.

Accurate and comprehensive data are the heart of information technology, and geographic location is a key feature of 80% to 90% of all government data. It is critical that as a nation we take the steps necessary to assure that strategic information assets relative to homeland security—particularly geospatial information assets—are created, and maintained for currency and accuracy, are readily available to those who need them, and are interoperable. Although homeland security requires much of the same basic, real-time spatial information needed for other uses and applications, we know from recent events that it must be immediately and comprehensively available.

The need had emerged to assure:

- Implementation of a comprehensive national spatial data infrastructure;
- Interoperability of the systems that process this information;
- Commonalities of the processes that collect, manage, and disseminate geospatial information.

A nationwide data publishing effort known as the National Spatial Data Infrastructure was established in the United States. The National Spatial Data Infrastructure is a physical, organizational, and virtual network designed to enable the development and sharing of the nation's digital geographic information resources. The National Spatial Data Infrastructure provides crosscutting mechanisms for organizations of many types, affiliations, and responsibilities to be able to collaborate in assuring that geospatial data and systems are in place, ready for use. The data, technology, and associated intergovernmental and government-private mechanisms forged in this effort will be invaluable to intelligence, law enforcement, and other national security-related elements, as well as to local communities, in dealing with terrorism and other major threats to public safety and welfare.

Presently, there are gaps that should be filled to achieve assurance of data and technology accessibility and interoperability. Examples are:

- National data standards still need to be developed for a number of framework and other data themes to provide data that is immediately useful in homeland security events.
- National Spatial Data Infrastructure Framework themes are not yet complete.
- Emergency wireless location capabilities are limited by the lack of consistent, standardized road data across the nation, preventing true interoperability between all levels of government.
- Current and accurate information about the nation's critical infrastructure is not consistently available or shareable among relevant agencies and the private sector which owns and maintains over 85% of these assets, leaving the nation unable to effectively plan for modern terrorist activities.

The National Spatial Data Infrastructure has established certain standards, processes, and relationships that serve to advance homeland security, including:

- Well-established relationships with federal, state, local, and tribal governments and ongoing coordination mechanisms.
- A multinode geospatial information clearinghouse network that can be extended to promote rapid discovery, sharing, and protection of critical geospatial information.

- Access to industry and international standards bodies and programs to advance standards that promote data consistency and interoperability of spatial technologies.

Because of the ubiquity of geospatial data within the United States government (the Government Accountability office states "about 80 percent of all government information has a geospatial data component"), there is a great deal of overlap of this data. Many agencies maintain the same data, but for different purposes.

Symbols and their definitions have been developed for incidents, natural events, operations, and infrastructures at a level to provide immediate and general understanding of the situation. Although these symbols do not include all emergency management features, they provide an initial framework for emergency management and emergency responder mapping and communication. Detailed attribution for any feature can be included by the user. Both a category structure and a damage-operational status hierarchy were developed using color and frame shapes with line patterns. The symbology was designed for use in digital and paper map products.

8.1.7 Organization for the Advancement of Structured Information Standards

The Organization for the Advancement of Structured Information Standards (OASIS) is a not-for-profit consortium supporting the development, convergence, and adoption of open standards for the global information society. OASIS was founded in 1993 under the name Standard Generalized Markup Language (SGML), and was open as a consortium of vendors and users devoted to developing guidelines for interoperability among products that support SGML. OASIS changed its name in 1998 to reflect an expanded scope of technical work, including the Extensible Markup Language (XML) and other related standards.

OASIS has more than 5,000 participants representing over 600 organizations and individual members in 100 countries. Members work to:

- Reduce the costs and risks of developing information-sharing situations;
- Identify requirements that would benefit from standardization;
- Articulate credible business use cases;
- Review proposed new standards;
- Collaborate on standards development;
- Promote best practices;
- Recommend the adoption of well-defined policies;
- Promote royalty-free licensing terms.

OASIS particularly values its transparent governance and operating procedures. Members set the OASIS technical agenda, using a lightweight process expressly designed to promote industry consensus and unite disparate efforts. Completed work is ratified by open ballot. Governance is accountable and unrestricted. Officers of both the OASIS board of directors and Technical Advisory Board are chosen by democratic election to serve two-year terms. Consortium leadership is based on individual merit and is not tied to financial contribution, corporate standing, or special appointment.

The public sector participates in OASIS to:

- Lower the costs and risks of developing Internet-based government services;
- Influence software vendors and systems integrators to develop open, standards-based solutions for the public sector;
- Improve the competitive landscape by helping to create global, interoperable standards;
- Ensure their own interests are represented as requirements and business cases are defined;
- Monitor the marketplace adoption of open standards through access to members-only information;
- Ensure their staff are kept up-to-date on key standards issues and developments;
- Learn from other agencies, industries, and companies how to implement and adopt open, standards-based, best practices.

OASIS is the source for incident- and emergency-related standards for data interoperability. Work includes the Common Alerting Protocol for alerting and event notification applications, the Emergency Data Exchange Language (EDXL) for a wide range of emergency data exchange operations, including for example, tracking and communicating hospital bed availability, status, and capacity. Within the United States OASIS emergency management standards are used by a growing body of agencies, including the U.S. Centers for Disease Control, the U.S. Department of Homeland Security, the U.S. Department of Justice, the National Weather Service, and the U.S. Geological Survey.

8.1.8 Common Alerting Protocol

Common Alerting Protocol (CAP) [14] is a simple but general format for exchanging all-hazard emergency alerts and public warnings and event notification over diverse types of communications networks. CAP allows a consistent warning message to be disseminated simultaneously over many different warning systems, thus increasing warning effectiveness while simplifying the warning task. CAP also facilitates the detection of emerging patterns in local warnings of various kinds, such as might indicate an undetected hazard or hostile act. CAP also provides a template for effective warning messages based on best practices identified in academic research and real-world experience maintained through OASIS work groups.

CAP is currently in version 1.1, and it functions both as a stand-alone protocol and as a payload for EDXL messages. EDXL is a broad initiative coordinated under the OASIS work groups to create an integrated framework for a wide range of emergency data exchange standards to support operations, logistics, planning and finance:

- EDXL Distribution Element (EDXL-DE);
- EDXL Hospital Availability Exchange (EDXL-HAVE);

- EDXL Resource Message (EDXL-RM);
- EDXL Reference Information Model (EDXL-RIM).

Among the principles that guided the design of the CAP alert message were:

- *Interoperability*—First and foremost, the CAP alert message should provide a means for interoperable exchange of alerts and notifications among all kinds of emergency information systems.
- *Completeness*—The CAP alert message format should provide for all the elements of an effective public warning message.
- *Simple implementation*—The design should not place undue burdens of complexity on technical implementers.
- *Simple XML and portable structure*—Although the primary anticipated use of the CAP alert message is as an XML document, the format should remain sufficiently abstract to be adaptable to other coding schemes.
- *Multiuse format*—One message schema supports multiple message types (e.g., alert, update, cancellations, acknowledgements, or error messages) in various applications (actual, exercise, test, or system message).
- *Familiarity*—The data elements and code values should be meaningful to warning originators and nonexpert recipients alike.
- *Interdisciplinary and international utility*—The design should allow a broad range of applications in public safety and emergency management and allied applications and should be applicable worldwide.

For the emergency management mission, CAP provides an open, nonproprietary digital message format for all types of alerts and notifications. It does not address any particular application or telecommunications method. The CAP format is compatible with emerging techniques, such as Web services, as well as existing formats, including the Specific Area Message Encoding used for the United States' National Oceanic and Atmospheric Administration Weather Radio and the Emergency Alert System, while offering enhanced capabilities that include:

- Flexible geographic targeting using latitude/longitude shapes and other geospatial representations in three dimensions;
- Multilingual and multiaudience messaging;
- Phased and delayed effective times and expirations;
- Enhanced message update and cancellation features;
- Template support for framing complete and effective warning messages;
- Compatible with digital encryption and signature capability;
- Facility for digital images and audio.

Key benefits of CAP include reduction of costs and operational complexity by eliminating the need for multiple custom software interfaces involved in all-hazard warning. The CAP message format can be converted to and from the "native" formats of all kinds of sensor and alerting technologies, forming a basis for scaling-up to a technology-independent, international "warning internet" [15]. As of 2009, the U.S. Federal Emergency Management Agency was adopting an alerting protocol in conformance with Common Alerting Protocol 1.1 as the standard for its Integrated Public Alert and Warnings System. The OASIS CAP standard is also an International Telecommunications Union Recommendation, meaning it is of direct value in multinational alerting. The OASIS committee seeks local and national participants worldwide to broaden adoption of its work.

8.1.9 Summary

In the aggregate, there are several national and international information-sharing initiatives to advance interoperability across a range of operational needs, including public health data and environmental geospatial data. In every instance, it is self-evident that the disciplined methodology inherent in the enterprise architecture could have enhanced our ability to define, in the broadest possible manner, the data life cycle and interoperability with derivative users. Advanced analytic tools exploit data in ways not envisioned only a few years ago, and one organization's noise and outlier data may be another agency's *data revelation*. This revelation could provide life-saving awareness in one or more of world's communities in the next moment.

8.2 Emerging IT Capabilities to Mitigate Deficiencies

This book has shown that human deficiencies are the source of failed information sharing in our public safety and emergency response IT/communication systems. From our worst-case World Interoperability Failure Model, we have postulated not only the human cause of lost opportunities and failed initiative, but also the faulty planning of the systems and systems-of-systems used by our responders-receivers and other stakeholders. We have described best practices for minimizing, mitigating, and ultimately eliminating these deficiencies from the planning of such systems. In this section, we describe emerging technologies and their value—if properly developed and planned—in mitigating and eliminating these deficiencies from their best-practices implementation going forward.

8.2.1 Beyond "Publish and Subscribe" to "Sense and Respond"

8.2.1.1 The Publish and Subscribe Environment

We have described the development of composite, federated, and even integrated architecture to share information and achieve holistic interoperability across disparate systems. Emerging information formats as described in Section 8.1 are providing standards needed to enable interoperability despite proprietary developments, understanding that an essential "key to interoperability and information sharing is to ensure projects adhere to standards and 'off the shelf' commercially available solutions rather than custom or proprietary solutions." In this regard, the World Wide Web provides a federated architecture for information sharing that spans all governance levels and stakeholders.

In state-of-the-art OPCENs, many Web sources of information—whether private or public—are displayed in an integrated situational awareness common operational picture (COP) [16]. This COP is dependent upon a publish and subscribe environment (P&SE), in which "each system *should* publish information to a platform that can be seen by other stakeholders in the region and those stakeholders *should* be able to subscribe to the information that is most relevant to them [italics added]" [17]. Much of the published information is available from a variety of online sources, such as weather map overlays, traffic and security cameras, infrastructure overlays (e.g., buildings, roads, bridges, waterways, fire hydrants, utility poles, telecommunications conduit), CBRNE sensors, among others, and predictive algorithms used to forecast and assess consequences. Figure 8.1 depicts a representative COP capability. This capability is made possible by the evolving data standards described in Section 8.1.

Figure 8.2 shows the COP located in the idealized OPCEN outlined in Chapter 6. Tailored information uploaded from these Web sites for COP presentation can then be relayed (see Chapter 6) to responders-receivers and other stakeholders in real time during crisis conditions. This technology allows information to be readily available to all who subscribe to the P&SE system-of-systems despite their otherwise disparate technology, forming a federated architecture. The specific items and views displayed can be tailored to the command elements shown in the figure.

The basis for this Web-based technology had arguably been envisaged by 1941, it had emerged in OPCENs worldwide before 9/11 and was well deployed by 2005 for Katrina. Nevertheless, despite the benefit of this publish-and-subscribe technology for information sharing, human initiative still failed. People not only had to have the same proprietary "publish" or "subscribe" capability, but they also had to understand how and when to publish information. What is arguably more important, they had to believe and act upon subscribed information to achieve true interoperability. Thus, the P&SE provides the available information in one or more selected common operational pictures for the human command elements to view, with perhaps directed audio and video feeds as well.

If Lt. Tyler had seen the same "large blip" witnessed by the Opana Point radar operators (see Chapter 2) via a publish-and-subscribe capability, would he have supplied the proper warning to the Navy, interceptor squadrons, and antiaircraft artillery? Would it have mattered, given the many other command, control, and communication deficiencies that existed at Pearl Harbor that December 7th morning? Civilian and military air traffic authorities knew where hijacked and military aircraft were on their single-purpose displays, but the "fog of war" prevented an interception. Clearly, the Louisiana and New Orleans authorities saw images of Katrina in the Gulf with New Orleans and the region in the center of the probability swath, but delayed mandatory evaluation until too late. The risk that these deficiencies of leadership and planning will become evident again increases with more important, larger incidents, and more disparate stovepipe organizations that are involved. Thus, the success of P&SE in achieving holistic (true) interoperability requires human initiative, and it is failed initiative—typically at and above the command level, as we have seen—that often determines the tragic outcome. The responders-receivers are, as always, prepared for their sacrifices necessitated, as usual, by the Pearl Harbor–9/11–Katrina deficiencies.

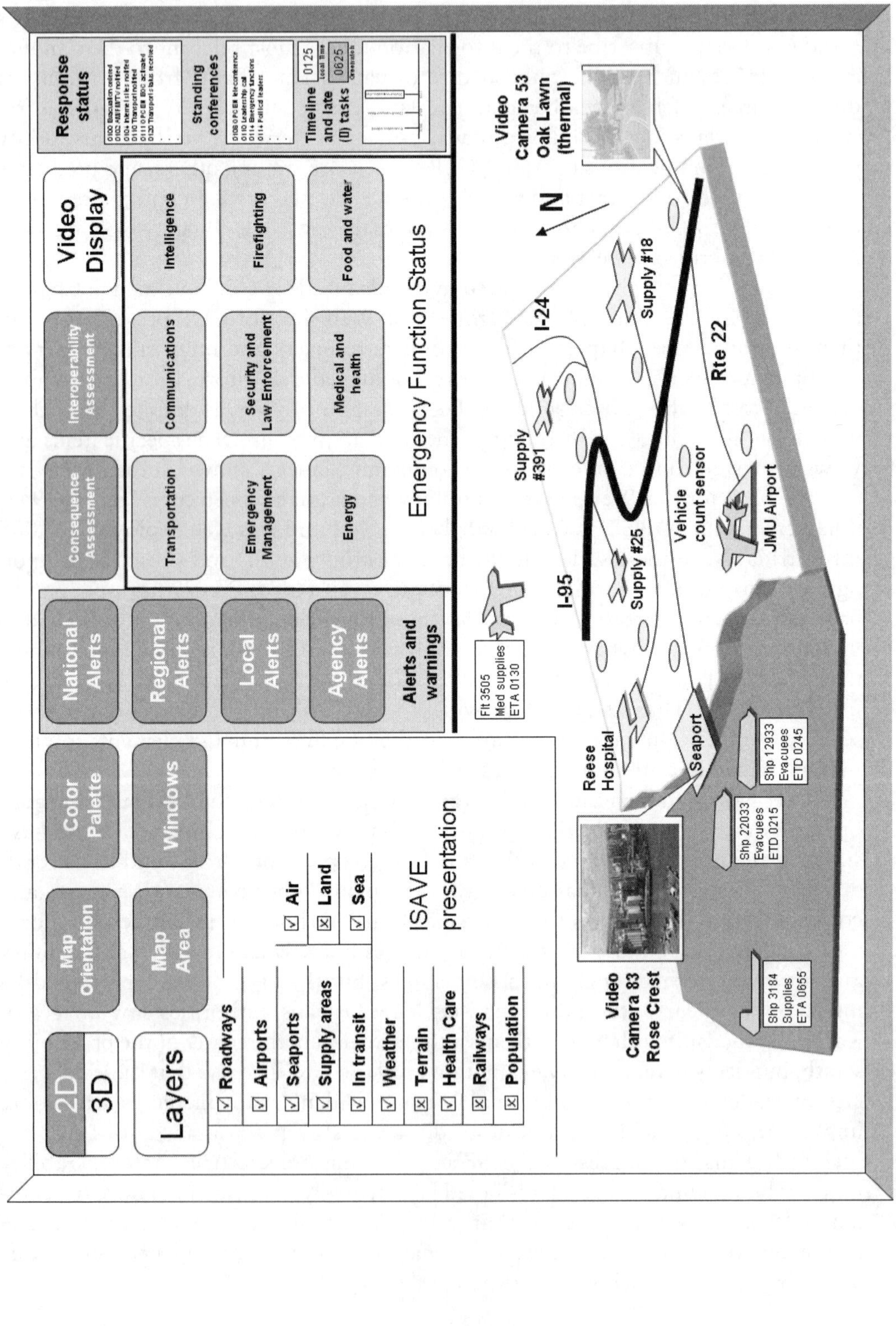

Figure 8.1 Common operational picture for shared situational awareness.

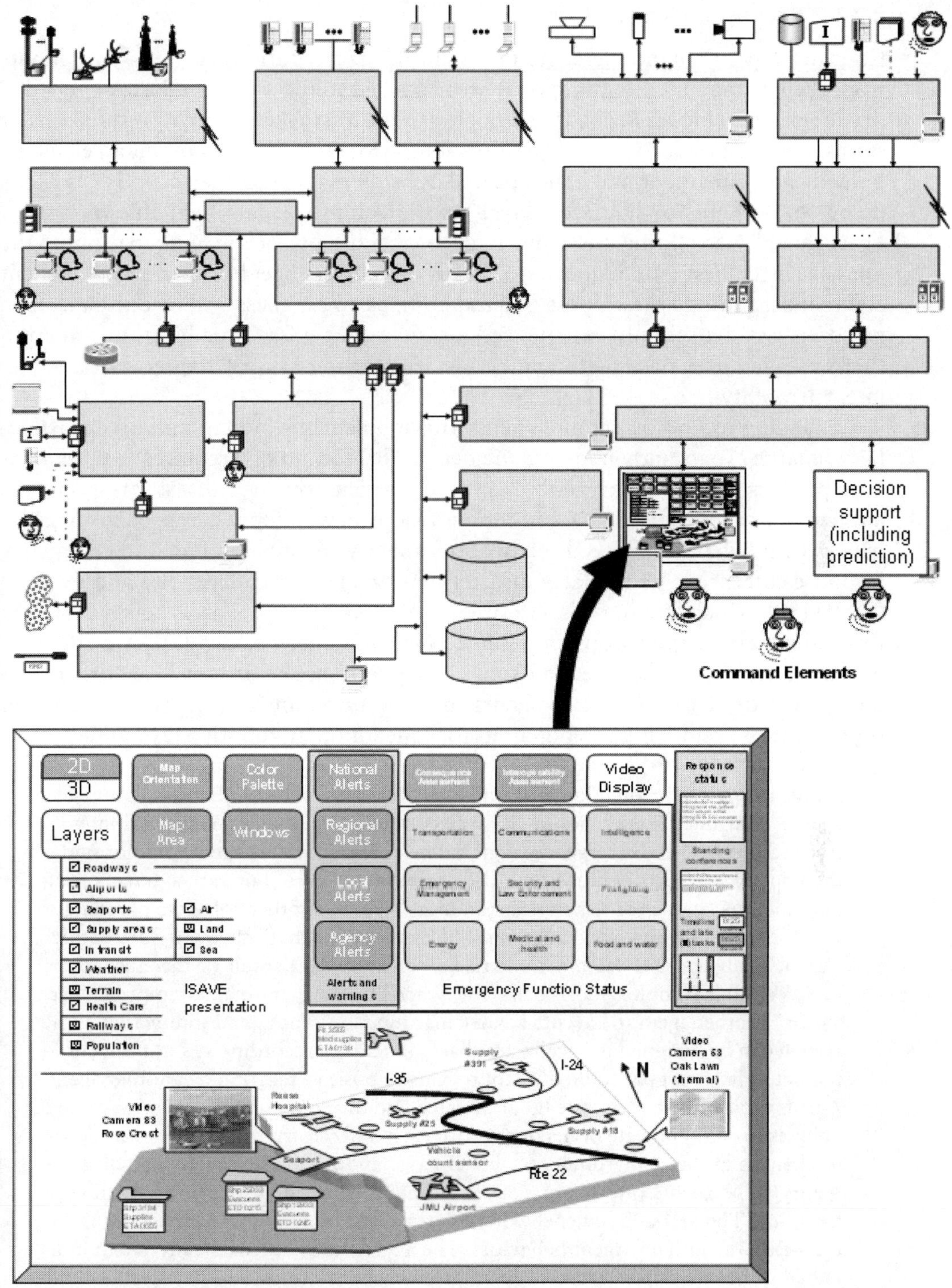

Figure 8.2 COP provides situational awareness.

8.2.1.2 Sense and Respond

Consider the command elements in Figure 8.1 face continuously evolving inputs from each of the OPCEN functional areas pushed to the situational awareness display depicted in Figure 8.2. These inputs span the intelligence, threat, response asset status, even responder-receiver qualifications [18]. How can the command elements or incident commander make the optimal decision in the next few seconds, and keep doing so, perhaps for hours or days? As the volume of data available mounts, a "best guess" accounting for what is deemed most important information at the moment is the best a human can do. If this human is a deeply experienced incident commander, despite the volume of data, perhaps a best guess will be optimal. If the person is less than deeply experienced, or the expert misses one bit of information that was available, then there will be historical material for Chapter 2 of a future interoperability book.

Emerging technology complements and supplements human initiative, particularly in large events and wide-area incidents. This technology requires that we visualize public safety and emergency response, particularly large disaster response, as an "enterprise" architecture (see Section 4.5). A new, adaptive approach for complex system behavior that is directly applicable to the public safety and emergency response enterprise, particularly disaster management, is called sense and respond (S&R) [19]. Much like the observe-orient-decide-act loop, its basic actions tailored for commercial applications, are *sense, interpret, decide, act* (SIDA), and it is well-suited to the public safety and emergency response environment. More importantly, it is driven by software agents that act autonomously at various levels of sophistication, removing the potential for failed human initiative [20] and

> localizing not only functionality and data but also control. Entities known as agents, controlled by decision-making algorithms, can execute many local interactions resulting in the emergence of global behaviors. That is, agents are not under the control of the program—they do not follow orders, but rather have strong autonomous characteristics that distinguish them from other software paradigms. Each agent retains its own functions and methods, so that other software entities cannot freely cause those functions to be executed. As a result of this autonomy, agents exhibit complex social behavior, whereby one agent may attempt to "persuade" another agent to execute a particular function. The agents interacting in the system can be designed to receive feedback and adapt accordingly, so one agent in pursuit of its own goal could attempt to cause a positive reaction by another agent. Agents typically are proactive (goal-directed and thus intentional) or reactive, have abilities to communicate or negotiate with each other, learn from experience, adapt to changes in their environments, make plans, and reason (e.g., via logic or game theory). The agents often interact in a self-adaptive, nonlinear manner with each time step. This self-adaptive behavior creates a vast number of variables and facilitates research into emergent behaviors. The aggregate effects of the myriad of individual decisions can be studied, for a given scenario, to assess the effects on the whole system.

The design of an S&R capability in itself forces planners to consider the many scenarios and potential outcomes that it must address (see Sections 4.5 and 4.8). Consider the specific elements of S&R logistics (S&RL) system, designed to adapt the provisioning of supplies in a changing environment. The technical elements of this S&RL capability, disaster-response situational awareness monitoring and recommendation/decision-support capability would include [21]:

- Passive and active tagging, instruments, and sensors that provide location status, diagnoses, prognoses, and other information relative to operations space entities, especially for conditions and behavior that affect ... capabilities management, logistics, and sustainment.
- Intelligent software agents that represent operations space entities, conditions, and behaviors, provide a focus for control of action or behavior, or act as monitors.
- S&RL knowledge bases oriented toward ... capabilities management, logistics, and sustainment.
- S&RL reference data, again focused on ... capabilities, assets, and resources related to force capabilities management, logistics, and sustainment.
- S&RL rule sets, which govern the operations and organization of S&RL functions, activities, and transactions.
- S&RL cognitive decision support tools uniquely supporting ... capabilities management, logistics, and sustainment.
- Unique [federated] S&RL processes, applications, portals, and interfaces.

S&RL offers advantages in the management of a large complex enterprise, such as the logistics management of disaster relief or catastrophe evacuation. Its use would provide a means of optimizing the deployment of available resources where the need is greatest. In addition, it would automatically adapt to the dynamic environment of the threat, the status of victims, and the associated evolution of emergency functions, such as transportation, communications, firefighting, security, food and water, and their sustainment. Software agents associated with each of these functions would negotiate with other agents to optimally apply these capabilities to their respective roles in the response. In this way, all known information would be used in the assignment of assets and resources to different needs emerging in the response, and they would do so automatically.

We will demonstrate the use of a conceptual S&R system using our catastrophe evacuation example from the DoDAF/PSAF OV examples provided in Section 4.5.5.2. We will make the assignment of S&R software agents and federated subsystems to the facilities and assets in Figure 8.1 and other units not shown there. These agents and subsystems operate in service-oriented architecture as described in Chapter 6, in which diverse agents work together to optimize the overall response. The software agents in the legend are depicted as polygons, where the increasing intelligence and computational load is indicated by an increasing number of sides of the polygon [22]. The circular agent is the most intelligent—effectively becoming a circle. It is this highest level-agent that produces recommendations to the command elements when required, but may be authorized to make many automatic decisions, such as in multimodel traffic routing for optimal flow of evacuees.

Figure 8.3 shows the catastrophe evacuation scenario with interoperating S&R software agents assigned to (see corresponding numbered circles on figure):

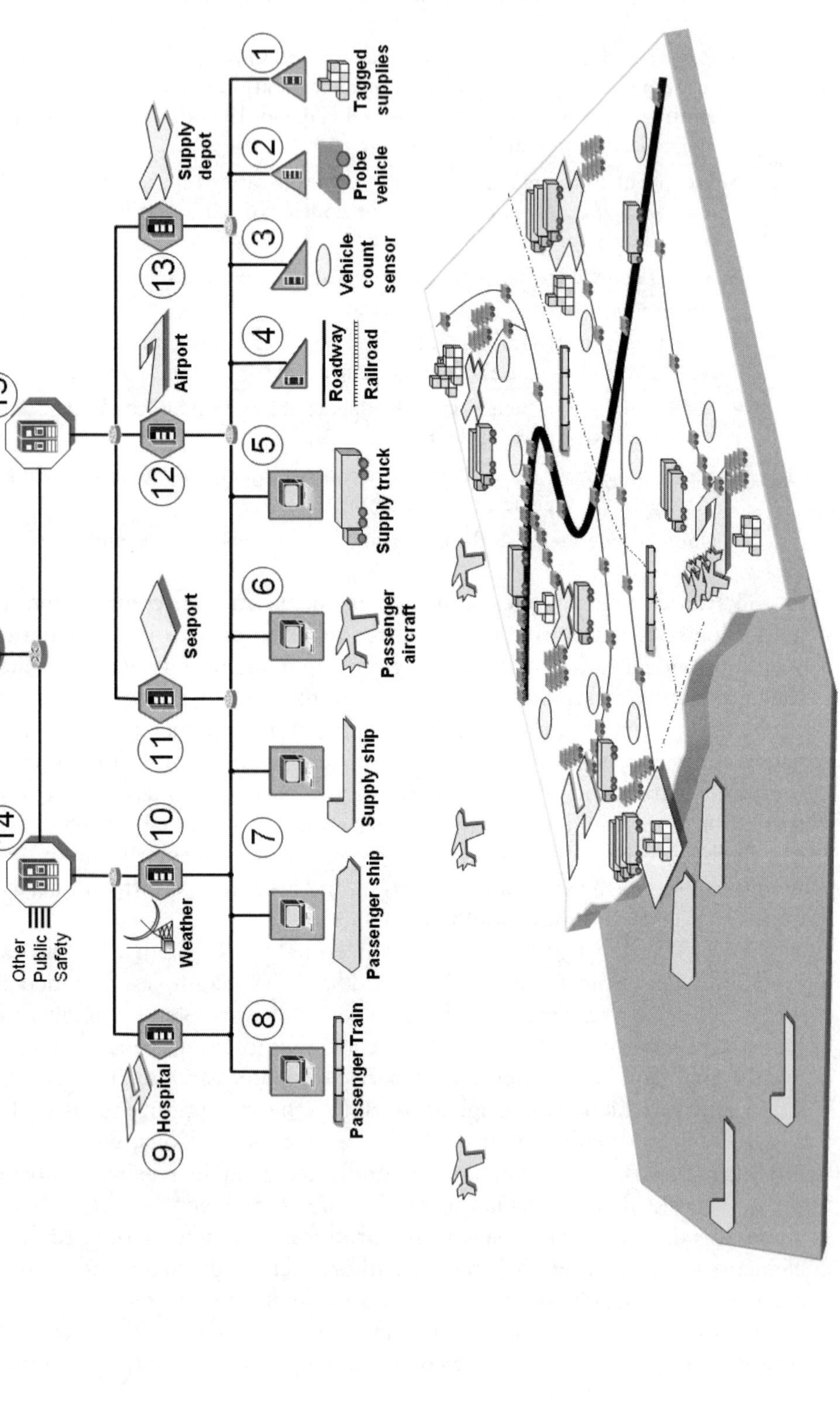

Figure 8.3 Sense and respond architecture for catastrophe evacuation management.

1. Supply tags, where the software agents are the RFID tags for item, box, or pallet tracking during multimodel transportation.
2. Roadside vehicle count sensors, designed to transmit the number of dual axle "hits" per unit time, including intelligence to differentiate vehicle types.
3. Road and railroad condition sensors, to detect black ice or other roadway safety issues and rail discontinuities indicating breaks in the line or other safety issues, respectively.
4. Cell phone users, who provide position reports from *probe* vehicles (civilians or others) by transmitting their geographic locations in standardized text message or email—these position reports from many such vehicles provide vehicular traffic flow rates when combined with roadside monitors.
5. Truck ID/tracking agents, typically using satellite communications and providing location of in transit supplies or empty trailers, with two-way communication to support truck routing optimization (also could be employed by passenger busses, not shown).
6. Aircraft ID/tracking capability agents—similar in sophistication to truck/bus tracking systems—to identify and route specific passenger (and cargo) aircraft in support of the evacuation.
7. Passenger and cargo ship ID/tracking agents, providing a similar function of location and status plus cargo details, such as people or containers, pallets, boxes, items, respectively.
8. Passenger trains ID/tracking with a smart agent as with other conveyances, reporting location, passengers, cargo, and so forth.
9. Smart agents at hospitals (and other responder-receiver locations) providing status, such as available beds, care specialties, available medications, blood supplies, supply needs, and all other relevant (standard) health-care information sharing.
10. Smart weather agents, providing a full range of climate and weather-related information, perhaps including consequence assessment for areas likely to flood, or experience freezing temperatures.
11. Smart seaport agents, providing ID/status, passenger and cargo throughput, supply needs, scheduling, and other functions—railheads at the seaport would supply similar information (rail stations not shown separately).
12. Smart airport agents, providing ID/status, passenger and cargo throughput, supply needs, scheduling, railhead connections, and other functions.
13. Smart agents operating at supply depots, which may be government, volunteer, or commercial facilities, providing available materials and assets as well as facility status and needs.
14. Intelligent public safety agents designed to integrate and develop ISAVE input as well as decision options related to all aspects of public safety, including the need for law enforcement (and its optimal deployment), firefighting/HAZMAT response, EMS and public health, utilities (e.g., electric and gas), transportation (e.g., repair) and others.
15. Intelligent transportation agents that optimizes operation of all facilities and conveyances, roadways, seaways, and airways, to maximize evacuee flow and simultaneous supply.

16. Overall intelligent agents that produces recommendations to command elements on all public safety and transportation deployment and redeployment needs based on all available information form lower order software agents supporting evacuation.

Finally, Figure 8.4 shows where the highest level S&R function (circle numbered 16 in Figure 8.3) is located in the OPCEN described in Chapter 6.

Applying properly designed (see Chapter 4) S&R architecture to major emergency management efforts such as catastrophe evacuation can mitigate or eliminate the Pearl Harbor–9/11–Katrina deficiencies by mitigating failed human initiative.

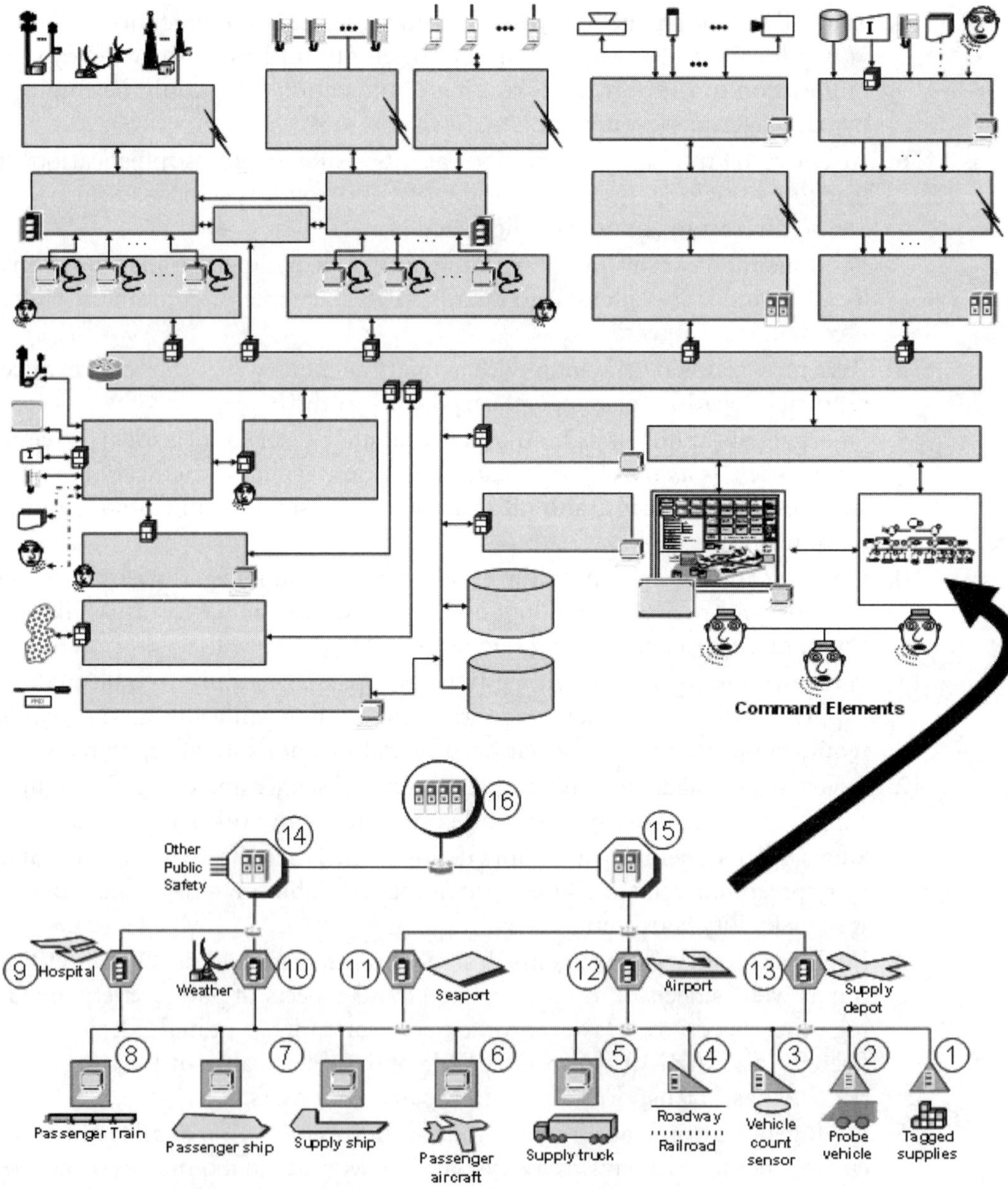

Figure 8.4 OPCEN sense and respond decision support integration.

8.2.1.3 S&R Reference Pilot Architecture

Working prototypes or reference pilot architecture (RPA) complements the ACE methodology described in Chapter 5 to improve capability rapidly. Referring again to the S&R architecture for an OPCEN supporting evacuation management, an RPA has been developed through cooperation among industry partners and government stakeholders. This reference pilot architecture demonstrates the effectiveness of S&R.

The RPA supporting catastrophe evacuation has captured the lessons learned from eight years of evolution and incorporates integrated system-of-systems patterned after the observe-orient-decide-act loop. For example, the RPA includes monitoring tools that sense and capture data related to vehicle flow rates, resource (shelter, fuel, medical, food) consumption rates, and disruption (observe/sense). It mediates in a service-oriented architecture data layer for concurrent usage by a correlation, course-of-action inference engine, and situational awareness picture. The correlation tools include simulators that anticipate chokepoints and bottlenecks (orient/sense), triggering threshold events to both the OPCEN operator and to a class of decision-support tools that support automated response recommendations (decide/respond). For catastrophe evacuation, this capability includes not only a geospatial presentation of projected backlogged route, color-coded in red, but autoquery of surrounding aid stations, shelters, and augmenting resources. OPCEN planning and response action is supported by automatically generated resource and support unit requests for those resources and units screened by the course-of-action tool as feasible to support corrective response (act/respond).

8.2.2 Advanced Incident Management

Sense-and-respond operational control requires an intelligent agent to make recommendations and support decisions with a volume of information which would overwhelm a human being, but with little to no time to make order of the best possible action. This subsection discusses hybrid nonmonotonic reasoning techniques and architectures for information fusion and for the integration of common operational understanding. We give an example of an application developed for incident commanders that uses these techniques along with specialized COP visualizations to give these commanders understanding of their situation as well as course-of-action recommendations [23].

8.2.2.1 Nonmonotonic Reasoning and Hybrid Algorithms

Nonmonotonic reasoning techniques. Nonmonotonic reasoning is a method of mimicking human thought patterns, like neural nets but with one important difference. It accounts for assumptions being made by listing those assumptions inside the rules. A *nonmonotonic rule* of inference takes the form $r = a_1, \ldots , a_n{:}b_1, \ldots , b_m/\varphi$ read as "if $a_1, \ldots , a_n$ are known to be true and $b_1, \ldots , b_m$ are not known to be true then conclude φ." Using this format, nonmonotonic reasoning can take advantage of assumptions made during the decision-making process, and even call them to the attention of the user. Nonmonotonic techniques also allow us to reason with conflicting and missing information, which is often a characteristic of the public safety and emergency response environment.

> Leadership requires decisions to be made even when based on flawed and incomplete information. Too often during the immediate response to Katrina, sparse or conflicting information was used as an excuse for inaction rather than an imperative to step in and fill an obvious vacuum. Information passed through the maze of departmental operations centers and ironically named "coordinating" committees, losing timeliness and relevance as it was massaged and interpreted for internal audiences [24].

Hybrid nonmonotonic approaches. Each available artificial intelligence method brings with it certain advantages. Blending two or more approaches as in Table 8.1 has been shown to strengthen the overall method and bring to the table the best abilities of each method. Each technique allows us to assist, analyze, or mimic human reasoning. These techniques advance the ability of computers to "think" and to help the command elements to make the optimal decision in the face of overwhelming amounts of information and seconds or minutes to make the best-possible decision. More advanced computer technology with higher processing speeds and immediately available memory has allowed the implementation of all of these techniques to be computationally feasible.

Transforming data into action. To take data and extract from it an understanding of the best actions to take next requires a process of assimilating the data to produce information, situational awareness, and finally optimal response. These levels of assimilation relate to the level of decision-making that is needed:

- *Data*—Data in this context are all the inputs to the COP (see Section 8.2), such as outputs of various sensors, observations of various witnesses, status of mobile assets and responders-receivers in the field, blueprints, schematics, and so forth.
- *Information*—Information in this scenario includes such statements as "Major roadway out or region is clogged," "Seaport open to all traffic," and so forth.

Table 8.1 Blended Approaches

Approach	*Benefits*
Probability with nonmonotonic reasoning	This blending returns the probable decision and gives information about why that decision was made, how to encourage that decision, and how to counteract a decision in the opposite direction. This method involves attaching statistical values to the elements of nonmonotonic formulas describing the decision.
Neural networks [25] with nonmonotonic reasoning	This method returns a justification along with the answer. Handles conflicting information and greater uncertainty easily. This method is known as "formula augmented networks" and involves attaching nonmonotonic formulas to the nodes of a traditional neural network.
Custom algorithms	These approaches can be used layer by layer in the decision-making process. Customizing each layer to the least complex method needed may reduce computation time and make the overall tool more efficient.

- *Knowledge*—Manifested as situational awareness, where knowledge incorporates the information gained in the last stage with temporal information to create a three-dimensional picture in the mind of the user.
- *Understanding*—Manifested as optimal response, means knowing what to do next and why.

8.2.2.2 The Advanced Incident Response Application and Common Operational Understanding

Given data and information, intelligent agents can be designed to assist the progression of a decision. There are many approaches to creating decision-support agents, depending on the type of data to be used and the level of decision to be made. A simplified application of these techniques has been developed and named the Advanced Incident Response System (AIRS) [26], intended to demonstrate the feasibility of aiding incident commanders during homeland security events using advanced expert-system technology. The Advanced Incident Response System intelligent agent takes information input from multiple disparate sources, translates it into a DAML+OIL (a language for describing an ontology [27]) schema, and then executes nonmonotonic algorithms on the information in near-real time. This application then provides command elements with a three-dimensional COP that conveys understanding of the situation along with a window that suggests courses of action, justifies those courses of action, and gives the command elements the ability to control the event with absolute confidence and complete understanding. DAML+OIL was chosen as a convenient and appropriate storage and retrieval structure. These nonmonotonic rules make automated decisions that define the best courses of action, and then AIRS displays the recommended actions.

The vision behind such intelligent agents as AIRS is the development of common operational (actionable) understanding, where various sources of information from all sources (i.e., observe) are collected and brought into central repositories. Various software agents (S&R) or people (P&S) tie into these repositories through a service-oriented architecture to use tools designed precisely for their job, allowing allow them to understand (i.e., orient) in the environment, select a course of action (decide), and even automate execution (act). The available decision-support tools can range from near total automation to near total collaboration. To mitigate bandwidth and computability issues, algorithms and filtering are distributed across the system. This approach ensure that the resulting intelligent agents are of modular design, and thus better able to integrate with existing and developing COP tools, remote smart agents, and evolving communication systems such as mesh networks. Figure 8.5 shows AIRS applied to the catastrophe evacuation scenario and situational awareness COP.

8.3 Emerging Communications Capabilities to Mitigate Deficiencies

Two evolving communications capabilities provide the human or smart agent interaction needed to increase media options for immediate push-to-talk voice communications.

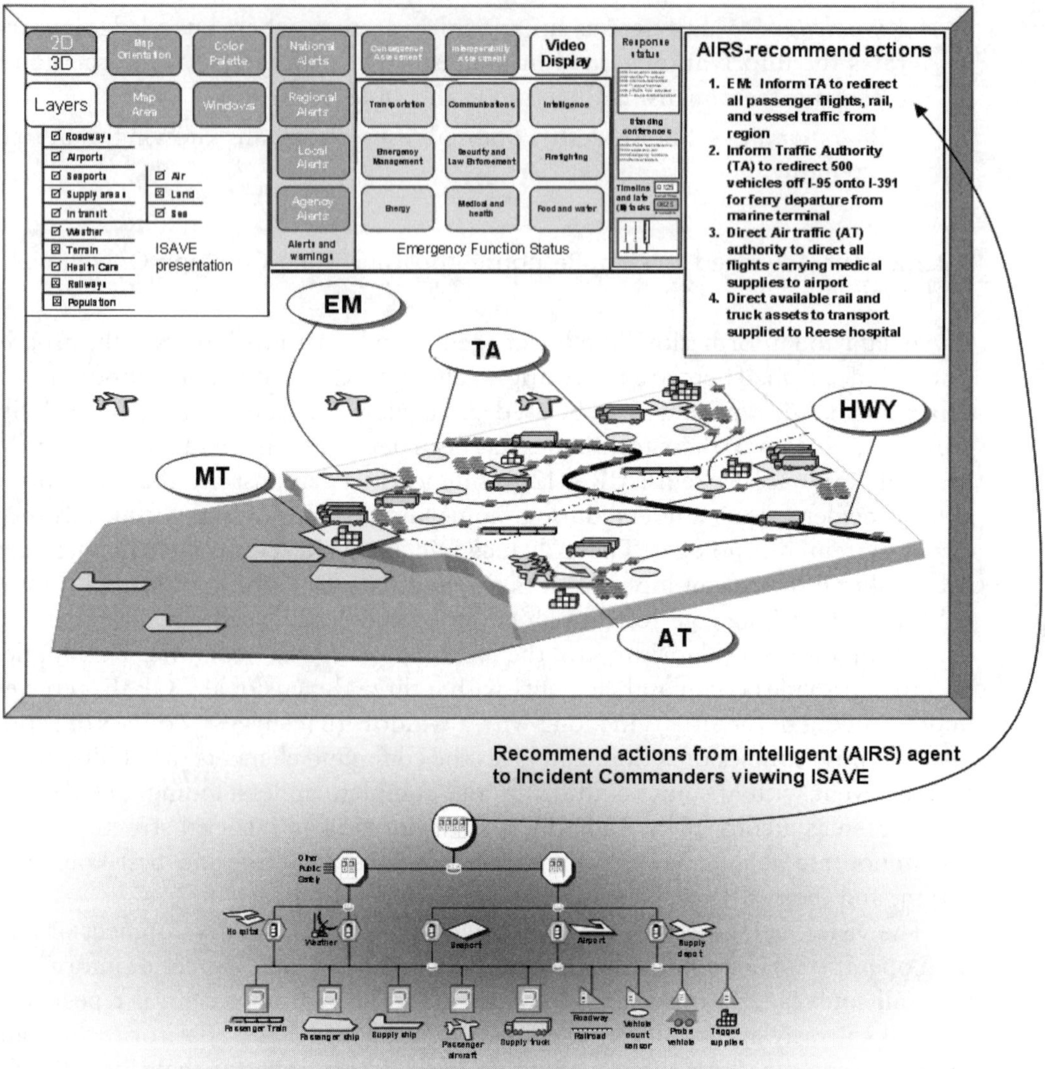

Figure 8.5 Integration of AIRS with the situational awareness COP.

8.3.1 Interoperability for Push-to-Talk Wireless Communications

Initially called the digital network management system (DNMS) [28], the wireless inter-system interface (WI-SI) [29] is intended to be a nonproprietary push-to-talk (PTT) interface permitting high-speed translation among all PTT-capable wireless systems, including land mobile radio (LMR), terrestrial wireless systems, and satellite systems.

8.3.1.1 Terminology and Abbreviation

The explanation of WI-SI requires the definition of some basic terms and abbreviations.

- *Network*—A collection of nodes. A network can also be known as a subsystem (in the terminology of Project 25 (P25) [30] or a subnet (in the terminology of the Internet Engineering Task Force).
- *Client network*—A network, such as P25 or push-to-talk over cellular, that is to be bridged by the WI-SI.
- *Client system*—This is a more generalized perspective and refers to a collection of client networks (e.g., a collection of different P25 networks integrated through the use of the inter-RF subsystem interface).
- *Edge nodes*—These are nodes that are placed at the edge of an administrative boundary, which have interface(s) connected to edge node(s) in other administrative boundaries. These nodes may be routers or servers.
- *Floor control*—This is the discipline of managing which users are allowed to transmit data/voice via half-duplex to the other users of a group.
- *Unicast*—This is a term most commonly used in today's Internet. It refers to communication (be it in the forms of packets, cells, frames) that is exchanged between two end-nodes. It is also refers to point-to-point or one-to-one communication.
- *Multicast*—Multicast is an extension to unicast and refers to point-to-multipoint or one-to-many communication. It is generally accomplished at the network layer, wherein the nodes of the network (e.g., routers) replicate the data to the various downstream receivers/users. Since the network does the replication, instead of dedicated servers, multicast makes very efficient use of transmission resources.

Important abbreviations in what follows include:

- CAI: Common Air Interface
- DHCP: Dynamic Host Configuration Protocol
- DSCP: Differentiated Services Code Point
- FQDN: Fully Qualified Domain Name
- FS: Fixed Station
- FSI: Fixed Station Interface
- IETF: Internet Engineering Task Force
- IMBE: Improved Multiband Excitation
- IP: Internet Protocol
- ISSI: Inter-RF Subsystem Interface
- MANET: Mobile Ad Hoc Networking
- MIP: Mobile IP Protocol
- MSC: Mobile Switching Center
- MIB: Management Information Base
- OMA: Open Mobile Alliance
- PCM: Pulse Code Modulation
- PoC: Press-to-talk over Cellular
- PTT: Push-to-Talk

- P25: Project 25
- RFSS: Radio Frequency Subsystem
- RSVP: Resource Reservation Protocol
- RTCP: Real-Time Transport Control Protocol
- RTP: Real-Time Transport Protocol
- SDP: Session Description Protocol
- SG: Subscriber Group
- SGID: Subscriber Group Identifier
- SIP: Session Initiation Protocol
- SNMP: Simple Network Management Protocol
- SU: Subscriber Unit
- SUID: Subscriber Unit Identifier
- TIA: Telecommunications Industry Association
- UDP: User Datagram Protocol
- WI-SI: Wireless Intersystem Interface

8.3.1.2 Objectives

The primary objective of the WI-SI architecture is to bridge disparate wireless communications systems through the use of open systems architecture not subject to proprietary designs or intellectual property. A snapshot of today's systems centered on the PTT communication model shows efforts by industry and standards groups to provide a measure of interoperability. However, these efforts tend to be narrow in scope and at times are subject to proprietary control. As an example, ISSI is an open architecture standardized by the Telecommunications Industry Association (TIA) to bridge Project 25 Radio Frequency Subsystems (RFSS) from multiple vendors. The advantage of having the Inter-RF Subsystem Interface is that it provides a focal point where various vendors of Project 25–based LMR systems can interoperate with each other. Prior to the Inter-RF Subsystem Interface, bridging of multiple RFSSs could only be accomplished if the component infrastructure of each RFSS came from the same vendor.

The emergence of PTT over cellular (PoC), as well as Internet Engineering Task Force standards used over the Internet, fosters a need for a bridging architecture that extends beyond P25 communications and can interact with more than one communication architecture. This need influences and has led to the design of the WI-SI as presented in the following sections.

8.3.1.3 Design Approach

The approach used in designing the WI-SI is based on a modular design and evolves in a chronologically phased manner. By this we mean that the WI-SI comprises a foundation of function and protocol elements that can be appended with new or additional elements as the need arises and capabilities become available. This approach is further refined in terms of phases; phase 1 being the point of origin of the WI-SI.

As a starting point, Project 25 and PoC have been used as target candidates for interoperability through the baseline design of the WI-SI, but many other LMR technologies [31] and commercial technologies [32] are also relevant for WI-SI application. Given that both of these digital radio communication architectures are migrating to IP, WI-SI is based on IP and Internet Engineering Task Force protocols to the greatest possible extent. We qualify this last statement because of the realization that TIA and the Third Generation Partnership Project (3GPP) have defined extensions to Internet Engineering Task Force protocols that are outside of the design and architectural philosophy of the Internet Engineering Task Force.

Phase 1 of the WI-SI is in the form of a backbone infrastructure, with edge nodes that have interfaces to the back-end infrastructures of various client systems. Intra-WI-SI communication is over IPv4 (IP version 4) [33] connectivity using either unicast or multicast packets. In the example shown in Figure 8.6, the WI-SI backbone bridges various Project 25 subsystems through the Inter-RF Subsystem Interface, PoC, and MANET systems based on routing protocols defined by the Internet Engineering Task Force. All of this is accomplished through classic [34] SIP-based communications over the Internet.

It is important to note that phase 1 of the WI-SI is not a complete solution. Open issues remain in terms of auto-configuration, registration, and exchange of presence information between various systems. In certain cases, phase 1 of the WI-SI articulates recommended extensions to existing systems that will automate the seamless exchange of information in order to avoid manual configuration of information.

There may also exist proprietary or nonstandard aspects unique to a given client network that will need to be supported by an edge node of the WI-SI. These addi-

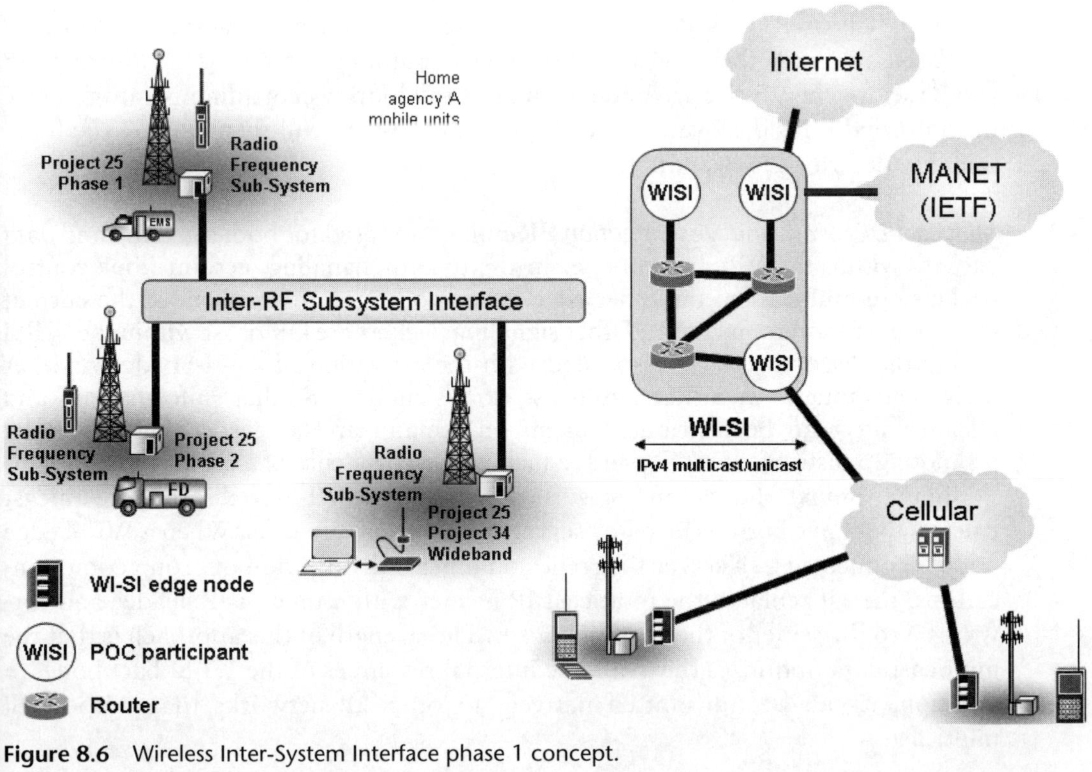

Figure 8.6 Wireless Inter-System Interface phase 1 concept.

tional features will be addressed in a separate supplemental document in order to maintain the open design and architecture of the WI-SI.

8.3.1.4 Features

There are several features that distinguish the WI-SI from other interoperability efforts, such as the Inter-RF Subsystem Interface design of P25. Below, we briefly describe the WI-SI capabilities that facilitate the bridging of disparate systems supporting wireless communication systems that support PTT, and in some cases, push-to-X applications (where X is voice, data, video, any media). Unless stated otherwise, all functionality listed in what follows is accomplished by WI-SI nodes, either internally as intranetwork communication or by edge interfaces connecting the WI-SI to client networks.

DHCP Auto-Discovery The Dynamic Host Configuration protocol (DHCP) [35] is a standard approach for on-demand allocation of IP addresses. In the context of the WI-SI architecture, we augment this capability by deploying RFC-4702 (DHCP Client Fully Qualified Domain Name option) at the edge WI-SI nodes with an interface that is connected to other networks. This RFC is a supplemental standard that specifies how DHCP clients can obtain the domain name established for the corresponding DHCP server.

The WI-SI edge node (a DHCP client) employs this supplemental standard to obtain the locally configured domain name associated with the interface connected to the client network. The inclusion of the domain name allows the WI-SI system and all of its other edge nodes to automatically discover which interface is associated with a given client network.

If DHCP or RFC-4702 are not supported by a client network (e.g., Project 25–based systems), then configuration files containing the necessary information will be set for the WI-SI edge node by man-in-the-loop system administrators. Once this information is discovered or configured, the WI-SI will then employ the functionality described in Section 4.2.

Multicast Discovery and Dissemination IP multicast is used for both signaling and data packets within the WI-SI backbone. In the case of signaling, certain floor control packets are multicast to the other edge WI-SI nodes informing them of the current state of group communication. Other signaling packets are multicast within the WI-SI during the discovery phase associated with the activation of a WI-SI edge node. A design advantage in using multicast for signaling is that information and responsibility can be distributed in an on-demand and as-needed basis, thereby promoting a distributed service and reducing the dependence of a client-server design.

In the context of data, multicast is used within the WI-SI to disseminate unicast packets that have been received by servers from other networks. When a WI-SI edge node is connected to a server that relies on unicast distribution of group communications, then it replaces the multicast IP header with a unicast IP header and forwards it to the server of the client network. The strength of this approach is that the multicast data optimizes the available internal resources of the WI-SI backbone. In addition, it can be transmitted natively to other IP networks that support IP multicast.

Internal Floor Control Both Project 25 and PoC use floor control for group communication. In the case of Project 25, additional headers are appended to RTP data packets to convey specific control primitives (such as request, grant, start, etc.). PoC uses RTCP control packets as carriers of its control primitives, and has a similar, although not isomorphic, set of commands to achieve floor control.

The WI-SI architecture defines its own set of floor control primitives in order to easily map and interoperate the same functionality found in Project 25 and PoC. The following commands are a subset of those used for Project 25 and only used internally within the WI-SI (note that some commands are multicast):

- WI-SI Push-to-Talk (W-PTT) Request—*multicast*
- W-PTT Grant—*unicast*
- W-PTT Deny—*unicast*
- W-PTT Start—*multicast*
- W-PTT End—*multicast*

The use of multicast for floor control has the added benefit of reinforcing a peer-to-peer model of control and in turn staying away from the nonoptimal third-party client-server architecture found in some of today's floor control designs. This approach should help reduce delays in bridging data between client networks.

Transcoding and Redundancy Research has shown that converting compressed digital voice streams from one encoding to another ("transcoding") degrades voice intelligibility [36]. There may exist a need to transcode digital voice streams in the WI-SI as they are forwarded from one client network to the next if the sets of codecs supported by those client networks do not overlap. If transcoding is needed, it shall be done by an egress WI-SI edge node. This ensures that bridged voice traffic suffers, at most, one transcoding.

Within the WI-SI network, media flows are encoded in both the native format from the ingress edge node and a standard, uncompressed encoding (e.g., pulse code modulation, or PCM, for audio). This approach is necessary for extensibility: egress nodes need not be updated as new encodings are added; they can transcode the standard encoding if a native encoding is unrecognized. By carrying the native as well as the standard encoding within the WI-SI backbone, *no transcoding is required* among WI-SI–connected systems that use the same native encoding (unlike alternative backbone approaches that convert all streams to a common encoding).

Redundancy is a technique used to help support a level of quality of service (QoS) by sending multiple packets of the same data. End-to-end support, from the perspective of origin SU to destination SU, may not support redundancy, thus requiring intervention by the WI-SI to compensate for this difference in capabilities. As in the case of transcoding, action taken on redundant encodings is taken at the egress WI-SI edge node.

Prioritization Both Project 25 and PoC incorporate prioritization in the upper layer signaling of calls. Project 25 redefines the structure and values of the Priority [37] field of SIP. PoC adds a priority value to the session description protocol (SDP). And to add to the complexity, the Internet Engineering Task Force has defined the Resource-Priority header to contain priorities associated with intermediary nodes.

In phase 1 WI-SI development, the WI-SI does not map these priorities within its internal backbone. Local policies are defined per deployment to obtain the "best" matching of upper layer priorities at the edge WI-SI nodes. Phase 2 of the WI-SI may map the upper layer priorities into network layer priorities through the Resource Reservation protocol (RSVP) and a related work-in-progress defining a new policy element for emergency priority.

Multiple Deployment Models The WI-SI has two deployment models. Phase 1 of the WI-SI focuses on a backbone model, with points of attachment that are placed at the backend of client networks. Typically, this would be the wire line infrastructure, which may extend beyond the wireless footprint of users and their SUs. It is anticipated that this model will be least intrusive to existing client networks in terms of requesting additional functionality to support the WI-SI.

The second deployment model, to be accomplished in phase 2, is termed *WI-SI in-a-box* [38], functioning much like the many deployable audio/IP gateways. In this case, the bridging functionality of WI-SI is self-contained in a local physical environment within the wireless footprint of different (and probably overlapping) client networks. The advantage of this approach is that it complements client networks that may have been established in an ad hoc manner with little, if any, back-haul connectivity to the wire-line networks due to damaged infrastructure. Since this model of the WI-SI would deploy emulated servers outside of their normal locations, minor augmentations will probably be needed of P25 and PoC systems. This model may impose limits on the functions and number of talk groups bridged between systems.

A goal for WI-SI is publication as an open standard, which would result in a third model: commercial offering of WI-SI interface cards that plug into the backplane of RFSS, Mobile Switching Center (MSC), and similar switching or console equipment, providing direct connection from a system into the WI-SI virtual backplane. These may be offered as options when ordering the equipment or sold as after-market enhancements to fielded systems.

Network Monitoring Network monitoring will be done at the edges of the WI-SI to maintain a near real-time database of information, particularly the connectivity status of various nodes. In the passive mode, RTCP packets will be examined to update the current list of active participants of all groups. This applies to those that are only observing or listening to active PTT flows.

Active mode monitoring involves several techniques. The first is internal to WI-SI, where all edge nodes join an internal well-known multicast group and send out periodic RTCP reports to each other. Depending on the local policies of client networks, WI-SI edge nodes will issue periodic ICMP echo-request messages, known as "pings" to server(s) of client networks to determine their connectivity. The results of this information is stored in management information bases (MIB) and accessed through the Simple Network Management Protocol (SNMP).

8.3.2 Netcentric Tactical Communication Architecture

Responders-receivers and other public safety and emergency response personnel rely on push-to-talk voice communications for tactical operations [39], and increasingly employ data communications for situational awareness and command coordination. Furthermore, the S&R systems described in Section 8.2 require automatic communications, presumably over IP, to feed the smart and intelligent software agents providing automatic decision support with recommendations for command elements. Instead of independent stovepipe systems, resilient converged multimedia communications with preferential push-to-talk voice traffic is an emerging capability needed to support S&R and well as P&S operations. By using the integrated best practices form Chapter 4, all relevant responders-receivers and other stakeholders for the chosen threat scenarios (see Chapter 7) would be included.

Consider a first-spiral National Tactical Communications Architecture for netcentric exchange of high-speed push-to-talk voice, short message service data, and (ultimately) imagery/video clips within workgroups (e.g., talk groups for push-to-talk voice) sharing a common operational picture using all private (e.g., land mobile radio) and commercial terrestrial, skywave (HF radio) [40], and space communication systems-free of intellectual property rights—a critical need for a resilient public safety and emergency response communication network of networks.

The intention of this architecture is to rapidly configure and maintain push-to-talk voice, short-message service data, and imagery/video for all those authorized to do so at the right time and right place. Figure 8.7 shows a high-level operational connectivity diagram (an OV-2 in DoDAF, see Section 4.5.5.2) showing a notional tactical connectivity to be implemented by the envisioned National Tactical Communications Architecture to optimally support homeland security

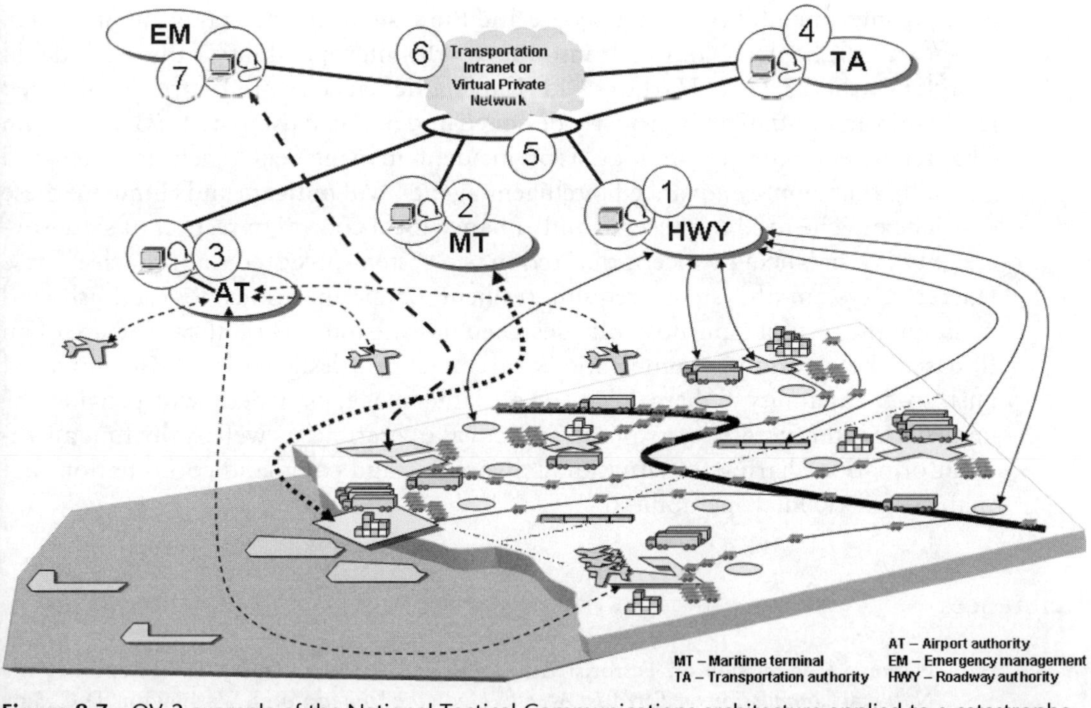

Figure 8.7 OV-2 example of the National Tactical Communications architecture applied to a catastrophe evacuation scenario.

action stages of "protect, detect, protect, alert, respond, and recover" for a major emergency, disaster, or incident, and return to normalcy.

The interfaces and acronyms in the figure are defined as follows:

- Um—Standardized air interface;
- G—Standardized interface to (heterogeneous) peer systems;
- Et—Standardized interface to telephone network;
- Ed—Standardized interface to data hosts and networks;
- En—Standardized interface to network management facilities;
- WI-SI—Wireless Inter-System Interface;
- HLR—Home Location Register;
- VLR—Visitor Location Register.

As shown in the figure, the NTAC would employ trained operators at each governance level and for each geographic region to manage the services-oriented architecture domain, adding governance levels, jurisdictions, and disciplines to provide situational awareness in advance of requiring assistance. In this way, the responders-receivers and other stakeholders will be able to plan ahead in the event their support will be requested. This planning is an innate part of the interoperability this emerging capability will enable.

8.4 Summary

In this book we have provided the definition tools and guidance needed to build effective interoperability among people and the systems in which they are included. As shown in Chapter 2 and built into the World Interoperability Failure Model in Chapter 3, the 25 Pearl Harbor–9/11–Katrina deficiencies are the result of failed leadership and initiative. If properly designed and planned using best practices from Chapter 4, emerging technologies for incident management (such as sense and respond), that employ advanced intelligent agents, will mitigate and eliminate these deficiencies. The need to preserve and enhance local control in respect of state sovereignty is a balancing concern that requires a system-of-systems approach. These federated system-of-systems require resilient communications to establish and maintain the resulting purposefully designed holistic interoperability. The use of an IP-based netcentric communications architecture designed with a universal push-to-X capability (where X is voice, data, image, or video) will provide the immediate interoperability required in tactical operations as well as the fundamental information sharing for situational awareness and command coordination that enables holistic interoperability.

References

[1] National Commission on Terrorist Attacks Upon the United States, *Final Report of the National Commission on Terrorist Attacks Upon the United States*, Washington, D.C., July 2004, p. 585.

[2] Office of the Director of National Intelligence (ODNI), Program Manager, Information Sharing Environment (ISE), *Information Sharing Environment Implementation Plan*, November 2006, Washington, D.C., p.186, available from http://www.ise.gov.

[3] Executive Office of the President, "Successes and Challenges in Improving Terrorism-Related Information Sharing," *National Strategy for Information Sharing*, Washington, D.C., October 2007, p. 2.

[4] U.S. Department of Homeland Security and U.S. Department of Justice, *Fusion Center Guidelines Developing and Sharing Information and Intelligence in a New Era Guidelines for Establishing and Operating Fusion Centers at the Local, State, and Federal Levels*, Washington, D.C., 2006, p. 104.

[5] U.S. Department of Justice, Office of Justice Programs and the Global Justice Information Sharing Initiative, *The National Criminal Intelligence Sharing Plan*, October 2003, Washington, D.C., p. 9.

[6] Wormeli, P., Executive Director, IJIS Institute, *National Information Exchange Model (NIEM)—Executive Introduction*, November 8, 2006, Washington, D.C., p. 16.

[7] Rodier, G, et al., *Global Public Health Security. Emerg Infect Dis* [serial on the Internet], Oct. 2007–November 2008, available at http://www.cdc.gov/EID/content/13/10/1447.htm.

[8] Kitler, M. E., P. Gavinio, D. Lavanchy, "Influenza and the Work of the World Health Organization," *Vaccine,* 2002, May 15; 20 Suppl 2, S5–14.

[9] Layne S. P., "Human Influenza Surveillance: the Demand to Expand," *Emerg Infect Dis,* April 2006, 12(4), pp. 562–568.

[10] Stohr, K., "The Global Agenda on Influenza Surveillance and Control," *Vaccine,* 2003, May 1, 21(16), pp.1744–1748.

[11] U.S. Centers for Disease Control & Prevention, *Progress in Public Health Informatics*, Atlanta, GA, Nov 2008, p. 4, available at http://www.cdc.gov/phin/spotlight/pdf/Progress_And_Public_Health_Informatics.pdf.

[12] http://www.cdc.gov/phin/library/documents/pdf/111759_aboutPHIN_title%20Page_c.pdf.

[13] See http://www.cdc.gov/phin/resources/architecture.html.

[14] OAISIS, *OASIS Common Alerting Protocol, v. 1.1, OASIS Standard CAP-V1.1*, October 2005, p. 35. Accessed on February 8, 2009 at http://www.oasis-open.org/apps/org/workgroup/emergency/download.php/14205/emergency-CAPv1.1-Committee%20 Specification.doc.

[15] Ibid.

[16] In this context, the term "common" implies the same COP is shared by multiple OPCENs.

[17] Contestabile, J, C. McGeorge, *Review of 2008 Urban Area Security Initiative [UASI] Projects*, unpublished report, November 20, 2008, p. 2.

[18] For example, the assignment of HAZMAT personnel experienced in managing exactly these substances detected at an incident scene.

[19] Tripp, Robert S., et al., *Sense and Respond Logistics: Integrating Prediction, Responsiveness, and Control Capabilities*, The RAND Corporation monograph series, RAND URL: http://www.rand.org/, 2006, p. 45.

[20] Ibid., p. 41.

[21] Ibid., p. 32.

[22] Abbott, E. A., *Flatland: A Romance of Many Dimensions*, Kindle, 1883.

[23] Vanderbilt, A. K. C. S., and Desourdis, R. I., Jr., "Information Fusion for Common Operational Understanding," *Visualisation and the Common Operating Picture* Derived with permission from *Proceedings (RTO-IST-MP-043) of the 2004 NATO RTO Workshop*, NATO IST-043-RWS-006, Toronto, 2004.

[24] *Katrina Report*, p. 360.

[25] Neural networks are a form of expert system.

[26] Science Applications International Corporation Advanced Incident Response, Independent Research & Development Project, January 2004.

[27] Description of conceptualized objects or agents and their interrelationships.

[28] Desourdis, R. I., Jr., et al., *Emerging Public Safety Wireless Communication Systems*, Norwood, MA: Artech House, November 2001, pp. 533-534.

[29] Science Applications International Corporation, *Development Scenario for the National Tactical Communications Architecture (NTCA)*, Nactional Communication System, unpublished report, June 13, 2007, pp. 26-46.

[30] Desourdis, R. I., et al., pp. 278-316.

[31] Ibid., Chapter 5, pp. 249-372.

[32] Ibid., Chapter 6, pp. 415-506.

[33] Future versions of the WI-SI are expected to add support for IPv6.

[34] We use the term "classic" in reference to protocols as defined by the Internet Engineering Task Force, but with no additional changes in semantics or augmentations as defined by other standards organizations or proprietary designs.

[35] Internet Engineering Task Force RFC-2131.

[36] Redding, C. et al., *Voice Quality Assessment of Vocoders in Tandem Configuration*, NTIA Report 01-386, April 2001. DoD studies have also shown significant degradation when transcoding between military and LMR vocoders.

[37] The Internet Engineering Task Force defines this field for use by end-users only, and not intermediate nodes. Hence, the redefinition by Project 25 is in direct conflict with the Internet Engineering Task Force version of SIP.

[38] Also known as WI-SI in-a-Van.

[39] Desourdis, R. I., et al., pp. 9-16.

[40] Johnson, E. E.,et al., *Advanced HF Communication Systems*, Norwood, MA: Artech House, March 1997.

Selected Bibliography

Andrus J. K., et al., "The Global Interrelatedness of Disease Control," *Lancet Infect Dis.*, 2007,7:176.

Baker, M.G., and D. P. Fidler, "Global Public Health Surveillance Under the New International Health Regulations," *Emerg Infect Dis.*, 2006, 7:1058–65.

Connell, II, Thomas W., and Michael A. Morin, "First Responder Credentialing Interoperability: A White Paper on the Necessity of Interoperability between Credentialing Systems," Gardner: MA, Vuance Inc., 2007.

Connell, Thomas, "Rapid Access of Privileges and Training for Operational Readiness (RAPTOR)," *Project Overview,* Gardner, MA, November, 2008.

Executive Office of the President, Homeland Security Presidential Directive 12 (HSPD-12), "Policy for a Common Identification Standard for Federal Employees and Contractors," August 27, 2004, Washington, D.C., http://www.whitehouse.gov/news/releases/2004/08/20040827-8.html .

Executive Office of the President, "Successes and Challenges in Improving Terrorism-Related Information Sharing," *National Strategy for Information Sharing*, Washington, D.C., October 2007, 48 pp.

IJIS Institute, *Strategic Plan for the IJIS Institute: FY2009-2011*, Ashburn, VA: IJIS Institute, July 30, 2008, 24 pp.

Kitler ME, Gavinio P, Lavanchy D. "Influenza and the Work of the World Health Organization," *Vaccine*, May 15, 2002, 20 Suppl 2:S5–14.

Layne, S.P., "Human Influenza Surveillance: The Demand to Expand," *Emerg Infect Dis.*, April 2006, 12(4):562–568.

Morin, Michael A., *National Responder Credentialing: The impact of the "Implementing Recommendations of the 9/11 Commission Act of 2007" on State, Local, and Tribal Governments*, Gardner, MA: Vuance Inc., 2007.

National Commission on Terrorist Attacks Upon the United States, *Final Report of the National Commission on Terrorist Attacks Upon the United States*, Washington, D.C., July 2004, 585 pp.

Office of the Director of National Intelligence (ODNI), Program Manager, Information Sharing Environment (ISE), *Information Sharing Environment Implementation Plan*, November 2006, Washington, D.C., 186 pp.

Rodier, G., et al., "Global Public Health Security," *Emerg Infect Dis.*, [serial on the Internet], October 2007–30 November 2008, http://www.cdc.gov/EID/content/13/10/1447.htm.

Stohr, K., "The Global Agenda on Influenza Surveillance and Control," *Vaccine*, May 1 2003, 21(16): 1744–1748.

U.S. Centers for Disease Control & Prevention, *Progress in Public Health Informatics*, Atlanta, GA, Nov 2008, 4 pp., http://www.cdc.gov/phin/spotlight/pdf/Progress_And_Public_Health_Informatics.pdf.

U.S. Department of Homeland Security and US Department of Justice, *NIEM Concept of Operations*, Global Justice Information Sharing Initiative, October 2005, Washington, D.C.

U.S. Department of Homeland Security, Federal Emergency Management Agency, *Resource Typing Initiative Continues* (Fact Sheet), Washington, D.C., http://www.fema.gov/pdf/emergency/nims/nims_resource_typing_initiative.pdf, April 2004.

U.S. Department of Homeland Security, Federal Emergency Management Agency, *Resource Typing*_(Fact Sheet), Washington, D.C., http://www.fema.gov/pdf/emergency/nims/resource_typing_qadoc.pdf , July 2005.

U.S. Department of Homeland Security, Federal Emergency Management Agency, *National Emergency Responder Credentialing System*_(Fact Sheet), Washington, D.C., October 26, 2005.

U.S. Department of Homeland Security and U.S. Department of Justice, *Fusion Center Guidelines.*

Developing and Sharing Information and Intelligence in a New Era: Guidelines for Establishing and Operating Fusion Centers at the Local, State, and Federal Levels, Washington, D.C., 2006, p. 104.

U.S. Department of Justice, Bureau of Justice Assistance and the Global Justice Information Sharing Initiative, *Findings and Recommendations of the Suspicious Activity Report (SAR) Support and Implementation Project*, October 2008, Washington, D.C.

U.S. Department of Justice, Office of Justice Programs and the Global Justice Information Sharing Initiative, *The National Criminal Intelligence Sharing Plan*, October 2003, Washington, D.C., 54 pp., http://www.it.ojp.gov/documents/NCISP_Plan.pdf .

U.S. Government Accountability Office, *Geographic Information: Challenges to Effective Data Sharing*, GAO-03-874T, June 2003, Washington, D.C., 20 pp., http://www.gao.gov/new.items/d03874t.pdf.

United States Senate, Intelligence Reform and Terrorism Prevention Act (IRTPA) of 2004, S.2845, Public Law 108-458, Washington, D.C., December 17, 2004, 236 pp.

Wormeli, Paul, Executive Director, IJIS Institute, *National Information Exchange Model (NIEM)—Executive Introduction*, November 8, 2006, Washington, D.C., 16 pp.

Douglas Cenzer, Jeffery B. Remmel, and Amy K. C. S. Vanderbilt, "Locally Determined Logic Programs," *Proceedings of the 5th International Conference on Logic Programming and Nonmonotonic Reasoning (LPNMR99)*, Springer-Verlag.

Douglas Cenzer, Jeffery B. Remmel, and Amy K. C. S. Vanderbilt, "The Complexity of the Set of the Extensions of Nonmonotonic Rule Systems," Artificial Intelligence and Mathematics conference paper, Fort Lauderdale, FL, January 2000, 10 pp.

Douglas Cenzer, Jeffery B. Remmel, and Amy K. C. S. Vanderbilt, "Characterizing the Stable Models of a Locally Determined Logic Program," *Proceedings of the 6th International Symposium on Artificial Intelligence and Mathematics, Annals of Mathematics and Artificial Intelligence*, March 2004,40 (3-4): 225–262.

Douglas Cenzer and Amy K. C. S. Vanderbilt, *Common Derivations in Locally Determined Logic Programs,* Artificial Intelligence and Mathematics conference paper, Fort Lauderdale, FL, January 2002, 10 pp.

[E-ARCH] EADS Inter-RF Subsystem Interface White Paper, Standards and regulatory EADS Secure Networks, July 2006.

[I-ARCH] APCO Project 25 Inter-RF Subsystem Interface Architecture Section V3e, TIA Committee TR8 Private Land Mobile Radio, November 2005.

[O-ARCH] Push to talk Over Cellular Architecture, Open Mobile Alliance, OMA-AD_PoC-V1_0_1-20061128-A, November 2006.

[rfc2131] Dynamic Host Configuration Protocol, RFC-2131, Internet Engineering Task Force, March 1997.

[rfc2205] Resource Reservation Protocol, Version 1, RFC-2205, Internet Engineering Task Force, September 1997.

[rfc2474] Definition of the Differentiated Services Field, RFC-2474, Internet Engineering Task Force, December 1998.

[rfc3246] An Expedited Forwarding PHB, RFC-3246, Internet Engineering Task Force, March 2002.

[rfc3261] Session Initiation Protocol, RFC-3261, Internet Engineering Task Force, June 2002.

[rfc3344] Mobile IP Protocol, RFC-3344, Internet Engineering Task Force, Aug 2002.

[rfc3411] An Architecture for Describing SNMP, RFC-3411, Internet Engineering Task Force, Dec 2002.

[rfc3550] A Transport Protocol for Real-Time Applications, RFC-3550, Internet Engineering Task Force, July 2003.

[rfc4702] DHCP Client Fully Qualified Domain Name Option, RFC-4702, Internet Engineering Task Force, October 2006.

Acronyms

3GPP	Third Generation Partnership Project
ACE	architecture-centric engineering
AEAS	Automated Exercise Assessment System
AFCEA	Armed Forces Communications-Electronics Association
AHC	All Hazards Consortium
AIC	Architecture and Infrastructure Committee
AIRS	advanced incident response system
ALI	automatic location identification
ANI	automatic number identification
ANSI	American National Standards Institute
AOR	areas of responsibility
APCO	Association of Public Safety Communications Officials
ASOCC	area security operations command and control
AV	all views
AVL	automatic vehicle location
AWS	Aircraft Warning System
BCA	business case analysis
BORTAC	border tactical
BPEL	Business Process Execution Language
BPMN	business process modeling notation
BSI	bridging system interface
CAC	common access card
CAD	computer aided dispatch
CAI	common air interface
CAP	Common Alerting Protocol
CAP Project 25	Compliance Assessment Program
CAT-EVAC	catastrophe evacuation

CBA	cost benefit analysis
CBRNE	chemical, biological, radiological, nuclear, explosive
CDC	Centers for Disease Control
CG	Coast Guard
CHUID	card holder unique identifier
CIA	Central Intelligence Agency
CIO	chief information officer
CNO	Chief of Naval Operations
COA	course of action
COBR (or Cobra)	Cabinet Office Briefing Room
CONOPS	concept of operations
COP	common operational picture
CRASOC	Coalition Rear Area Security Operations Command and Control
CSSI	console subsystem interface
DAML+OIL	DARPA Agent Markup Language + Ontology Web Language
DCO	defense coordinating officer
DHCP	Dynamic Host Configuration Protocol
DHS	Department of Homeland Security
DIO	defensive information operations
DMAT	disaster medical assistance team
DMV	Department of Motor Vehicles
DNMS	digital network management system
DoD	Department of Defense
DoDAF	Department of Defense Architecture Framework
DoJ	Department of Justice
DSCP	differentiated services code point
EA	enterprise architecture
EAS	Emergency Alert System
EDXL	Emergency Data Exchange Language
EMAC	Emergency Management Assistance Compact
EMS	Emergency Medical Services
EOC	emergency operations center
ESF	emergency support function
ESU	emergency service unit
EVDO	evolution data optimized

FAA	Federal Aviation Administration
FAC	Federal Advisory Committee
FAO	Food and Agriculture Organization
FBI	Federal Bureau of Investigation
FCC	Federal Communications Commission
FCO	Federal Coordinating Office
FD	Fire Department
FDNY	Fire Department of New York
FEAF	Federal Enterprise Architecture Framework
FEMA	Federal Emergency Management Agency
FGDC	Federal Geographic Data Committee
FIPS	Federal Information Processing Standards
FISA	Foreign Intelligence Surveillance Act
FiXs	Federation for Identity and Cross Credentialing Systems
FOS	family of systems
FQDN	fully qualified domain name
FRAC	first responder access card
FRS	Fire and Rescue Services
FS	fixed station
FSI	fixed station interface
GAO	Government Accountability Office
GHz	gigahertz
GIS	geographic information system
GJXDM	Global Justice Extensible Data Model
GLT	government liaison team
GO	government offices
GPHIN	Global Public Health Intelligence Network
GSA	General Service Administration
GUI	graphical user interface
HAZMAT	hazardous material
HF	high frequency
HIT	health information technology
HLR	home location register
HM	Her Majesty
HSOC	Homeland Security Operations Center
HSPD	Homeland Security Presidential Directive

HUMINT	human sources intelligence
IA	information assurance
IAP	incident action plan
IC	incident commander
ICE	Immigration and Customs Enforcement
ICEPIC	ICE pattern analysis and collection system
ICMP	Internet Control Message Protocol
ICP	incident command post
ICS	incident command system
IDP	incident demobilization plan
IEA	interoperability enterprise architecture
IEEE	Institute of Electrical and Electronics Engineers
IERs	information exchange requirements
IETF	Internet Engineering Task Force
IJIS	integrated justice information system
ILP	intelligence-led policing
IM	immediate message
IMBE	improved multiband excitation
IMP	incident medical plan
INCOSE	International Council on Systems Engineering
IP	Internet Protocol
IPAWS	Integrated Public Alert and Warnings System
IPR	intellectual property rights
IPT	integrated product teams
IRTPA	Intelligence Reform and Terrorism Prevention Act
ISA	information sharing architecture
ISAVE	integrated situational awareness visualization environment
ISE	information sharing environment
ISE EAF	ISE enterprise architecture framework
ISE PAIS	ISE profile and architecture implementation strategy
ISE-SAR	ISE suspicious activity reporting
ISSA	information sharing segment architecture
ISSI	inter-RF subsystem interface
IT	information technology
ITP	incident traffic plan
ITU	International Telecommunication Union

IWG	Industry Working Group
JFO	Joint Field Office
JTTF	Joint Terrorism Task Force
LEO	Law Enforcement Online
LMR	land mobile radio
LRF	Local Resilience Forum
M & S	modeling and simulation
MACA	military aid to civil authorities
MACS	multiagency coordination system
MANET	mobile ad hoc networking
MDOT	Maryland Department of Transportation
MHz	megahertz
MIB	management information base
MIP	Mobile IP Protocol
MOU	Memorandum of Understanding
MSC	mobile switching center
NATO	North Atlantic Treaty Organization
NCR	national capital region
N-DEx	NIEM National Data Exchange
NDMS	national disaster medical systems
NEADS	North East Air Defense Sector
NEPLOs	Navy emergency preparedness liaison officers
NFPA	National Fire Protection Agency
NGO	nongovernmental organization
NIEM	national information exchange model
NIMS	national incident management system
NISG	National Information Sharing Guidelines
NIST	National Institute for Standards and Technology
NLETS	National Law Enforcement Telecommunications
NMCC	National Military Command Center
NOAA	National Oceanic and Atmospheric Administration
NOC	network operations center
NORAD	North American Aerospace Defense Command
NPHI	National Public Health Institute
NRCC	National Response Coordination Center
NRF	national response framework

NRP	national response plan
NSDI	national spatial data infrastructure
NSIS	national strategy for information sharing
NTCA	net-centric tactical communication architecture
NTIA US	National Telecommunications and Information Administration
NWS	National Weather Service
O&M	operations and maintenance
OASIS	Organization for the Advancement of Structured Information Standards
OCD	operational concept description
OD&E	organizational development and effectiveness
OEA	organizational enterprise architecture
OEC	Office of Emergency Communications
OIC	officer in charge
OMA	Open Mobile Alliance
OMB	Office of Management and Budget
ONCRC	Office of National Capitol Region Coordination
OODA loop	observe, orient, decide, act loop
OPCEN	operations center
OSI	Open System Interconnection
OV	operational view
P25	Project 25
PAPD	Port Authority Police Department
PCM	pulse code modulation
PFO	principal federal official
PHD	Pearl Harbor deficiencies
PHIN	Public Health Information Network
PIN	personal identification number
PIV	personal identity verification
PM	project management
PMBOK	Program Management Body of Knowledge
PMC	pulse code modulation
PMI	Program Management Institute
PM-ISE	Presidential Memorandum Information Sharing Environment
PoC	push to talk (PTT) over cellular
ProMED	Program for Monitoring Emerging Diseases

PS	public safety
PSAF	public safety architecture framework
PSAP	public safety answering point
PSIC	public safety interoperable communications
PSTN	public switched telephone Network
PSWAC	Public Safety Wireless Advisory Committee
PSWN	public safety wireless network
PTT	push to Talk
PTX	push-to-X
QoS	quality of service
RADM	Rear Admiral
RAPTOR	rapid access of privileges and training for operational
RCCC	Regional Civil Contingencies Committee
RFID	RF identification device
RFSS	radio frequency subsystem
RISS	regional information sharing systems
ROW-B	radio over wireless broadband
RPA	reference pilot architecture
RSS	regional sharing service
RSVP	Resource Reservation Protocol
RTCP	Real-Time Transport Control Protocol
RTP	Real-Time Transport Protocol
S&R	sense & respond
S&RL	sense & respond logistics
SA	situational awareness
SAME	specific area message encoding
SAR	suspicious activity report
SARS	Severe Acute Respiratory Syndrome
SCG	Strategic Coordination Group
SCIPs	statewide communications interoperability plans
SCO	state coordinating officer
SDLC	systems development life cycle
SDO	Standards Development Organization
SDP	Session Description Protocol
SDR	software defined radio
SE&I	systems engineering and Integration

SELC	systems engineering life cycle
SEP	systems engineering plan
SFLEO	Senior Federal Law Enforcement Office
SG	subscriber group
SGID	subscriber group identifier
SGML	Standard Generalized Markup Language
SIDA	Sense, Interpret, Decide, Act
SIEC	State Interoperability Executive Committee
SIP	Session Initiated Protocol
SLC	system life cycle
SMART	specific, measurable, agreed upon, realistic, time-bound
SME	subject matter expert
SMS	short message service
SNMP	Simple Network Management Protocol
SOA	service oriented architecture
SOP	standard operating procedure
SoR	Statement of Requirements
SoS	system of system
SU	subscriber unit
SUID	subscriber unit identifier
SV	systems view
SWOT	strengths, weaknesses, opportunities, threats
TACSTACK	tactical stack
TETRA	terrestrial trunked radio
TIA	Telecommunications Industry Association
TRD	Technical Requirements Document
UASI	Urban Area Security Initiative
UC	unified command
UDP	User Datagram Protocol
UML	Universal Modeling Language
UN	United Nations
US	United States
VLR	visitor location register
VoIP	Voice over IP
WBS	work breakdown structure
WHO	World Health Organization

WI-SI	wireless intersystem interface
WMD	weapons of mass destruction
W-PTT	WI-SI push-to-talk
XML	Extensible Markup Language

Glossary

Applications	A class of software components characterized as delivering functionality directly related to end-user mission capability.
Application interoperability	The capability of different software applications to exchange data via a common set of exchange formats, to read and write the same file formats, and to use the same protocols.
Architecture	The configuration of operational, system, or technology components and their interactions in an enterprise.
Archive	The repository for data retained on a longer-term basis, such as completed incidents, recorded radio traffic, incident photographs, and incident video useful in judicial and operational proceedings, training and exercises, and so fourth.
As-is	The current state of an architecture or architectural component. From an engineering architecture perspective, it is the systems baseline that is in current operations. This perspective provides the basis to identify areas for improvement.
Audio processing	In land mobile radio, audio processing includes trunking and simulcast system control functions, such as assignment of remote site assets to radio calls, removal of radio system signaling for extraction of the intelligence, vocoding, and signal voting. For data, this configuration includes removal of radio system signaling for extraction of the intelligence, formatting as required by the radio system, routing of data to the appropriate radio channels, and packet or message switching services. Audio processing may involve encryption and decryption as well.

Backplane A printed circuit board with slots into which other cards are plugged. A backplane is typically accessed by one or more connectors and does not usually have many active components.

Baseline (project management) An established and agreed upon point in a project or engineering effort to which changes are managed and controlled.

Bridging System Interface Standards based on Session Initiation Protocol (SIP) to ensure that IP-based gateway implementations developed by different vendors do not employ proprietary hardware and software rendering them uninteroperable.

Bridging with Legacy Systems New systems implemented with software-defined radio technology which can be programmed for technical interoperability with external systems using virtually any technology.

Caches of radio equipment Equipment prepositioned for use in emergencies or disasters, which can readily support communication with disparate radio systems. This includes access to the various public safety mutual aid channels in various bands.

Class An abstract categorization of related objects. Often used as the basis of static modeling for object-oriented approach to engineering.

Composite mobile communications architecture The use of a gateway or other reduced functionality interface between disparate IT and communication systems.

Cross-system features (radio interoperability) Radio system capabilities/controls that can be carried through the interoperability interface between radio systems. Audio gateways do not carry proprietary signaling, such as talk-group identification information, across the gateway interface.

Data processing (radio systems) This configuration includes removal of radio system signaling for extraction of the intelligence, formatting as required by the radio system, routing of data to the appropriate radio channels, and packet or message switching services. Data processing may involve encryption and decryption. In general, it is any automated or computerized process that converts data into information or knowledge.

Data standards Documented agreements on representations, formats, and definitions of common data. Data standards improve the quality and share ability of environmental data by increasing data compatibility, improving the

	consistency and efficiency of data collection, and reducing data redundancy.
Decision gate or decision milestone	A point within a project or engineering effort where it is agreed that moving forward is within acceptable risk limits and that new activities are not pursued until current activities are completed satisfactorily.
Decision support (including prediction and forecasting)	This function uses actionable knowledge from sensor, intelligence, public, responder-receiver, and external information-sharing networks to simplify and prioritize the decision-making process for OPCEN command elements. In addition, predictive tools that assess theoretical consequences of hypothetical or real events (e.g., release of a toxic plume) can be used in combination with other data (e.g., prevailing winds, interposed terrain) to predict impact contours pushed to become an overlay to the GIS-based situational awareness display.
Deficiency	One of 25 failures of leadership or planning identified by the congressional Pearl Harbor investigative report that have been enhanced with the addition of corresponding failures identified by the congressional 9/11 and Katrina reports, which when combined explain continued holistic interoperability failures.
Deployment and build	The phase of a project or engineering effort where procurement, development, and manufacturing takes place.
Digital interfaces	Physical communication linkages between disparate systems, including radio systems that retain digital representation of voice and other information media. These linkages enable computing devices and/or software applications to communicate, control, and synchronize with each other.
Evolution-Data Optimized (EVDO)	Mobile phone technology for high speed internet access.
External network interface	A physical & logical network linkage or gateway for communicating/connecting a computer, or computer network, with one or more computers or computer network(s).
First Receivers	Healthcare workers at a medical facility receiving contaminated victims from exposure to potential biological pathogens, radiation, or other contaminating agents. They can be regarded as a subset of first responders, with the caveat that they typically perform at a healthcare facility, rather than at the site where the hazardous material release occurred.

First Responders Those individuals who in the early stages of an incident are responsible for the protection and preservation of life, property, evidence, and the environment. This includes firefighters, law enforcement, HAZMAT teams, and ambulance service personnel as well as emergency management, public health, clinical care, public works, and other skilled support personnel that provide immediate support services during prevention, response, and recovery operations.

Framework An abstract bounding of conceptual approach, notation standards, and engineering artifacts to provide discipline for software-intensive systems engineering.

Fusion Center A collaborative effort of two or more agencies that provide resources, expertise, and/or information to the center with the goal of maximizing the ability to detect, prevent, apprehend, and respond to criminal and terrorism activity.

Geographic Information System based situational awareness This function converts actionable information into a format for effective and efficient presentation to command elements using location-based display, which shows the location of incidents and operation centers, and other assets in terrain, built-up areas, or within structures in a spatial or geographic format.

Governance Mechanism to provide stakeholders a means or process to control the planning and implementation of interoperability in critical IT and communication systems for joint use, usually established and maintained to transition from an "as-is" system current baseline to a "to-be" system target baseline and sustains that baseline.

Holistic interoperability A measure of shared, trusted, understanding that drives predictable collaborative action to achieving a common goal, and which must be intentionally planned and designed in the specification of critical IT and communication systems.

Home infrastructure accommodation A local operations center employs technologies that accommodate the diverse capabilities of the visiting agency subscriber devices, enabling federated mobile architecture. More generally, the local area employs diverse interface capabilities to accommodate the holistic interoperability of visiting organizations.

Information sharing The planned and vetted distribution of sensor data, alerts, and warnings; tactical command and control status and directives; assembly and sharing of situational awareness using the common operational pic-

	ture, and administrative/logistics orders and delivery tracking.
Information-sharing systems architecture	The configuration of integrated IT and communications system elements that ensures "information sharing" defined immediately above.
Infrastructure gateway	An interconnection appliance that permits the translation of information between two or more disparate systems, where connecting systems through the gateway is automatically or manually controlled.
Integrated logistics support	This function includes monitoring of the system architecture for status, including impending or actual equipment failures, and provides all sustainment functions including facility/equipment operations and maintenance (O&M), including operator and maintenance training, critical parts ordering/sparing, and so forth.
Integration and verification	The phase in a project or engineering effort where items developed are assured to meet specification and where individually developed items are combined to form the whole system.
Intelligence collection	The gathering of information from a variety of sources, including archived data, the Internet, electronic surveillance, reports (from other agencies), media outlets, casual observation, prediction forecasts, and human sources (HUMINT), among others, needed to build an understanding of individual or organization intentions and capabilities, plans, and timelines.
Intelligence fusion	The received information content from one or more sources is analyzed collectively to determine capability, intent, timing, and so forth in the context of other available information (e.g., sensor data, responder-receiver reports, ongoing investigations, Public Safety Answering Point calls).
Internal network	The information sharing capabilities within a physical facility, such as an operations center.
Interoperable RF systems	Mobile voice, data, and video communications among home and visiting responder-receivers and other stakeholders and their dispatchers through the OPCEN backbone to support tactical interoperability.
Interoperability	A measure of shared, trusted understanding that drives predictable collaborative action towards a common goal.
Internet backbone	The main trunk connections of the Internet made up of a large collection of interconnected commercial,

	government, academic, and other high-capacity data routes and core routers that carry data across the countries, continents, and oceans of the world.
Interoperability architecture	The proper application of the principles of integrated, composite, aligned, and federated architecture in order to assure that new systems satisfy the need to interoperate with a wide range of existing and future systems. Interoperability architecture addresses the need for future systems to interoperate with aging legacy systems, for different systems within a jurisdiction to interoperate crossfunctionally with other systems in that jurisdiction, for systems in adjacent jurisdictions to interoperate laterally with one another, and for systems to interact vertically with other systems at all other levels of governance.
Interoperability Assessment Checklist	A list of 25 categories of questions derived from the 25 Peal Harbor-9/11-Katrina deficiencies (failures of leadership and planning), which can be used to identify potential sources of failed holistic interoperability.
Interoperability channels	Radio frequencies and associated signal bandwidth that are reserved for multiorganizational events, such as natural disasters or terrorist activities. In the United States, dedicated interoperability channels are not licensed to any particular agency, but are a general resource to support multiagency interoperability.
Interoperability network interface	A linkage or gateway between two or more disparate voice and data communications systems, records management systems, computer aided dispatch systems, wireless networks, and other critical IT and communication systems.
Interoperability system architecture	Combines elements of network, platform, system, data, messaging, and application architecture and focuses on the degree to which the user at one operations center is able to share information with and access information from other operations centers.
IT and communication system architecture	Defines the general hardware and software tools that enable the information sharing needed for public safety and emergency response interoperability.
"Katrina Report"	The published findings of the U.S. Congressional Commission on the aftermath of Hurricane Katrina that investigated the emergency preparedness and response.
Lateral disparity	The inability of systems in adjacent administrative divisions of the same level to interoperate. For exam-

ple, two counties that share a border may not be able to share data or talk with one another.

Local database This database provides storage and retrieval of data resident to the local operations center, which typically includes a mobile unit dispatch capability, records management, and location-based information and processes. Much data of potential value available in remote databases such as criminal, license, hydrant, and hazard data as well as building floor plans may be stored in this local database.

Mobile gateway An internetworking system capable of joining together two or more mobile networks that use different base mobile protocols. This appliance is similar to an infrastructure gateway, except that it is operated most often as a hands-free device in a ground, air, or water-mobile platform searching for planned signals. For voice, a mobile gateway is simply an array of audio interconnect circuits permitting multiple voice radios to share the same reversed audio (i.e., all microphones connected to all speakers) from all radios. For broadband data (and VoIP voice) applications, the mobile gateway serves as a "best-effort router" to find and maintain the best performing packet data communications path for visitor mobile data communications with the home (local) operations center.

Mobile emergency call service The processing and routing of mobile (wireless) phone calls from the public for emergency service, including routing of calls to available call-takers (potentially using automatic call distribution).

Mutual aid channels Interagency-planned mutual aid channels are licensed for the exclusive use of the participating agencies and are commonly organized through county, regional, or state planning organizations (or managing authorities) that exercise general supervision and disciplinary control.

Maturity In this context, the term applies to the benefits of longevity of use of IT and communication systems. These benefits often mean that system defects have been revealed and corrected, many system users have been trained and become familiar with trusted system operations, and there are sustainment capabilities in place for operations and maintenance as well as improvement cycles.

Media supported The source and end state formats of information as collected by various information sources or as utilized

	by various information recipients, meaning voice, data, or video and its variations, such as sensor data and thermal-imaging video.
Modus operandi	This Latin expression identifies the specific means the method or mode of operation.
National Information Exchange Model (NIEM)	An interagency initiative that leverages established XML schema efforts such as GJXDM to provide a coordinated set of interoperable XML standards across domains. Currently, the seven NIEM domains are emergency management, immigration, intelligence, infrastructure protection, international trade, justice, and screening.
Network management	The activities, methods, procedures, and tools that pertain to the *operation*, *administration*, *maintenance*, and *provisioning* of networked IT and communication systems.
Noise and outlier data	Statistical anomalies in observed data points.
Nonmonotonic reasoning	A process or technique of expert systems that does not necessarily rely on static and complete information, but can build powerful inferences in environments with incomplete and changing information.
Operational agreements and procedures	In this context, these documented agreements define common (agreed and authorized) operational concepts and standard operating procedures for cooperative use of shared information (perhaps shared assets, such as critical IT and communication systems). The training, discipline, and foresight to adhere to these agreements and procedures are an essential part of their content.
Operational architecture	The interworking of people, process, and tools needed to perform actions in the form of documented, vetted, and practiced tasks and activities associated with achieving a common goal.
OPCEN backbone	Set of networks and supporting infrastructure services that interconnect multiple operations centers of all types, capabilities, disciplines, jurisdictions, and levels of governance.
OPCEN system architecture	The generic configuration and interface technical components integrated to achieve a common capability or action-based objective in an operations center, where the center could be one of many types, capabilities, disciplines, jurisdictions, and levels of governance.
Orchestration (orchestrated services)	The capability within an enterprise of system-of-systems to guide the operational workflow of human-system and system-system interaction to support and

	deliver improved mission capability when compared with ad hoc system interaction based on manual procedure.
Outlier	An observation that lies outside the overall pattern of a behavior, where often such behavior is statistically quantified.
Phased implementation	The best practice of implementing changes or upgrading, as well as replacing operational or system architecture in progressive stages over a period of time. In this way, rapid changes that could damage ongoing or near-term operations are minimized or avoided.
Ping-ponging	Feedback in repeater-based systems bridged with an audio switch can occur when the squelch tail (postvoice energy transmission) from a repeater keeps the repeater transmitting long enough to key radios (i.e., cause them to transmit) connected via the switch before the originating radio can transmit.
Project 25	An evolving suite of standards by the Association of Public Safety communications officials for digital land mobile radio communications for use by federal, state/province, and local public safety agencies. (See Chapter 5, *Emerging Public Safety Wireless Communication Systems*, Artech House, November 2001.)
Protocol conversion (radio systems)	If a radio system is analog, or if it is digital with an analog interface to a gateway, then protocol conversion means translation from analog to digital, coding the digitized data stream for compression, and the insertion of the resulting payload data into data packets. If digital, a means of translating from one bit stream to another must be employed.
Provide emergency call service	The provisioning of processing and routing of Public Switched Telephone Network calls by telephone company switches to emergency services, including the routing of calls from the public to available call takers (potentially using automatic call distribution technology), call forwarding, and retrieval.
Public Safety Answering Point (PSAP)	The telephone audio equipment, including headsets and attached microphones, switches, and other infrastructure, that enable trained emergency call-takers to speak directly to the caller provided in an environment conducive to one-to-one voice communications.
Radio-Over-Wireless Broadband (ROW-B)	The basic concept of a bridging system interface is expanded to include the interaction of voice between narrowband (i.e., 25-kHz or smaller channels) and broadband wireless systems. That is, the narrow band

voice or data communications carried in those channels is carried instead simultaneously over broadband channels.

Radio system interface An interconnection with a radio system that permits some or all of the functionality (e.g., control signaling as well as transmitted voice and data) of that radio system to be received and utilized among one or more other communication or IT systems.

Received call processing The extraction of Enhanced ANI/ALI (caller ID and location information) from an incoming voice call via the service provider, it is called Enhanced 9-1-1 or E9-1-1 in the United States, 999 in the United Kingdom and British commonwealth countries, and 112 in much of Europe.

Recording Analog or digital storage and archiving of received voice calls and the subsequent conversation at a Public Safety Answering Point (PSAP), or the recording of two-way radio call traffic between a dispatcher and mobile units. Minimal configurations of this capability store only PSAP calls for some weeks or months, accessible by standard playback to find specific calls. Maximal configurations will employ all digital recording with immediate playback features as well as long-term archival storage.

Regional planning In the present context, it is the use of best practices to gather and vet wide area threat-based information-sharing requirements drawn from all threat-relevant disciplines, geographic jurisdictions, stakeholders (including private as well as public) and levels of governance within the region.

Reliability/survivability Reliability is a measure of the probability that a system, subsystem, equipment, process, or procedure will maintain its functional capabilities under a specific set of conditions for an interval of time, while survivability refers to the probability that a system, subsystem, equipment, process, or procedure will continue to provide some specified reliability following a specified event.

Responder-receiver mobile communications The various wireless communications systems used by one or more agencies to communicate with the responders-receivers under its command or coordination.

Requirements elicitation and concept definition The phase in a project or engineering effort where operational needs and stakeholder expectations are captured.

Requirements and functional analysis	The phase in a project or engineering effort where system requirements are validated and assessed for feasibility, completeness, and verifiability; and where the activities of the system are defined, along with interfaces and information flow, and traced to requirements.
SAFECOM	A communications program of the Department of Homeland Security which provides research, development, testing and evaluation, guidance, tools, and templates on interoperable communications-related issues to local, tribal, state, and federal emergency response agencies.
Scalability	A measure of the upper bound on the expansion of a particular system, subsystem, equipment, process, or procedure to accommodate an increased geographic area or number of users.
Security	The leadership, planning, and activities associated with protecting information and information-sharing systems from unauthorized access, use, disclosure, disruption, modification, or destruction.
Secure communications	The use of one or more media to accurately communicate without information loss, spoofing, disruption, interception or decryption in its transmission from the information source to the intended recipient.
Security management	Protection of the availability, confidentiality, and integrity of the associated information systems with the multilayered, end-to-end prevention, detection, and vulnerability management capabilities.
Sensor array	The collection of raw environmental information at one or more locations, such as audio, imagery, video, airborne chemicals, radiation, temperature, and so forth.
Sensor array with processing	A set of intentionally positioned detectors of environmental conditions (e.g., audio, video, chemical, biological, radiological, thermal) in a sensor array with a corresponding IT subsystem that processes this array output to extract one or more characteristics from the array output. Possible process output includes frequency of an event, scope of an event, intensity of an event, geographic location of an event, reliability that an event has occurred, and more.
Sensor data (including video) processing	A single sensor (versus multiple sensors in an array) with an IT subsystem that compresses and decompresses (and sometimes encrypts or decrypts) sensor outputs and converts them into usable formats. Some

of the types of sensors monitored include intrusion sensors, weather sensors, RFID, seismic, heat detection, gunshot, and HAZMAT leakage.

Service Oriented Architecture (SOA) A framework for deployment of system-of-systems that embraces standards-based tools and communication protocols enabling rapid integration of loosely coupled, modular components to deliver user capability and leverage common crossfunctional services.

Shared systems In this context, the planned use of a single "make and model" of information-sharing capability by more than one organization, potentially spanning multiple disciplines, jurisdictions, and levels of governance. For optimal private land mobile radio systems, it consists of an integrated architecture of a single standards-based or proprietary wide-area system in which all user agencies have interoperability and programmed interoperability.

Stakeholder A group or individual affected by or in some way accountable for the outcome of an undertaking.

Success planning factors In the interoperability context, governance, resources, approach, architecture, performance, and acceptance are the key elements to be addressed in developing interoperability for critical IT and communication systems.

Sustainment and disposal The phase in a project or engineering effort where a capability is maintained in its operational environment, and where upgrades and replacement/removal are managed.

System An integrated set of elements, including—in this context—IT and communication elements or components that together achieve one or more stated objectives for a common purpose.

System lifecycle A framework for the technical transformation of user needs into an operational solution using decision gates and engineering phases.

Systems architecture The generic configuration and interfaces of technical components integrated to achieve a common capability or action-based objective.

Systems engineering An interdisciplinary approach and means to enable the realization of a system which balances stakeholder expectations with system performance as well as cost, schedule, and risk constraints.

Systems engineering plan A document that identifies how a systems engineering effort will be managed, how technical progress will be

	measured, and what baselines and milestones will be implemented.
Talkgroup	A subset or group of users in a given communication system, usually a radio system, with operational requirements to communicate among themselves. Radio system planning includes the identification of these groups and the programming of their identities into their individual subscriber devices for their ready access when called.
Taxonomy	Method of classification of objects or concepts, like that used by biologist to classify organisms.
Third-party network or parallel network affiliation	In the present context, home (i.e., local) and visiting stakeholders share information over a network owned and operated by a third organization, often private, and not involved in the communications itself. For wireless users, mobile subscribers affiliate to an independent or coexisting mobile communications capability, forming integrated mobile communications architecture.
To-be	An engineering architecture perspective of the systems desired future capability baseline. This perspective provides the basis to target and evaluate engineering alternatives to close dysfunctional gaps in current capability.
Token	An abstract object used in simulations and executable architectures to stimulate a representation of enterprise or system-of-system behavior. For example, they may represent such objects as vehicles in a catastrophic evacuation simulation.
Transcoding	The conversion from one form of information encoding to another, such as translating from one technical means of converting voice from analog sound to digital bits to another means of doing analog vice to digital data conversion.
Trunking	The means by which a communications system can provide common transmission media access to many clients simultaneously by sharing a set of wireline or radio/wireless frequencies/channels instead of providing them individually. The intent of this technique is to gain an economy of scale in the numbers of independent communication paths required to service a specific number of users. Since all users in multiple talkgroups do not communicate simultaneously, the

system can be sized probabilistically to accommodate the expected traffic load for day-to-day activities or perhaps (with greater expense) special events or even major disasters. This universal approach is analogous to the structure of a tree with one trunk and many branches out to the individual leaves, where each leaf would be a communications user. Examples of this include telephone systems and nonconventional radios commonly used by public safety agencies in metropolitan areas.

Validation and deployment The phase in a project or engineering effort where a capability is transitioned to its operational environment and stakeholders accept its functionality.

Vertical disparity The inability of systems of nested administrative divisions at different levels to interoperate. For example, a statewide system may not be able to share information with a city within that county. The county may not be able to interoperate with a statewide or national system even though the county lies within the state or that country, and so on.

Visiting mobile affiliation Visiting agencies are granted access to home mobile communications infrastructure as if the visiting subscriber devices were operating as home devices.

Vocoding This process typically resides in an IT or communication system to convert voice into digital bits for use in a computer or transmission using digital communication systems. Technically speaking, it is an analysis/synthesis system, mostly used for speech in which the input is passed through a multiband filter, each filter is passed through an envelope follower, control signals from the envelope followers are communicated, and the decoder applies these (amplitude) control signals to corresponding filters in the (re)synthesizer.

Work plan A plan which outlines staffing, tasks, and resources required to achieve community interoperability at any governance level.

World Interoperability Failure Model The worst case worldwide information-sharing scenario where the leaders, planners, bureaucracies, consultants, and vendors fail to establish interoperability for their critical IT and communication systems. Reasons for this failure include uninvolved, uniformed, or indifferent leadership; individual/organizational jealousy or profit maximization; lack of problem awareness; lack of knowledge of optimal integrated best practices to employ; fear of attempting sweeping change despite necessity; belief that existing capabilities are good enough; and other factors captured by the 25 Pearl Harbor-9/11-Katrina deficiencies (see Chapters 2 and 3).

World Interoperability Success Plan A proposed approach to achieve the desired future state of holistic interoperability, in which planners, leaders, and stakeholders collaborate to bring about a dramatically improved situation in which interoperability is the norm. In this achievable success model, the required systems are in place, cross organizational processes have been defined, and people are trained and ready. This situation will be characterized by a very high level of shared trusted understanding that drives predictable coordinated action among responders-receivers collaborating to achieve a common public safety and emergency response goal. The World Interoperability Success Plan can be realized through systematic application of the principles, tools, checklists, and lessons provided in this book.

About the Authors

Robert I. Desourdis, Jr. is vice president for technology and senior systems architect for Science Applications International Corporation (SAIC) as well as a member of the SAIC Technology Fellows Council. Desourdis led the development of a statewide communications interoperability plan for IT and communications interoperability and has lead local efforts to develop a consolidated dispatch plan. He created the SAIC Public Safety Integration Center (PSIC), a scenario-based solution-oriented integration center for public safety systems, and is currently extending the PSIC concept to university environments. He served as technical manager for a major nationwide IT and communication system supporting search and rescue operations. He has led and supported the development of strategic wireless plans for several states and a U.S. federal law enforcement agency.

Desourdis has worked for the development of interoperability in critical IT and communication systems since 1996. During that year, he served as the spectrum analysis lead for the inaugural year of a U.S. nationwide project to plan public safety wireless network interoperability. Prior to 1996, Desourdis worked for 16 years as a radio systems engineer, including the planning of assured radio communications for strategic missile forces in a postnuclear attack environment and wireless communications for intelligent transportation, including meteor burst communications and high frequency radio.

In addition to this book, he has coauthored four previous books entitled *Preferential Emergency Communications: From Telecommunications to the Internet*, with Dr. Kenneth G. Carlberg in 2003; *Emerging Public Safety Wireless Communication Systems*, with Dr. David R. Smith in 2001; *Advanced High-Frequency Radio Communication Systems*, with Dr. Eric E. Johnson in 1996; and *Meteor Burst Communication: Theory and Practice*, with Dr. Donald L. Schilling in 1993. He has also published over 20 papers on meteor burst communications as well as ground-wave and sky-wave radio propagation experiments.

Desourdis has a B.S. in mathematics and an M.S. in electrical engineering from the Worcester Polytechnic Institute, and an M.S. in technology and policy from the Massachusetts Institute of Technology.

Desourdis conceived of *Achieving Interoperability in Critical IT and Communication Systems*, proposed and received the book contract, assembled the coauthors and contributors, managed manuscript development, wrote initial drafts of Chapters 1, 2, 3, 7, and 8, conceived of the integration of best practices in Chapter 4, cowrote Chapters 5 and 6, and edited all chapters. He also created most of the book's artwork.

Peter J. Rosamilia initiated and created the Interoperability Solutions practice for Science Applications International Corporation (SAIC). As vice president and general manager for the Interoperability Solutions practice, Rosamilia is responsible for all aspects of strategy, operations, technical service delivery, funding, and business development. Rosamilia leads program teams that create state-of-the-art interoperable communications and information sharing systems for federal, state, local, and commercial clients. Recent clients include state and federal emergency management agencies, public safety, public health, and healthcare organizations. Rosamilia has expertise in networks, communications, systems, security, and data interoperability as it applies to critical systems in public safety, healthcare, and energy.

Rosamilia initiated and ran the development of the Law Enforcement Justice Information System, a first-of-its-kind data interoperability and information-sharing solution. It is based on the Global Justice XML Data Model (GJXDM) and the National Information Exchange Model (NIEM). This cutting-edge project has provided real-time data interoperability among competing records management systems (RMS) and computer-aided dispatch (CAD) systems. Now available statewide, this solution delivers data interoperability among dozens of police departments, district attorney offices, sheriffs' offices, and state and county operations centers (OPCENS). The system is fully integrated with the statewide public key infrastructure (PKI).

Previously, Rosamilia led the development and operation of data interoperability solutions for public health, energy, and healthcare clients, including the Beth Israel Deaconess Medical Center, other integrated delivery networks, large health maintenance organizations in the United States, and a provincial government in Canada.

Rosamilia was chief operating officer (COO) of the Advanced Network eXchange (ANX), the high-performance, high-security network interoperability service that reliably interconnected U.S. auto manufacturers with their largest suppliers. As COO and founding member of ANXeBusiness, Rosamilia led the commercialization and expansion of the ANX network. He managed relationships among competing telecommunications companies that served the auto industry as certified providers of the interoperable ANX service. Rosamilia managed company operations through startup and wrote the business plan that resulted in a $20 million capital raise.

Rosamilia holds an M.S. in technology management from the Sloan School of Management at the Massachusetts Institute of Technology (MIT) and a B.S. in chemical engineering from MIT.

Rosamilia wrote the IT system architecture descriptions in Chapter 6 and added key figures. He wrote the first two sections of Chapter 6, developed the interoperability systems architecture concepts depicted in Figures 6.1 through 6.5, and contributed to all sections of Chapter 6. He also supported the final 48-hour manuscript production phase, reviewing and verifying each chapter. He provided editorial input for the entire book.

Christopher P. Jacobson serves as Science Applications International Corporation's (SAIC's) chief architect for cyber command and control, as well as several programs involving command and control (C2), incident response, homeland security, aerospace operations, and chemical, biological, radiological, nuclear, and high-yield explosive (CBRNE) across multiple federal agencies and departments.

After graduating from the Air Force Academy, Jacobson began a 28-year career in software-intensive systems engineering for critical command, control, communications, and computers (C4I) systems, beginning as a software engineer at the North American Aerospace Defense Command (NORAD), working with space computational software systems. During the first half of his career, Jacobson applied structured analysis techniques as program manager for the Space Defense Operations Center (SPADOC 4B), Interim Iceland Air Defense System (IIADS), ROCC/AWACS Digital Information Link (RADIL) and was selected as Program Manager of the Year for Strategic Systems at the Electronics Systems Center.

Jacobson served at the Pentagon as acquisition program element monitor (PEM) for Western Hemisphere air defense systems, and later as Joint Staff (J6) technical lead for the Global C2 System and collaboration technology. He also served as military executive assistant to the newly formed Defense Information Systems Agency (DISA), where he continued to apply structured analysis techniques in working business process reengineering (BPR) for the U.S. Department of Defense's information technology core competencies. Today, Jacobson continues to refine an object-oriented approach to architecture-centric engineering for his programs and was asked by the Office of the Secretary of Defense (OSD) to present his views as keynote speaker at the Software and Systems Technology Conference 2008.

Jacobson earned a B.S. from the United States Air Force Academy, an M.S. in public administration from the University of Colorado in Colorado Springs, and is a Ph.D. candidate in information technology at George Mason University. His PMP accreditation is from the Project Management Institute.

Jacobson contributed to Chapter 5, Operational Architecture, writing Sections 5.3 and 5.7. He also provided architecture material for Chapter 4, RPA material for Chapter 8, and case study material for the annex..

James E. Sinclair is currently vice president for Public Safety Systems at Young Enterprise Systems and has over 40 years of experience, which place him among the technical leaders in the public safety community. In this position, he has provided major public safety integrators with consulting services related to land mobile radio systems and interoperability solutions.

In addition to public safety wireless systems and communications center operations, Sinclair's background includes radio frequency communications from very low frequencies to satellite communications, terrestrial broadband communications, and development of data-link and network-level protocols. Further, he has applied all of the systems engineering disciplines relevant to evaluating, selecting, designing, installing, testing, and delivering these systems.

While with TRW (now Northrop Grumman), he led the systems engineering effort on one of the largest statewide public safety systems developed to date: Ohio Multi-Agency Radio Communications System (MARCS). This included the development, installation, integration, and testing of a statewide trunked voice and conventional data land mobile radio system with 201 remote tower sites. The project included in-building coverage for government buildings and correctional institutions. Also featured were one major and 67 remote dispatch facilities.

Sinclair's experience with public safety radio system interoperability includes Ohio MARCS, the District of Columbia Unified Command Center (UCC) Project, the Silicon Valley Regional Interoperability Project, the Texas Communications Interoperability Project (CIPP), and the Navy Enterprise Land Mobile Radio (ELMR) System.

Sinclair's education includes a B.S. in electrical engineering from Ohio University and an M.S. in electrical engineering from the University of Illinois, where he taught undergraduate courses while pursuing his graduate degree. He has been a registered professional engineer in the state of California since 1968, is a life member of the Institute of Electrical and Electronics Engineers (IEEE), and is a longtime member of the Association of Public-Safety Communications Officials (APCO).

Sinclair was a major contributor to all sections of Chapter 6 and related material in the Web site annex that accompanies this book.

James R. McClure is a senior enterprise architect with Science Applications International Corporation's (SAIC's) Homeland & Civilian Services business unit. He supports the Department of Homeland (DHS) Security chief architect developing the Homeland Security Enterprise Architecture (HLS EA). The HLS EA has at its core DHS's mission, business, performance, systems, investments, and technologies but its scope encompasses the whole U.S. emergency response community, including federal, state, tribal, and local governments as well as private industry. McClure formerly led the Enterprise Architecture and Systems Engineering team in developing the Emergency Communications Architecture (ECA), based on the National Response Plan (NRP) and the 15 Emergency Support Functions (ESFs). The ECA specifically focused on communication and communications failures in the wake of Hurricane Katrina.

Prior to joining SAIC, McClure spent almost 20 years with GTE and then Verizon in many technical and managerial roles, ending as manager of Technology & Architecture Strategy. McClure designed and developed information technology systems and products for GTE and provided technical and strategic integration for wireless voice and data services for Verizon.

McClure holds a B.S. from the University of North Carolina at Charlotte.

McClure wrote Section 4.5 and provided editorial inputs for the entire book.

About the Contributors

Kenneth G. Carlberg has over 20 years experience in the design and development of communication systems. In the area of emergency-related communications, he is currently a consultant to U.S. national authorities for the investigation and definition of Internet Engineering Task Force (IETF) protocols to support preferential treatment of IP-based communications.

In this capacity, Carlberg has coauthored several requests for comment (RFCs) on the topic of emergency telecommunications service for the IETF. He has been the lead author of a book on emergency communications, taught a course on the subject at Johns Hopkins University, and has been an invited speaker for the International Telecommunication Union and EMTEL workshops on disaster communications.

Carlberg earned a B.S. and an M.S. in computer science from Loyola College, and a Ph.D. in computer science from University College London.

Carlberg wrote Section 4.9.1 and contributed to Section 8.3.1.

Robert J. Coullahan is the president of Readiness Resource Group, Inc. (RRG), a veteran-owned small business serving homeland security, protection, preparedness, and enterprise resilience programs. Board-certified in emergency management and security management (Certified Emergency Manager and Certified Protection Professional), Coullahan is also a Certified Business Continuity Professional (CBCP). He served over a 25-year period at Science Applications International Corporation (SAIC), where he earned the appointment of senior vice president and operations manager for Homeland Security Support.

Coullahan also served five years as vice president and director of Government & International Programs for an award-winning federally funded university consortium based in Washington, D.C. The consortium was designated the World Data Center-A for Human Interactions in the Environment, based on a rigorous scientific peer-review. He is a certified emergency medical technician (EMT) and served nine years of active and reserve duty in the United States Army.

His professional experiences have been at the nexus of interagency, intergovernmental, and interdisciplinary collaboration. His work includes leadership roles as director of the Crisis Management Development Center and program manager of the White House Y2K National Information Coordination Center. His diverse command, control, communications, computers, and intelligence (C4I) and emergency management experiences have yielded a comprehensive understanding of data interoperability, decision support, situational awareness, and the elusive but precious common operational picture.

Coullahan attended Rutgers University and is a graduate of the University of California (B.A.) and The George Washington University, where he earned a M.S. in telecommunications and an M.A. in security management–forensic sciences.

Coullahan wrote Section 8.1.

Ezmeralda Khalil de Raet is an assistant vice president and organizational effectiveness (OE) practice manager at Science Applications International Corporation (SAIC). The OE practice focuses on delivering innovative, holistic business solutions to improve client effectiveness and efficiency. Clients include federal (civilian and military), state, and commercial organizations.

De Raet serves as a senior consultant and advisor, supporting clients from development of initial strategy through implementation and successful institutionalization of major initiatives and transformations. Employing collaborative decision-support tools and techniques, she has developed actionable strategic plans and implementation guidance for federal and state organizations. For federal, state, and commercial organizations, she has reengineered major processes to increase the quality of products and services while eliminating waste. She has also worked to align organizational structures to enable effective operations and better client service within the federal government.

She has supported development of SAIC methodologies in process improvement, organizational alignment, and change management while also creating staffing models to maximize the efficiency of an organization's current staff while forecasting staffing needs based on changing requirements. Prior to joining SAIC, De Raet served as an analyst at Accenture and an applications engineer at Boland Trane.

De Raet earned a B.S. in mechanical engineering from the University of Maryland and an M.S. in engineering management from The George Washington University.

De Raet contributed to Sections 4.3 and 4.4. These sections focus on best practices in planning for, implementing, and institutionalizing an entity's interoperability mission and vision while increasing stakeholder trust, buy-in, and adoption.

Holger Hinsch is a senior economist for Science Applications International Corporation (SAIC). He has 15 years experience in cost-benefit analysis, strategic planning, risk analysis, transportation system analysis, and economic impact assessments. He has acted as project manager for a number of cost-benefit analysis, transportation, economic, financial modeling, and engineering efforts for both public and private sector clients. Some of these efforts include valuation and mergers and acquisitions analysis for clients in the biotech and telecommunications sectors, the development of privatization strategies for the Hamburg Airport, numerous cost-benefit analyses for federal agencies' IT and communications systems investments, as well as supporting the Massachusetts Executive Office of Public Safety, its designated interoperability coordinator, and the standing State Interoperability Working Committee to design and produce a five-year strategic plan for achieving optimal statewide communications interoperability in Massachusetts. The Massachusetts Statewide Communications Interoperability Plan (SCIP) is a requirement of the Department of Homeland Security, Science, and Technology Directorate's SAFECOM Program. Hinsch led stakeholder interviews and facilitated focus groups sessions. He further was responsible for the analysis of the financial viability of the proposed investment into technology to be included in the strategic plan.

Hinsch has published a number of transportation research-related articles and a book, *The Impact of Liberalizing International Aviation Bilaterals*, written with David Gillen, Jonathan Harvey, MKmetric, and Institut für Weltwirtschaft, and published by Ashgate Publishing in February 2001.

Hinsch has an M.B.A. from the University of Maryland and a B.S. in engineering from the University of Michigan.

Hinsch wrote Section 4.6.

Eric Johnson is a professor of electrical and computer engineering at New Mexico State University, where he is also head of the Signals and Systems Group at the Physical Science Laboratory. His teaching and research areas include wireless network protocols and standards, and network security.

For more than two decades Johnson has led a joint government/industry/academic team in developing and standardizing technologies for interoperability in high-frequency radio communications in NATO, the U.S. Department of Defense, and civilian government. He regularly offers short courses in high frequency (HF) radio through the Armed Forces Communications-Electronics Association (AFCEA).

In addition to this book, Johnson coauthored *Advanced High-Frequency Radio Communication Systems* with Robert I. Desourdis, and is the author of over 100 technical papers and reports on wireless communications and computer architecture.

Johnson earned B.S. degrees in electrical engineering and physics, an M.S. in electrical engineering from Washington University in St. Louis, and his Ph.D. in electrical and computer engineering from New Mexico State University.

Johnson wrote Sections 4.9.2 and 8.3.1. He also wrote the case study on HF radio standardization on the book Web site annex.

Karen S. Kelly is a senior systems engineer at Science Applications International Corporation (SAIC), where she has over 10 years of systems engineering experience applied to a broad spectrum of information technology problems. Her customers have included the Department of State (DoS), the Federal Emergency Management Agency (FEMA), the Department of Homeland Security (DHS), and the Defense Information Systems Agency (DISA). She has focused on areas of systems engineering management and the system life cycle, requirements and functional analysis, modeling, and decision analysis. She has developed systems engineering plans, tailored processes and products, composed system requirements and managed traceability, and implemented risk management practices.

Kelly has extensive experience in data analysis and modeling, and in the implementation of simulation models for network performance assessment. In addition, she has applied her skills to areas of emergency operations center communications, procedures, and exercise simulation. She has helped her customers focus their organizational goals and mission statements, ensuring that initiatives dovetail with strategic planning.

In 2008, Kelly achieved the International Council of Systems Engineering (INCOSE) Certified Systems Engineering Professional (CSEP) designation. She also serves as an instructor for SAIC's four-day Systems Engineering 101 course.

Kelly earned her B.A. in mathematics from Boston College and her M.S. and Ph.D. in systems engineering from the University of Virginia.

Kelly wrote Section 4.8.

Karen M. Krause is the focus area lead for Homeland Security Initiatives at Raytheon, Integrated Defense Systems, Civil Security and Response Programs.

Krause has 15 years of funding strategy and program management experience in the homeland security, transportation, public safety, environmental, and energy sectors. Currently Krause is developing strategic plans for homeland security initiatives at Raytheon, an international defense contractor. Krause is the former cofounder and managing director of the Public Safety Strategies Group, a niche public safety management consulting firm, where she secured more than $1.6 billion in federal contracts and grants for U.S. clients. She also served as a Harvard Business School research associate in operations management and a former project manager with Camp Dresser & McKee, an international environmental engineering firm.

Krause earned her B.S. in environmental systems technology from Cornell University and her M.B.A. in entrepreneurship from the F.W. Olin Graduate School of Business, Babson College.

Krause wrote Section 4.7.

Suzanne J. MacBain is a senior project manager at Science Applications International Corporation (SAIC) focused on interoperability and intercommunications in public safety and emergency response departments of local, state, and federal government agencies. She has more than 15 years experience consulting to public and private sector entities, providing program management, financial analysis, and strategic planning.

Much of MacBain's experience and leadership in program management and interagency communication stems from her years supporting the directorate-level Strategic Plans & Programs office of a key U.S. military organization. This experience, combined with her participation in regional and state interoperability initiatives for government, commercial, and nonprofit institutions, provides her with specialized insight to the varied communications needs and interoperability solutions necessary to support first responders and their respective agencies.

MacBain was educated at Vassar College in Poughkeepsie, New York, and the Sorbonne University in Paris, France, where she received her dual degree in economics and French literature. She also completed graduate level studies at Johns Hopkins University while working in Washington, D.C. as a corporate liaison at the International Telecommunication Union (IEEE) Computer Society. She received her Project Management Professional (PMP) accreditation from the Project Management Institute.

MacBain contributed to and edited several chapters. She reduced the original extensive Chapter 2 content on Pearl Harbor to the essential facts, and merged the draft Chapter 3 material on the 9/11 and Katrina reports into Chapter 2. In addition, MacBain managed the collection of materials for the book Web site annex, particularly the collection and integration of vendor offerings.

William R. Michalson is a professor in the electrical and computer engineering department at the Worcester Polytechnic Institute (WIP), where he performs research and teaches in the areas of navigation (including indoor and non-GPS-based navigation systems), communications, and computer system design. He supervises the WPI Center for Advanced Integrated Radio Navigation (CAIRN), where he is developing a Public Safety Integration Center focused on the integration of communications, navigation, and information technologies for public safety applications. His research focuses on the development, testing, and evaluation of systems for both civilian and military applications, with a special emphasis on techniques focused on indoor, underground, or otherwise Global Positioning System (GPS)-deprived situations. Prior to joining the faculty at WPI, Michalson spent approximately 12 years at the Raytheon Company, where he was involved with the development of embedded computers for guidance, communications, and data processing systems for spaceborne and terrestrial applications.

Michalson has been involved with GPS-based positioning and navigation systems for nearly 17 years, and with radio communications systems for over 30. He has been on the faculty of the electrical and computer engineering department at the

Worcester Polytechnic Institute since 1991. In addition, he has background in the coverage and information technology issues associated with network-based radio communications and has extensive background in public safety communications.

Michaelson has a B.S. in electrical engineering from Syracuse University (1981), an M.S. in electrical engineering from the Worcester Polytechnic Institute (WPI) (1985), and a Ph.D. in electrical engineering from WPI (1989).

Michalson provided input on radio network usage statistics and other interoperability technologies appearing in the book Web site annex.

Jelena Roljevic is the director of the Center for Enterprise and Solutions Architecture at Science Applications International Corporation (SAIC). She has over 16 years of successful service delivery experience in both federal government and industry markets. She has led numerous enterprise architecture development and information technology integration projects. She has extensive hands-on experience and knowledge of systems engineering and integration, enterprise architecture planning, policy and standards development, capital planning and investment control, business process improvement, technology deployment, and program management.

Roljevic earned a B.S. in electrical engineering from the University of Belgrade in Yugoslavia, and an M.S. in systems engineering and management from The George Washington University.

Roljevic wrote Section 4.2.

David R. Smith, beginning in 1967, held positions as an electrical engineer and manager within the Department of Defense for both the Navy Department and the Defense Information Systems Agency. Since 1970, he has held adjunct, visiting, research, and regular faculty positions at Georgia Tech, The George Washington University, and George Mason University. Smith is currently a professor in the Department of Electrical and Computer Engineering of The George Washington University (GWU), Washington, D.C., director of Graduate Programs at the GWU Virginia Campus, and deputy director of the Space and Advanced Communications Research Institute. He also consults with numerous companies in wireless and digital communications. His current interests include teaching, research, and consulting in wireless telecommunications, public safety communications, telecommunications networks, digital communications, propagation modeling, and computer simulation and modeling of communication systems.

His publications include over 20 Institute of Electrical and Electronics Engineers (IEEE) articles, the books *Digital Transmission Systems* published by Kluwer and *Emerging Public Safety Wireless Communication Systems* (coauthor) published by Artech House with Robert Desourdis, along with chapters contributed to two other books.

Smith received a B.S. in physics from Randolph-Macon College, an M.S. in electrical engineering from Georgia Tech, and a Ph.D. in electrical engineering and computer science from The George Washington University.

Smith contributed significant technical material to Chapter 6 with much of this detailed material placed in the book Web site annex.

William R. Stewart, Jr. is a principal systems engineer with Science Aplications International Corporation (SAIC), responsible for architecting command, control, and communication solutions for public safety, emergency response, surveillance, and homeland security at all federal, state, and local levels. Key activities include interoperability planning and design, with a primary life-cycle focus in enterprise architecture and systems engineering of systems requiring situational awareness, dispatch, wireless communications (e.g., WiMAX, Land Mobile Radio, NAIS), sensor management (i.e., radar, cameras, microwave intrusion detection systems, etc.) and multiorganizational collaboration. Key projects include U.S. Armed Services P25 enterprise land mobile radio architecture, large municipal WiMAX public safety command and control networking, Fusion Center concept exploration and architecture, critical infrastructure protection system architecture and design, and others. Other major systems/software engineering projects include command and control systems engineering, space station systems/software engineering, and advanced fighters, to name a few.

Stewart earned both his B.S. degrees in computer science and engineering technology from Old Dominion University.

Stewart was primarily involved in integrating, editing, and cowriting sections of Chapter 4.

Index

Recent Titles in the Artech House Mobile Communications Series

John Walker, Series Editor

3G CDMA2000 Wireless System Engineering, Samuel C. Yang

3G Multimedia Network Services, Accounting, and User Profiles, Freddy Ghys, Marcel Mampaey, Michel Smouts, and Arto Vaaraniemi

802.11 WLANs and IP Networking: Security, QoS, and Mobility, Anand R. Prasad, Neeli R. Prasad

Achieving Interoperability in Critical IT and Communication Systems, Robert I. Desourdis, Peter J. Rosamilia, Christopher P. Jacobson, James E. Sinclair, and James R. McClure

Advances in 3G Enhanced Technologies for Wireless Communications, Jiangzhou Wang and Tung-Sang Ng, editors

Advances in Mobile Information Systems, John Walker, editor

Advances in Mobile Radio Access Networks, Y. Jay Guo

Applied Satellite Navigation Using GPS, GALILEO, and Augmentation Systems, Ramjee Prasad and Marina Ruggieri

Artificial Intelligence in Wireless Communications, Thomas W. Rondeau and Charles W. Bostian

Broadband Wireless Access and Local Network: Mobile WiMax and WiFi, Byeong Gi Lee and Sunghyun Choi

CDMA for Wireless Personal Communications, Ramjee Prasad

CDMA Mobile Radio Design, John B. Groe and Lawrence E. Larson

CDMA RF System Engineering, Samuel C. Yang

CDMA Systems Capacity Engineering, Kiseon Kim and Insoo Koo

CDMA Systems Engineering Handbook, Jhong S. Lee and Leonard E. Miller

Cell Planning for Wireless Communications, Manuel F. Cátedra and Jesús Pérez-Arriaga

Cellular Communications: Worldwide Market Development, Garry A. Garrard

Cellular Mobile Systems Engineering, Saleh Faruque

The Complete Wireless Communications Professional: A Guide for Engineers and Managers, William Webb

EDGE for Mobile Internet, Emmanuel Seurre, Patrick Savelli, and Pierre-Jean Pietri

Emerging Public Safety Wireless Communication Systems, Robert I. Desourdis, Jr., et al.

The Future of Wireless Communications, William Webb

Geographic Information Systems Demystified, Stephen R. Galati

GPRS for Mobile Internet, Emmanuel Seurre, Patrick Savelli, and Pierre-Jean Pietri

GPRS: Gateway to Third Generation Mobile Networks, Gunnar Heine and Holger Sagkob

GSM and Personal Communications Handbook, Siegmund M. Redl, Matthias K. Weber, and Malcolm W. Oliphant

GSM Networks: Protocols, Terminology, and Implementation, Gunnar Heine

GSM System Engineering, Asha Mehrotra

Handbook of Land-Mobile Radio System Coverage, Garry C. Hess

Handbook of Mobile Radio Networks, Sami Tabbane

High-Speed Wireless ATM and LANs, Benny Bing

Interference Analysis and Reduction for Wireless Systems, Peter Stavroulakis

Introduction to 3G Mobile Communications, Second Edition, Juha Korhonen

Introduction to Communication Systems Simulation, Maurice Schiff

Introduction to Digital Professional Mobile Radio, Hans-Peter A. Ketterling

Introduction to GPS: The Global Positioning System, Ahmed El-Rabbany

An Introduction to GSM, Siegmund M. Redl, Matthias K. Weber, and Malcolm W. Oliphant

Introduction to Mobile Communications Engineering, José M. Hernando and F. Pérez-Fontán

Introduction to Radio Propagation for Fixed and Mobile Communications, John Doble

Introduction to Wireless Local Loop, Second Edition: Broadband and Narrowband Systems, William Webb

IS-136 TDMA Technology, Economics, and Services, Lawrence Harte, Adrian Smith, and Charles A. Jacobs

Location Management and Routing in Mobile Wireless Networks, Amitava Mukherjee, Somprakash Bandyopadhyay, and Debashis Saha

Mobile Data Communications Systems, Peter Wong and David Britland

Mobile IP Technology for M-Business, Mark Norris

Mobile Satellite Communications, Shingo Ohmori, Hiromitsu Wakana, and Seiichiro Kawase

Mobile Telecommunications Standards: GSM, UMTS, TETRA, and ERMES, Rudi Bekkers

Mobile Telecommunications: Standards, Regulation, and Applications, Rudi Bekkers and Jan Smits

Multiantenna Digital Radio Transmission, Massimiliano Martone

Multiantenna Wireless Communications Systems, Sergio Barbarossa

Multipath Phenomena in Cellular Networks, Nathan Blaunstein and Jørgen Bach Andersen

Multiuser Detection in CDMA Mobile Terminals, Piero Castoldi

OFDMA for Broadband Wireless Access,

Personal Wireless Communication with DECT and PWT, John Phillips and Gerard Mac Namee

Practical Wireless Data Modem Design, Jonathon Y. C. Cheah

Prime Codes with Applications to CDMA Optical and Wireless Networks, Guu-Chang Yang and Wing C. Kwong

QoS in Integrated 3G Networks, Robert Lloyd-Evans

Radio Engineering for Wireless Communication and Sensor Applications, Antti V. Räisänen and Arto Lehto

Radio Propagation in Cellular Networks, Nathan Blaunstein

Radio Resource Management for Wireless Networks, Jens Zander and Seong-Lyun Kim

Radiowave Propagation and Antennas for Personal Communications, Third Edition, Kazimierz Siwiak and Yasaman Bahreini

RDS: The Radio Data System, Dietmar Kopitz and Bev Marks

Resource Allocation in Hierarchical Cellular Systems, Lauro Ortigoza-Guerrero and A. Hamid Aghvami

RF and Baseband Techniques for Software-Defined Radio, Peter B. Kenington

RF and Microwave Circuit Design for Wireless Communications, Lawrence E. Larson, editor

Sample Rate Conversion in Software Configurable Radios, Tim Hentschel

Signal Processing Applications in CDMA Communications, Hui Liu

Smart Antenna Engineering, Ahmed El Zooghby

Software Defined Radio for 3G, Paul Burns

Spread Spectrum CDMA Systems for Wireless Communications, Savo G. Glisic and Branka Vucetic

Technologies and Systems for Access and Transport Networks, Jan A. Audestad

Third Generation Wireless Systems, Volume 1: Post-Shannon Signal Architectures, George M. Calhoun

Traffic Analysis and Design of Wireless IP Networks, Toni Janevski

Transmission Systems Design Handbook for Wireless Networks, Harvey Lehpamer

UMTS and Mobile Computing, Alexander Joseph Huber and Josef Franz Huber

Understanding Cellular Radio, William Webb

Understanding Digital PCS: The TDMA Standard, Cameron Kelly Coursey

Understanding GPS: Principles and Applications, Second Edtion, Elliott D. Kaplan and Christopher J. Hegarty, editors

Understanding WAP: Wireless Applications, Devices, and Services, Marcel van der Heijden and Marcus Taylor, editors

Universal Wireless Personal Communications, Ramjee Prasad

WCDMA: Towards IP Mobility and Mobile Internet, Tero Ojanperä and Ramjee Prasad, editors

Wireless Communications in Developing Countries: Cellular and Satellite Systems, Rachael E. Schwartz

Wireless Communications Evolution to 3G and Beyond, Saad Z. Asif

Wireless Intelligent Networking, Gerry Christensen, Paul G. Florack, and Robert Duncan

Wireless LAN Standards and Applications, Asunción Santamaría and Francisco J. López-Hernández, editors

Wireless Technician's Handbook, Second Edition, Andrew Miceli

For further information on these and other Artech House titles,including previously considered out-of-print books now available through our In-Print-Forever® (IPF®) program, contact:

Artech House
685 Canton Street
Norwood, MA 02062
Phone: 781-769-9750
Fax: 781-769-6334
e-mail: artech@artechhouse.com

Artech House
16 Sussex Street
London SW1V 4RW UK
Phone: +44 (0)20 7596-8750
Fax: +44 (0)20 7630-0166
e-mail: artech-uk@artechhouse.com

Find us on the World Wide Web at: www.artechhouse.com